BIOLOGY OF FRESHWATER POLLUTION

THIRD EDITION

BIOLOGY OF FRESHWATER POLLUTION

THIRD EDITION

C. F. MASON
DEPARTMENT OF BIOLOGY, UNIVERSITY OF ESSEX

Longman

Longman
Longman Group Limited
Longman House, Burnt Mill, Harlow
Essex CM20 2JE, England
and Associated Companies throughout the world

© Longman Group UK Limited 1982, 1991
This edition © Longman Group Limited 1996

First published 1981
Second edition 1991
Third edition 1996

British Library Cataloguing in Publication Data
A catalogue entry for this title is available from the British Library.

ISBN 0–582–24732–2

Library of Congress Cataloging-in-Publication data
A catalog entry for this title is available from the Library of Congress.

Set by 3 in 10/12 pt Times Roman
Produced by Longman Singapore Publishers (Pte) Ltd
Printed in Singapore

CONTENTS

PREFACE

Five years have elapsed since I completed the second edition of this book, the first edition having been written as long ago as 1980. The structure of the current edition remains much the same as its predecessor, but I have taken the opportunity to update throughout, much having been published on all aspects of pollution in the early 1990s. New sections have been added on neoplasms, biomarkers and global warming, while the final chapter has been revised to reflect new developments in water resource management.

Water quality remains a matter of considerable concern, for, in many parts of the world, wholesome water is still a scarce commodity and waterborne diseases are still rife. The problem is largely one of politics, not of science. The United Nations 1994 Development Report has estimated that a 12 per cent cut in military spending worldwide could provide safe drinking water and primary health care for everyone, while eliminating severe malnutrition. Nevertheless, 18 developing countries still spend more on the military than on health and education. Rapid economic development and industrialization in the past has left the developed world with a legacy of chronic water pollution, while accidents, causing considerable damage to freshwater life, are daily occurrences. The demand for the restoration of clean water flowing in natural channels is increasingly vociferous. The problem in the industrialized countries, however, is one of unsustainable patterns of consumption and production, leading to major environmental degradation. For example, it has been estimated that the environmental impact of 2.6 million newborn Americans each year will far exceed that of 34 million new Chinese and Indians. The long-term survival of our life-support systems depends, not only on curbing population growth, but also on reducing poverty and promoting equity, and on developing sustainable approaches to production and consumption. Otherwise, globally, aquatic resources will continue to deteriorate.

While this book is aimed primarily at advanced university students, I hope the style makes it accessible to a wider audience who may have an interest in the pollution problems besetting our freshwater habitats and in the techniques developed to combat them.

I would like to thank my wife, Sheila Macdonald, for her considerable help at all stages in the production of this book. Rita Bartlett and Judi Blackmore provided additional secretarial assistance. Dr Brian Whitton and Dr John Eaton provided a number of suggestions to improve the revision of the text. Alex Seabrook, Pauline Gillett and Tina Cadle at Longman provided their customary excellent support.

C. F. Mason
February 1995

ACKNOWLEDGEMENTS

We are grateful to the following for permission to reproduce copyright material:

Academic Press Inc. and the authors for fig. 2.11 (Higgins and Burns, 1975); American Chemical Society for fig. 5.6 (Havas *et al.*, 1984), fig. 11.3 (Foster and Bates, 1978), table 4.2 (Lee *et al.*, 1978) and table 6.2 (Giesy *et al.*, 1994), © 1978, 1984 and 1994 American Chemical Society; Board of Editors, *Annals of Applied Biology* for fig. 2.10 (Lloyd 1960); ASTM for fig. 11.7 (Westlake and Van der Schalie, 1977), © 1977 American Society for Testing and Materials; Birkhauser Verlag for fig. 4.10 (Hartmann, 1977); Blackwell Scientific Publications Ltd for figs 1.5 and 1.6 (Edwards *et al.*, 1984), fig. 2.5 (McCahon and Pascoe, 1990), figs 3.16 and 3.20 (Holland and Harding, 1984), fig. 5.10 (Fryer, 1980) and table 5.1 (Gee and Stoner, 1988); Cambridge University Press and the respective authors for figs 1.12 and 2.12 (Holdgate, 1979), fig. 12.2 (Porter, 1978), fig. 5.14 (Turnpenny, 1989) and fig. 5.15 (Brown and Sadler, 1989); *Canadian Journal of Fisheries and Aquatic Science* for fig. 2.4 (Sprague, 1964), fig. 4.12 (Schindler *et al.*, 1973) and 4.15 (Dillon and Rigler, 1975); Chapman & Hall Ltd for fig. 5.16 (Ormerod, J.J. and Tyler, J.J., 'Birds as indicators of change in water quality', in Furness, R.W. and Greenwood, J.J.D., eds, *Birds as Monitors of Environmental Change*, 1993) and figs 12.7 and 12.8 (Edwards, R.W., 'Predicting the environmental impact of a major reservoir development', in Roberts, R.D. and Roberts, T.M., eds, *Planning and Ecology*, 1984); Chartered Institution of Water and Environmental Management for fig. 3.7 (Cooper, P.F. *et al.*, *Journal of Water and Environmental Management* **3**, 60–74, 1989); Croom Helm Ltd for fig. 6.5 (Birkhead and Perrins, 1986); reprinted with kind permission from Elsevier Science Ltd, The Boulevard, Langford Lane, Kidlington OX5 1GB, UK, fig. 2.15 from *Biol. Conserv.* **7**, 79–118, Newbold, C., 'Herbicides in aquatic systems', 1975, figs 3.15, 7.3 and 7.4 from Hellawell, J.M., *Biological Indicators of Freshwater Pollution and Environmental Management*, 1986, fig. 3.18 from *Environmental Pollution* **5**, 1–10, Aston, R.J., 'Tubificids and water quality: a review', 1973, fig. 4.3 from 'Concept of stress and recovery in aquatic ecosystems', R.D. Gulati, in *Ecological Assessment of Environmental Degradation, Pollution and Recovery*, O. Ravera, ed., fig. 4.20 from 'Biomanipulation of aquatic food chains to improve water quality in eutrophic lakes', De Bernardi, R. in *Ecological Assessment of Environmental Degradation, Pollution and Recovery*, Ravera, O., ed., figs 5.7 and 5.12 from 'Air pollution effects on aquatic ecosystems and their restoration', Henrikson, A., in *Ecological Assessment of Environmental Degradation, Pollution and Recovery*, Rivera, O., ed., fig. 6.7 from *Environmental Pollution* **43**, 163–73, Venant, A. and Cumont, G., 'Contamination des poissons du secteur français du Lac Léman par les composes organochlorés entre 1973 et 1981', 1987, fig. 9.2 from Green and Trett, *The Fate and Effects of Oil in Freshwater*, 1989, fig. 9.5 from Brown, M., 'Biodegradation of oil in freshwaters' in Green, J. and Trett, M., eds, *The Fate and Effects of Oil in Freshwater*, 1989, fig. 10.4 from *Environmental Pollution* **58**, 55–70, Raven, P.J. and George, J.J., 'Recovery by riffle macroinvertebrates in a river after a major accidental spillage of chlorphyrifos', 1989, fig. 10.6 from *Environmental Pollution* **81**, 217–28, Rutt, G.P. *et al.*, 'The impact of livestock farming on Welsh streams: the development and testing of a rapid biological method for use in the assessment and control of organic pollution from farms', figs 12.4 and 12.6 from Mance, G., *Pollution Threat of Heavy Metals in Aquatic Environments*, 1987 and table 10.2 from *European Water Pollution Control* **3(4)**, 15–25, Wright, J.F. *et al.*, 'A technique for evaluating the biological quality in the UK'; The Eugenics Society for fig. 3.11 (Arthur, 1972); Dr I. Häkkinen for fig.6.2 (Häkkinen and Häsänen, 1980); *Holarctic Ecology* for table 6.1 (Särkkä *et al.*, 1978); Holt, Rinehart and Winston for fig. 3.23 (Warren, 1971) © 1971 Holt, Rinehart and Winston; Ellis Horwood for fig. 11.8 (Diamond *et al.*, 1988); The Institution of Water Engineers & Scientists for fig. 2.3 (Herbert, 1961); International Association for Great Lakes Research for fig. 2.8 (Fox, 1993); International Atomic Energy Agency for fig. 8.3 (Preston, 1974); IOP Publishing Ltd for figs 3.4 and 3.24 (Wood, 1982); Kluwer Academic Publishers for fig. 4.5 (R.V. Smith, 1993) and fig. 4.7 (Hussein and Mason, 1988); Longman Group UK Ltd for fig. 3.17 (Macan, 1959); Liverpool University Press for figs 3.12 and 3.14 (Hynes, 1960) and fig. 7.6 (Langford, 1983); Sir John Mason for fig. 5.1 (Mason, 1989); Ministry of Agriculture, Fisheries and Food for table 8.2 (Hunt, 1987); National Academy of Sciences for fig. 4.2 (Edmondson, 1969); National Technical Information Service for fig. 7.5 (McFarlane *et al.*, 1976); Norwegian Institute for Water Research for fig. 5.8 (Henriksen *et al.*, 1984); fig. 4.14 from *Eutrophication of Waters: Monitoring, Assessment and Control*, reproduced by permission of the OECD, © 1982 OECD; Pergamon Press PLC for fig. 2.6 (Sprague, 1970), fig. 2.7 (Calamari and Marchetti, 1973), fig. 2.9 (Lloyd and Orr, 1969), fig. 4.5 (Smith, 1993), fig. 9.4 (Vandermeulen, 1987), figs 11.5 and 11.6 (Payne, 1975), table 2.2 (Solbé and Cooper, 1976), table 2.3 (Williams *et al.*, 1984) and table 3.2 (Curtis and Curds, 1971); fig. 12.5 reprinted by permission from *The Chemical Industry. Friend to the Environment?* by Slater, D., Royal Society of Chemistry, Cambridge, 1992; The Royal Swedish Academy of Sciences for fig. 4.17 (Rast and Holland, 1988) and table 8.3 (Petersen *et al.*, 1986); E. Schweizerbart'sche for fig. 3.21

(Edward *et al.*, 1991) and fig. 11.2 (Mouvet, 1985); Springer-Verlag and the respective authors for fig. 6.3 (Czarnezki, 1985), and fig. 6.9 (Schuler *et al.*, 1985); UK Atomic Energy Authority for fig. 8.1; Water Environment Federation for fig. 11.1 (Cabelli, 1983); reprinted by permission of John Wiley & Sons, Inc. fig. 2.2 (*The Chemistry and Ectotoxicology of Pollution*, Connell, D.W. and Miller, G.J., 1984), fig. 3.25 ('Domestic and industrial pollution', Hammerton, D., in *The Freshwaters of Scotland*, Matthews, P.S., ed., 1994) fig. 4.1 (*Drinking Water Quality*, Gray, N.F., 1994), fig. 6.8 (*Ectotoxicology*, Ramude, F., 1987) © 1984, 1987 and 1994 John Wiley & Sons, Inc.; WRC for table 11.1 (Evans *et al.*, 1986).

While every effort has been made to trace the owners of copyright material, in a few cases this has proved impossible and we take this opportunity to offer our apologies to any copyright holders whose rights we may have unwittingly infringed.

Chapter 1

INTRODUCTION

The oceans hold some 97 per cent of the Earth's total resource of water, too salty to be used for drinking, irrigation or industry. Of the remaining 3 per cent, the vast majority is frozen in ice-caps and glaciers or is buried too deeply underground to be economically exploited. Only 0.003 per cent of the total volume of water is exploitable, though this is replenished by the hydrological cycle. Not only is fresh water essential to life, it is also clearly a relatively scarce resource, and is likely to become more so with the impacts of global warming and population growth: the human population, currently estimated at 5.66 billion, is predicted to rise to 10 billion by the year 2050.

Many freshwater resources are contaminated through human activities. Each day some 25 000 people die from their everyday use of water, dirty water being the greatest cause of death worldwide. Many millions more suffer from frequent and debilitating water-borne illnesses. About half of the inhabitants of developing countries do not have access to safe drinking water and 75 per cent have no sanitation, some of their wastes eventually contaminating their drinking supply. As well as causing much suffering, water-borne diseases also result in great economic loss. In India, for example, it is estimated that 73 million working days are lost each year, costing $600 million in lost production and health care (Lean *et al.*, 1990).

The United Nations launched the International Drinking Water Supply and Sanitation Decade in 1980. Its aim was to provide clean water and adequate sanitation for all by the year 1990. The improvement at the end of the decade was, at best, only marginal. Pollution must be seen as the gross misuse of an essential but scarce resource.

The dictionary variously defines pollution as the 'act of making dirty, defiling, contaminating, profaning, corrupting'. For our purposes such definitions are too broad to be useful. Streams flowing through deciduous woodlands receive large autumnal inputs of leaves which may deoxygenate the water and result in an impoverishment of the fauna. To the angler, wishing to catch trout, the leaves are pollutants, though the ecologist would consider deoxygenation as a normal seasonal feature in the dynamics of such streams.

To avoid the consideration of naturally stressed environments, definitions of pollution are often restricted to include only the effects of substances or energy released by humans themselves on their resources (Edwards *et al.*, 1975). The definition followed in the present book is that of Holdgate (1979):

The introduction by man into the environment of substances or energy liable to cause hazards to human health, harm to living resources and ecological systems, damage to structure or amenity, or interference with legitimate uses of the environment.

Pollutants may be derived from *point sources*, often discharges known to the authorities and readily amenable to abatement provided resources are available. Examples of point sources are discharges of effluent from sewage treatment works or of wastes from factories. Alternatively, sources of pollution may be *diffuse*, entering watercourses from run-off and land drainage. Fertilizers and pesticides applied to crops, and acid precipitation, provide examples of diffuse pollution.

Much pollution is *chronic* (or steady state), that is, the watercourse receives discharge continuously or regularly. Given the right legal framework and resources, such pollution can be reduced so that its impact on the aquatic ecosystem is acceptable. In much of the developed world, marked improvements in water quality have taken place over the last three decades. A greater problem now is that of *episodic* (or intermittent) pollution, which is unpredictable in both space and time. Heavy rainfall, releasing large amounts of acid from soils or causing sewerage systems to overflow, may result in pollution episodes. Accidents are a major cause, for example carelessness in handling wastes or the crashing of road tankers close to rivers (in one case concentrated orange juice from a crashed tanker caused a major pollution incident). Vandalism and the deliberate discharge of wastes also result in episodic events. Some recent examples from England that resulted in successful prosecutions in 1994 include the spillage of lime slurry by a gas company, of cyanide from a railway engineers, of raw sewage from a water company, of pig effluent from a farm and of caustic soda from a dairy. All resulted in large-scale loss of aquatic life.

Episodic pollution is of especial concern to water managers, for a single event can destroy years of careful, patient work in reducing the impact of pollution from known discharges. Of course, as chronic pollution is increasingly brought under control, the impact of single pollution events becomes much more apparent.

Four examples of water pollution

1. Pollution incident on the River Rhine

The River Rhine rises in the Swiss Alps and passes through Germany, for a time forming the border with France, and the Netherlands, to discharge into the North Sea some 1320 km from its source (Fig. 1.1). The river is navigable for much of its length and there are a number of major cities (Fig. 1.2). It re-

Fig. 1.1. River Rhine at Köln (photograph by The International Commission for the Protection of the Rhine against Pollution).

ceives several major tributaries, including the River Emscher which drains the industrial region of the Ruhr. The Rhine therefore receives many polluting discharges, but much effort has been put into reducing the level of contamination. Friedrich and Müller (1984) concluded of the Rhine that 'its ecosystem is on the way to recovery from very severe damage, but the possibility of future danger remains because of accidents associated with a river used so intensively'.

Such an accident happened on 1 November 1986. A fire broke out in a chemical warehouse near the Swiss city of Basel. Some 1300 t of chemicals were stored there, including 934 t of pesticide and 12 t of organic compounds containing mercury. Of this, some 30 t of chemicals, including mercury and organophosphorus pesticides, were washed into the Rhine during the fire-fighting operation. The slick of pollution also contained rhodamine, a red dye, which enabled its passage down the river to be monitored. Water intakes to German towns were closed and water had to be brought in by truck. The Rhine also serves two important drinking water reservoirs for the Netherlands which had to be protected, and the Dutch attempted quickly to channel as much as possible of the slick into the North Sea. In the Netherlands the slick was

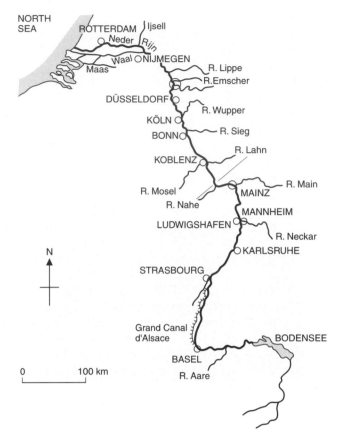

Fig. 1.2. Map of the River Rhine showing major tributaries and cities.

200 km long, with a mercury content of $0.22\ \mu g\,l^{-1}$, three times the normal level in Rhine river water.

Daphnia, an important planktonic component of the food chain in the river, were destroyed as the slick passed. Half a million fish were estimated to have died, including almost all the eels (*Anguilla anguilla*) for 400 km downstream of Basel. The river for 300 km downstream of the city was described as biologically dead.

In the same month 1100 kg of the herbicide dichlorophenoxyacetic acid leaked into the Rhine from a factory in Ludwigshafen (Fig. 1.2), requiring the shutting off of water intakes downstream, and up to 50 kg of the solvent chlorobenzol were spilled down a drain by another chemical works into the River Main. It was thought by environmental groups that such accidents were only being announced because of the Basel fire, and that they are a problem of regular occurrence in the river.

It was estimated at the time that it would take between three and ten years for the Rhine to recover from the Basel accident, though within a year the

International Commission for the Protection of the Rhine against Pollution (ICPR) announced that the river had practically recovered (see also Lelek and Köhler, 1990). Eel populations, however, were expected to take up to eight years to return to their former size. Pollutant residues in the sediments were also back to their former levels, which were, though, described as alarmingly high. The rapid recovery of the Rhine was made possible by the many side channels and tributaries of the river, which were unaffected by the pollution.

The Rhine represents an example of a chronically polluted river which suffers periodic episodes of pollution, frustrating long-term attempts at clean-up. The Basel accident resulted in the ICPR developing a Rhine Action Programme with four main aims to be achieved by the year 2000. Improvements in habitat and water quality must allow for the return of the native fauna, including the salmon (*Salmo salar*), once abundant but declining from the 1880s to extinction by 1940. The resident fish fauna is now quite diverse, though numbers and biomass are often low, and some 155 macro-invertebrate species have been recorded since 1989, though a number of species with specialized requirements are still absent (Tittizer *et al.*, 1994). The second aim is to guarantee a reliable supply of drinking water from the river. Thirdly, a substantial reduction in toxic chemicals, including in the sediments, must take place. Eels from the Rhine, for example, contain some of the highest PCB levels ever reported in fish (de Boer and Hagel, 1994). Finally, the North Sea must be protected from pollution. A summary of the action programme, the cost of implementation of which is huge, is provided by Schulte-Wülwer-Leidig (1995).

2. The Camelford accident

In contrast to the River Rhine, the River Camel, in Cornwall, southwest England, is small (Fig. 1.3). It rises on moorland and flows through woodland and agricultural land to enter tidal waters some 37 km from its source. Its main tributary, the River Allen, is 17 km long.

On 6 July 1988 a lorry arrived at Lowermoor Water Treatment Works with a load of 20 t of aluminium sulphate, which is used in the treatment process to flocculate suspended matter from water. The works was unmanned. The aluminium sulphate was destined for a storage tank, but the tanks were unmarked. The driver, who was unfamiliar with the site, opened an aperture with his key and pumped his cargo out of the lorry. The tank receiving the chemical was the wrong one and the aluminium sulphate went straight into the water main serving 22 000 people in the vicinity of Camelford.

In less than five hours complaints were being received over the quality of water coming out of the taps (Craig and Craig, 1989). Milk was curdling and the water stung lips and made fingers feel sticky. Over the next few days, before the cause of the problem was found, tap water was very acid (pH 4.2) and the concentration of aluminium was up to 4000 times the permitted level of

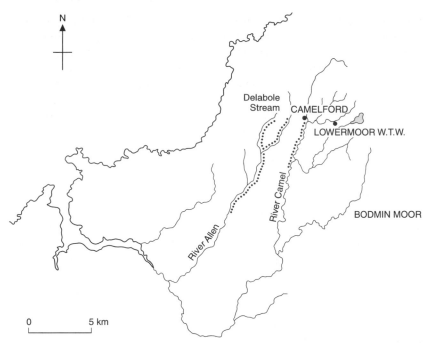

Fig. 1.3. River Camel catchment, Cornwall, showing the extent of the fish mortality (dotted line) in July 1988.

200 μg l^{-1}. Aluminium has been linked with diseases such as arthritis, brittle bone disease and Alzheimer's disease, a form of senile dementia. The acid water leached out metals such as lead, zinc and copper from the pipes. A large proportion of the population of the Camelford area complained of symptoms such as sore throats, nausea and vomiting, muscle cramps and joint pains, skin rashes and even of hair turning green! The water authority consistently under-played the seriousness of the accident, while at national level the relevant authorities remained inactive. The scientific periodical *New Scientist*, in an editorial on 21 January 1989, said of the Department of the Environment that 'its limp attitude to a major pollution incident was nothing short of a scandal'. It took almost six months for the Department of Health to set up a team of experts to investigate the long-term health effects of the incident, which were rightly causing great concern to those who had drunk the water. Their report (the Clayton Report) concluded that there would be no long-term health effects, though many who suffered symptoms remain sceptical.

In order to clear the public supply of polluted water, the authority flushed out the water distribution system overnight into the Rivers Camel and Allen. 'In the extreme and extended pressure of the incident, with immediate attention focussed on the water supply to customers, the potential effects on water-courses and aquatic life were not given any consideration' (Bielby, 1988). It was estimated that the average pH of the water was 4.5 and the aluminium concentration averaged $100 \, \text{mg} \, \text{l}^{-1}$, but was probably very variable. Many fish were found dead, covered with mucus and with bright red gills, characteristic of aluminium poisoning. Aluminium may be toxic to fish at concentrations as low as $0.1 \, \text{mg} \, \text{l}^{-1}$. The aluminium concentration in the gills of two fish sent for autopsy was 76 times higher than in control fish. The estimated total loss of salmon and brown trout (*Salmo trutta*) was between 43 000 and 61 000 and substantial losses of other, less important species also occurred. It was concluded, however, that the incident had no measurable effect on the invertebrate fauna of the river (Bielby, 1988). A detailed account of the development of this incident is provided by Rose (1990).

The Camelford example demonstrates how a single pollution episode can have a major impact on an otherwise clean river, as well as putting the health of the local population at risk. Unfortunately the River Camel suffered a further pollution episode in 1994 when farm slurry killed 100 000 trout, 500 in the river and the remainder in a fish farm fed by river water.

3. Recovery of the River Ebbw Fawr

The two examples above are of accidents, unpredictable events which had serious biological consequences for the rivers receiving pollution. This example presents a case of chronic pollution, with recovery once pollution control was enacted. In South Wales, the River Ebbw Fawr, some 40 km long (Fig. 1.4) flows through a large coalfield, with a major steelworks near its headwaters at Ebbw Vale (Fig. 1.5). The catchment is highly populated. The main river has been essentially lifeless for most of the past 100 years, though the biota has survived in the uppermost reaches of the river and in some tributaries. Three improvements to the river took place in the 1970s (Edwards *et. al.*, 1984; Edwards, 1989). The steelworks, which had been responsible for the acutely toxic conditions in the river, received a new and comprehensive effluent treatment system. Discharges of suspended solids to the river were reduced by improvements to the coal washeries and the provision of a large-capacity trunk sewer reduced discharges of untreated sewage during storm conditions.

Before 1972 there was no visible plant growth in the main river downstream of Ebbw Vale, and even sewage fungus (p. 76) was restricted to the middle and lower reaches. From 1972 areas of green algal growth were noticed in downstream sites, and diatoms, mosses and higher plants also appeared. In the most downstream sites some macroinvertebrates were present in the first year of the study (1970), especially oligochaetes (naidid and tubificid worms), chironomid midge larvae and mayflies, particularly *Baetis rhodani* (Fig 1.5). Although macroinvertebrates were abundant in the headwaters and could have colonized by drift, the site below Ebbw Vale had no invertebrates until 1973, when small numbers appeared. At no time during the study did the macroinvertebrate fauna downstream of the steelworks approach that of the headwater sites in diversity and abundance.

Fig. 1.4. An industrialized river, the Ebbw Fawr in South Wales (photograph by the author).

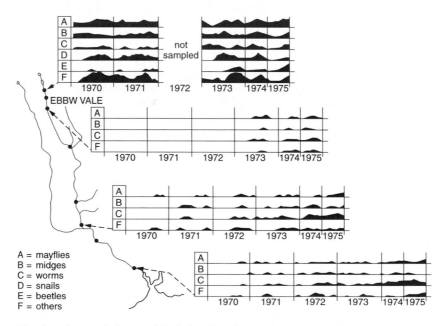

Fig. 1.5. Seasonal changes in relative abundance of macroinvertebrates in the Ebbw Fawr, 1970–1975 (from Edwards *et al.*, 1984).

By 1973 a marked improvement in oxygen concentration and pH enabled fish to colonize the lower reaches of the river, five species being present by 1975 (Fig. 1.6). Two more species had colonized by 1980 and further up-stream movements had occurred, and in 1986 eight species were present. Sticklebacks were recorded immediately below the steelworks, emphasising the improvement in water quality. Since the mid-1980s the river has been stocked with salmon and sea trout and there is evidence of spawning, though natural populations are low. Otters (*Lutra lutra*) were recorded in the lower catchment in 1991 for the first time in living memory. Fig. 1.7 illustrates the improvements in water quality in the Ebbw catchment over 20 years and shows what can be achieved in combatting chronic pollution, given co-operation between industry and the regulatory authorities and the provision of resources (over £22 million).

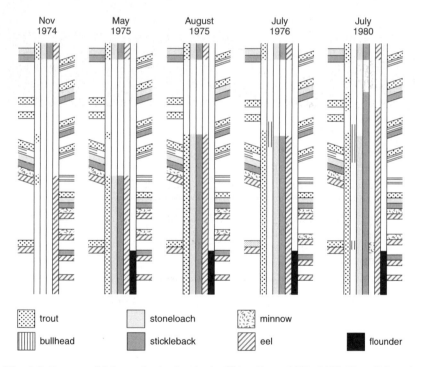

Fig. 1.6. Pattern of fish recolonization in the Ebbw Fawr, 1970–1980 (from Edwards *et al.*, 1984).

4. The decline of the otter

The otter (Fig. 1.8) is an amphibious, mammalian carnivore, feeding exten-sively on fish and therefore at the top of the aquatic food chain. The species was once widespread over most of Europe (Fig. 1.9a) but has declined over an extensive area, especially since the 1950s, and is now very restricted in range

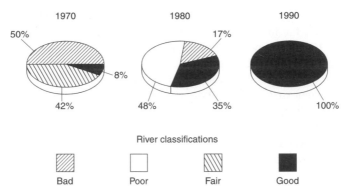

Fig. 1.7. River water quality classification of the River Ebbw catchment, Wales, 1970–1990 (from data provided by the National Rivers Authority).

Fig. 1.8. European otter (photograph by K. Bavinck, Otterstation *Aqualutra*, the Netherlands).

(Fig. 1.9b). Otters are now absent, for example, from much of southern and central England and most of central Europe. Even in Sweden, with its small human population and myriad rivers and apparently pristine lakes, the otter is almost extinct in the south and rare in the north. The best populations in Europe are closest to the Atlantic (coastal Norway, Scotland, Ireland, Portugal) and in east and southeast Europe. Otters are largely nocturnal and occur naturally at low densities, males patrolling home ranges of up to 40 km of waterway, so that the decline went largely unnoticed (Mason and Macdonald, 1986; Macdonald and Mason, 1994).

Several factors detrimental to the survival of otters, for example persecution and the destruction of waterside habitat (Mason and Macdonald, 1986), cannot

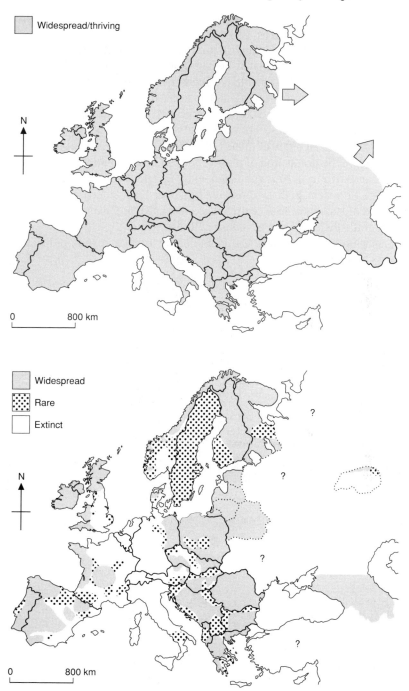

Fig. 1.9. Distribution of the otter in Europe (a) in 1950 (b) in 1992 (adapted from Macdonald and Mason, 1994).

explain the dramatic decline in range: the widespread introduction of a pollutant seems more likely. The 1950s saw the introduction of organochlorine pesticides, such as DDT and dieldrin, into agriculture, while during the same period the use of polychlorinated biphenyls (PCBs) in many industrial processes was increasing exponentially. Such compounds readily enter watercourses from agricultural run-off and industrial discharges to contaminate aquatic ecosystems. Organochlorines, especially PCBs, are also transported over wide areas by wind, to be washed into waterbodies with precipitation. Fig. 1.10 provides an index of industrial output for individual European countries, based on the production of plastics for 1983/84, and also shows the prevailing winds. If Fig. 1.10 is compared with Fig 1.9b it can be seen that otters are scarce or absent both in those regions with high industrial output *and* downwind of such industrial centres. The most intact populations of otters are upwind of the major centres of industry.

Organochlorines (pesticides and PCBs) are hydrophobic and lipid soluble,

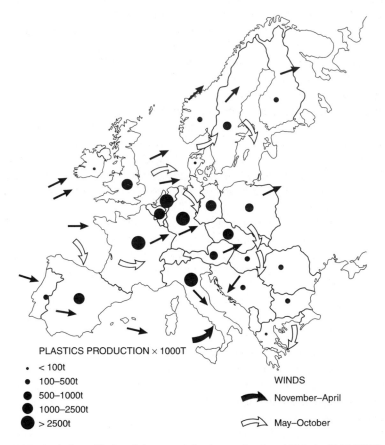

PLASTICS PRODUCTION × 1000T

- ·   < 100t
- •   100–500t
- ●   500–1000t
- ⬤   1000–2500t
- ⬤   > 2500t

WINDS

➤ November–April

➪ May–October

Fig. 1.10. An index of industrial output (plastics production, 1000 t in 1983/1984) and prevailing wind patterns over Europe.

so they rapidly pass from water to living tissue and biomagnify in the food chain. Concentration factors of PCBs from water to predators may be as high as ten million times (Tanabe *et al.*, 1984). Organochlorines are also highly persistent. In large concentrations they may directly kill otters under conditions of physiological stress, such as during periods of food shortage or pregnancy. In the long term, however, sublethal effects are probably more significant. In experiments with mammals it has been shown that PCBs cause disturbance of female reproductive function, affecting the reproductive tract, the neuro-endocrine system controlling puberty, oestrus and ovulation, and foetal and neonatal survival. Subsequent adult reproductive success may be affected by exposure in the uterus. There are also effects on male reproduction.

Experiments have been conducted on the effects of PCBs on the American mink (*Mustela vison*), a species in the same family as the otter and taking food from wetland habitats. Fewer young were born to females receiving 3.3 mg PCB kg^{-1} of food over 66 consecutive days and they weighed 28 per cent less than controls. Only 1.7 per cent of young survived to five days, compared with 82 per cent of controls. Females receiving 11 mg PCB kg^{-1} food produced no young at all. At death females taking the lower dose of PCB had tissue concentrations of 50 mg kg^{-1} in muscle fat. Further experiments have shown that reproduction in mink is inhibited on a daily intake of PCB of only 25.2 µg and a lower rate of reproduction is recorded on a daily intake of only 2.5 µg PCB (see review in Mason, 1989; Macdonald and Mason, 1994).

Figure 1.11 shows the concentration of PCBs in tissues of otters found dead in various regions of Europe, compared with that concentration known to cause reproductive impairment in mink. It is clear that animals from those populations that are declining or are endangered have mean PCB concentrations greater than 50 mg kg^{-1} tissue weight, though there is a wide variation between individuals. One animal from southern Sweden had nearly 1000 mg PCB kg^{-1} in muscle fat.

In eastern England a cub, born in the wild to a mother released as part of a restocking programme and killed by a lorry when not yet weaned at 11 weeks of age, had already accumulated 62 mg kg^{-1} in extractable fat in the liver. High levels of PCBs have also been found in otter spraints (faeces) in this region, indicating a disturbing general level of contamination with PCBs in the rivers (Mason and Macdonald, 1993). In the British Isles as a whole there is an inverse relationship between the regional distribution of otters (the number of sites found positive for the species in standard field surveys) and the average concentration of PCBs in spraints; high concentrations of PCBs are found in areas where otters are scarce and low concentrations where otters are widespread (Mason, 1995). In situations of PCB pollution we might expect a slow decline and contraction in range of otters as animals cease to reproduce, while themselves surviving to old age. This seems to be the case.

In seals in the Baltic Sea, which have exhibited poor breeding success over a number of years, many other symptoms have been described which are considered to be due to adrenocortical hyperplasia, resulting in hormonal

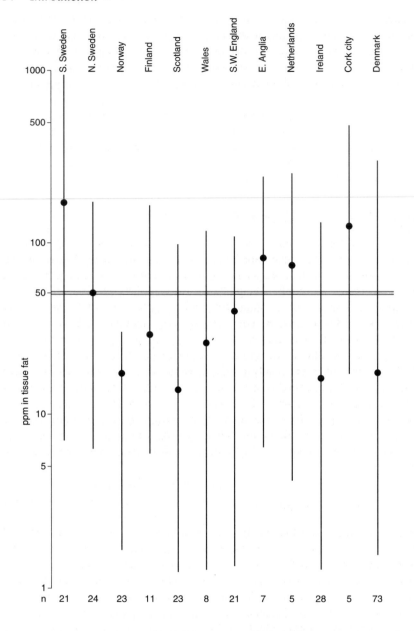

Fig. 1.11. Mean (and range) of PCBs (mg kg^{-1} in lipid) in tissues from otters from various European populations. The hatched line at 50 mg kg^{-1} indicates the tissue concentration of PCBs known to cause reproductive failure in mink. Numbers of samples are indicated at the bottom. Populations with means at or above the line are severely endangered, those below the line are stable or recovering following a decline (adapted from Macdonald and Mason, 1994).

imbalances and a breakdown of the immune system. The most likely cause is contamination with organochlorines, especially PCBs, and immuno-suppression facilitates the establishment of infections which would otherwise have no adverse effects. Similar symptoms have recently been described in otters from eastern England and Ireland (see review in Macdonald and Mason, 1994).

Those otters with high levels of PCBs were found in rivers where PCBs were not recorded by regulatory authorities using routine analytical methods on water. As far as the water industry is concerned there are no significant problems with PCBs, but it is quite clear that fresh waters are chronically polluted with PCBs over large areas of Europe.

Why need we be concerned about pollution?

From the examples given above it is obvious that pollution is important because humanity's resources are being damaged. When accidents occur the aquatic scientist can only assess the damage and suggest remedial action, such as restocking with fish, once the pollution has passed. The impact of accidents on the environment could often be minimized if hazard assessments were made of facilities dealing with dangerous chemicals and preventative measures taken. They rarely are.

As human activities cause pollution we should be able to control much of it. Pollution control is extremely costly, however, and the benefit in resource terms may be far outweighed by the cost of control. Furthermore, many rivers cross frontiers (as in example 1), and nation (or individual) 'A' might, by pollution, be damaging the resources of nation (or individual) 'B' rather than its own, and may be unwilling to reduce its profits to benefit its neighbours. Stringently applied laws, or a high degree of altruism, are required to control pollution.

Water supply and demand

In England and Wales the average domestic water consumption is about 140 l per person per day of which 32 per cent is used in toilet flushing, 17 per cent for baths/showers and 12 per cent in washing machines, the remaining 39 per cent being miscellaneous activities such as cooking, drinking, hand-washing, outside use in the garden, etc. Water is also used by industry and agriculture. Most of the rainfall in England and Wales falls in the hills of the west and north, and most of the residual rainfall (the difference between precipitation and evapotranspiration), which is potentially available for use, falls in the winter. Moreover, regions of high population and industry, for example London, the Southeast and the Midlands, are in areas of low rainfall and demand for water is greatest during the summer. Arable farming, with its requirements for irrigation, is also concentrated in the drier areas of the country. Water abstractions increased by 4 per cent in England and Wales between 1980 and 1990.

Abstractions for piped mains water increased by 13 per cent but the decline in industry and more efficient re-use of water partly offset this (Department of the Environment, 1992).

This discrepancy between water availability and water use means that resources may, at times, not be able to meet demand, as happened in the drought years of 1976, 1984 and 1989, when considerable restrictions on water use became necessary. There are similar problems of supply and demand in many other areas of the world: the Potomac River in Washington DC, for example, can no longer safely supply the demand for water in dry years. Evapotranspiration exceeds precipitation in large areas of the world, including many developing countries, where rational water management is of paramount importance.

The disparity between areas of high rainfall and areas where water is needed results in large quantities of water being abstracted for the public supply from lowland reaches of rivers. There is also considerable direct abstraction by industry and agriculture. *Such water must be of an acceptable quality.* Abstractions from the lower reaches of a river may be sustained by a controlled discharge from a reservoir, usually situated in the headwaters where rainfall is heavier, this being an economical way of transporting water to where it is needed. Water may also be transferred from one catchment to another which is short of water. Many rivers are being used as aquaducts in this way. Over-abstraction is an increasing problem and 40 rivers in England and Wales are suffering low flows, at least in part due to abstraction (Department of the Environment, 1992).

Effluent disposal

Domestic, industrial and agricultural users produce large quantities of waste products, and waterways provide a cheap and effective way of disposing of many of these. During dry weather, the flow of some rivers consists almost entirely of effluents. The effluents of some towns become the water supplies of other towns downstream. There is the well-known saying that the water coming from the taps in London has already passed through five sets of kidneys! It is therefore essential that the effluent discharged into a watercourse is of high quality and the degree of pollution is such that the self-purifying capacity of the river (p. 73) is not overloaded.

In addition to providing a source of water and a sink for effluents, freshwaters have an important amenity role, including such activities as boating, angling and wildlife studies. Some of these pastimes require water of a very high quality. The service industries associated with them are locally very valuable. In England and Wales, angling is by far the largest participation-sport in the country. There are also lucrative commercial fisheries for salmon, migratory trout and eels. It is also very desirable that the natural communities of animals and plants in freshwaters be maintained.

Finally, the resources of the seas are vast and most of the pollutants travelling down rivers will eventually end up there. Animals such as arctic seals and antarctic penguins are already loaded with pollutants, and we should not be complacent that the immense quantities of water in the oceans can absorb pollution indefinitely without effect.

Setting priorities for pollution control

There are, therefore, a number of reasons why the control of pollution is important. We do not, of course, have equal concern for all the components of our environment, and this is especially so when the enormous costs of pollution abatement are taken into account. The targets for pollutants can be ranked in order of decreasing concern to man (Holdgate, 1979):

Man $\rightarrow$ Domestic $\rightarrow$ Crops and $\rightarrow$ Most wildlife $\rightarrow$ Pests and
             livestock   structures   and amenity   disease
                                                                         vectors

Our prime concern is to reduce the risk of pollution affecting human health, whilst at the other extreme, if pollution kills organisms that are pests, then it might be considered positively beneficial. Some groups of organisms may be of fairly low priority in terms of human, or at least politicians', concern, but they may require water of an especially high quality. Nature reserves, for example, may need completely unpolluted water (though this is not necessarily so, it depends on what is being conserved), whereas grossly polluted water will suffice for shipping. Because of the efficiency of water treatment processes, water for potable supply need not be of the highest quality.

It would obviously not be economically feasible to clean all waters to such an extent that they would make pristine nature reserves, and economic considerations may make it unrealistic to improve the quality of some waters for recreation and fisheries. Increasingly the concept of maintaining water quality at a standard relating to the use to which that water is put is being adopted. Thus a classification system for water uses is constructed and water quality criteria for these uses are formulated (see p. 272).

What are pollutants?

In terms of the definition given at the beginning of this chapter almost anything produced by man can be considered at some time to be a pollutant. Indeed, to the farmer whose land is about to be lost under a new reservoir scheme, pure water is itself a pollutant in almost every sense of the definition given. Substances that are essential to life (e.g. copper, zinc) can be highly toxic when present in large amounts.

Some 1500 substances have been listed as pollutants in freshwater ecosystems, and a generalized list is given in Table 1.1. Some of the categories

Table 1.1   Categories of pollutants found in freshwater

Acids and alkalis
Anions (e.g. sulphide, sulphite, cyanide)
Detergents
Domestic sewage and farm manures
Food processing wastes (including processes taking place on the farm)
Gases (e.g. chlorine, ammonia)
Heat
Metals (e.g. cadmium, zinc, lead)
Nutrients (especially phosphates and nitrates)
Oil and oil dispersants
Organic toxic wastes (e.g. formaldehydes, phenols)
Pathogens
Pesticides
Polychlorinated biphenyls
Radionuclides

---

are not necessarily mutually exclusive. Domestic sewage, for example, may contain, in addition to oxidizable material, detergents, nutrients, metals, pathogens and a variety of other compounds.

Whether or not a compound will exert an effect on an organism or a community will depend on the concentration of that compound and the time of exposure to it (i.e. the dose). The effect of a pollutant on a target organism may be either acute or chronic. Acute effects occur rapidly, are clearly defined, often fatal and rarely reversible. Chronic effects develop after long exposure to low doses or long after exposure, and may ultimately cause death. Sublethal doses result in the impairment of the physiological or behavioural processes of the organism (e.g. it may grow poorly, or fail to reproduce). Its overall fitness is reduced.

At the community or ecosystem level it is unlikely that pollution will cause irreversible effects, except possibly in the case of radioactive pollution. The effects of pollution are recorded in the loss of some species, with possibly a gain in others, generally a reduction in diversity but not necessarily in numbers of individual species, and a change in the balance of such processes as predation, competition and materials cycling.

The generalized pathway of a pollutant from source to target is shown in Fig. 1.12. There are three important rate processes in the pathway: the rate of emission from the source of pollution; the rate of transport through the ecological system; and the rate of removal or accumulation of the pollutant in the pathway. The rate of transport will depend on the diffusion rate of the pollutant and on a variety of environmental factors as well as properties of transport within organisms in the pathway. The rate of removal or accumulation will depend on rates of dilution or sedimentation and on chemical and biological transformations. These determine the dose reaching the target organisms. Processes within the target will either transport the pollutant to where it exerts

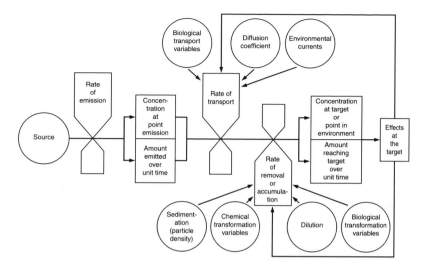

Fig. 1.12. A generalized pollutant pathway (from Holdgate, 1979).

an effect or will excrete the pollutant. We can see how this basic pathway can be repeated along a food chain. Also, with slight terminological changes, Fig. 1.12 can illustrate the basic pathway from entry point to site of action within a target organism.

The complexity of pollution

The following chapters will show the general characteristics and effects of various types of pollutants. Only rarely, however, is a single pollutant present in a watercourse. Normally an effluent will consist of a variety of potentially harmful substances. Most watercourses will receive a number of effluent discharges. The effects of these will often be difficult or impossible to disentangle.

Pollutants occurring together may act completely independently on a target, and the one exerting the greatest effect would then be the most important. One would not, for example, worry unduly about high levels of zinc in an effluent if the oxygen demand was so high that all life in the receiving stream was suffocated, but if the organic loading in the effluent was reduced such that the stream could support life, the concentration of zinc might then become important. The effects of pollutants might also be additive, antagonistic or synergistic (p. 29). These interactions will become apparent in later chapters.

It is worth remembering that the uses to which we put water (drinking water for ourselves and our livestock, irrigation, food processing, etc.) often have biological implications. The effects that are perceived in natural communities might be considered as an early warning system for the potential effects of pollutants on ourselves.

Chapter 2

AQUATIC TOXICOLOGY

A great variety of pollutants affect the majority of watercourses which receive domestic, industrial or agricultural effluents, and these complex situations become especially apparent when considering toxicity. Klein (1962) described the rivers draining the conurbations of Manchester and Liverpool in northwest England (Fig. 2.1) as containing 'waste waters from tanneries, fellmongers and leather dressers; food processing; rubber proofing; gas works; tar distilling; electro-plating; iron pickling; coal washing; sand washing; quarrying; oil and grease processing and refining; the scouring of cotton and wool; the bleaching, finishing and macerizing of cotton and rayon; the dyeing of cotton, wool, jute and rayon; piggeries; slaughterhouses; calico-printing; and from the manufacture of batteries, paint, light alloys, concrete, rubber, plastics, rayon, dyes, chemicals, glue, gelatine, size, paper pulp and paper'. It is small wonder that the watercourses in this area remain some of the most polluted in Britain and, in 1990, 1117 km of rivers in the northwest were still considered too polluted to support fish (National Rivers Authority, 1991). Some of the industrial origins of toxic compounds are listed above, but agriculture and forestry also add many toxic pollutants to freshwaters. These additions may be indirect, such as run-off of insecticides and herbicides applied to the land, while waste pesticides and their empty containers are frequently carelessly dumped into ponds or streams with unfortunate effects. Landfill sites and toxic waste dumps, many of them poorly managed and controlled, are a matter of great concern, especially with respect to groundwater contamination. Toxic chemicals are also used in the direct control of particular members of the freshwater community. The most widely used are herbicides to control water plants considered to be interfering with human use of freshwaters. Other organisms may be directly poisoned by the herbicide or may be indirectly affected by the change in community structure caused by the loss of plants. Insecticides are also applied directly to freshwaters, for instance to destroy larvae of mosquitoes, the vectors of malaria. In tropical Africa, DDT is added in large quantities to rivers to destroy larvae of the blackfly *Simulium damnosum*, the vector of the disease onchocerciasis or river blindness. Molluscicides (e.g. niclosamine) are widely used in the tropics to control the snail vectors of schistosomiasis (p. 56). In some countries piscicides are used to control fish. This

Fig. 2.1. Typical river landscape in the industrial northwest of England (photograph by the author).

may involve the elimination of entire fish communities (e.g. with rotenone), the elimination of selective groups of fish (e.g. with antimycin), or the control of particular species, such as larval sea lampreys *Petromyzon marinus* in the Great Lakes area of North America, using trifluoro methyl nitrophenol, TFM.

A generalized flow diagram of the effects of toxic pollutants on freshwater ecosystems is presented in Fig. 2.2.

Types of toxic pollutants

The major types of toxic pollutants can be listed as follows:

1. Metals, such as lead, nickel, cadmium, zinc, copper and mercury, arising from many industrial processes and some agricultural uses. The term

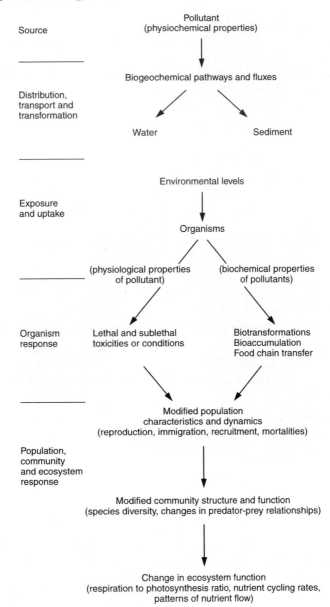

Fig. 2.2. Impact of pollutants on organisms and ecosystems (adapted from Connell and Miller, 1984).

'heavy metal' is somewhat imprecise, but includes most metals with an atomic number greater than 20, and excludes alkali metals, alkaline earths, lanthanides and actinides.

2. Organic compounds, such as organochlorine pesticides, herbicides, poly-

chlorinated biphenyls (PCBs), chlorinated aliphatic hydrocarbons, solvents, straight-chain surfactants, petroleum hydrocarbons, polynuclear aromatics, chlorinated dibenzodioxins, organometallic compounds, phenols, formaldehyde. They originate from a wide variety of industrial, agricultural and some domestic sources.
3. Gases, such as chlorine and ammonia.
4. Anions, such as cyanides, fluorides, sulphides and sulphites.
5. Acids and alkalis.

The most dangerous toxic compounds were listed by the European Community on a 'Black List', less dangerous substances forming the 'Grey List' (Table 2.1). The properties that influence the selection of 'Black List' chemicals include their toxicity, persistence and potential for bio-accumulation.

Some potentially toxic compounds, such as heavy metals, are continually

Table 2.1   'Black' and 'Grey' List compounds of the EC

List No. 1 ('Black List')

1. Organohalogen compounds and substances which may form such compounds in the aquatic environment
2. Organophosphorus compounds
3. Organotin compounds
4. Substances, the carcinogenic activity of which is exhibited in or by the aquatic environment
   (substances in List 2 which are carcinogenic are included here)
5. Mercury and its compounds
6. Cadmium and its compounds
7. Persistent mineral oils and hydrocarbons of petroleum
8. Persistent synthetic substances

List No. 2 ('Grey List')

1. The following metalloids/metals and their compounds:
   1. Zinc   2. Copper   3. Nickel   4. Chromium   5. Lead
   6. Selenium   7. Arsenic   8. Antimony   9. Molybdenum   10. Titanium
   11. Tin   12. Barium   13. Beryllium 14. Boron   15. Uranium
   16. Vanadium   17. Cobalt   18. Thalium   19. Tellurium   20. Silver
2. Biocides and their derivatives not appearing in List 1
3. Substances which have a deleterious effect on the taste and/or smell of products for human consumption derived from the aquatic environment and compounds liable to give rise to such substances in water
4. Toxic or persistent organic compounds of silicon and substances which may give rise to such compounds in water, excluding those which are biologically harmless or are rapidly converted in water to harmless substances
5. Inorganic compounds of phosphorus and elemental phosphorus
6. Non-persistent mineral oils and hydrocarbons of petroleum origin
7. Cyanides, fluorides
8. Certain substances which may have an adverse effect on the oxygen balance, particularly ammonia and nitrites

released into the aquatic environment from natural processes such as volcanic activity and weathering of rocks, and a number (e.g. copper, zinc) are essential, in small amounts, to life. Industrial processes have greatly increased the mobilization of many metals. Human-induced releases of tin, lead and mercury, for example, are respectively 110 times, 13 times and 2.3 times greater than geological mobilization. The rate of manufacture of organic compounds, such as pesticides, has increased exponentially since the 1950s. Global pesticide usage is currently in excess of 2.2×10^9 kg per annum.

Toxicity

Two general categories of toxic effect can be distinguished. *Acute* toxicity (a large dose of poison of short duration) is usually lethal, whereas *chronic* toxicity (a low dose of poison over a long time) may be either lethal or sublethal. There are a a number of words in regular use in the study of toxic effects:

acute coming speedily to a crisis
chronic continuing for a long time, lingering
lethal causing death, or sufficient to cause it, by direct action
sublethal below the level which directly causes death
cumulative brought about, or increased in strength, by successive additions.

There are also a number of terms which are used to express quantitatively the results of toxicity studies.

Lethal concentration (LC), where death is the criterion of toxicity. The results are expressed with a number (LC_{50}, LC_{70}), which indicates the percentage of animals killed at a particular concentration. The time of exposure is also important in studies of toxicity so that this must also be stated. The 48-hour LC_{50} is the concentration of a toxic material which kills 50 per cent of the test organisms in 48 hours.

Effective concentration (EC) is the term used when an effect other than death is being studied, for instance respiratory stress, developmental abnormalities or behavioural changes. The results are expressed in a similar way to lethal concentration (e.g. 48-hour EC_{50}).

Incipient lethal level is the concentration at which acute toxicity ceases, usually taken as the concentration at which 50 per cent of the population of test organisms can live for an indefinite time.

Safe concentration is the maximum concentration of a toxic substance that has no observable effect on a species after long-term exposure over one or more generations.

Maximum acceptable toxicant concentration (MATC) is the concentration of a toxic waste which may be present in a receiving water without causing harm to its productivity and its uses.

It is necessary to determine what concentrations of particular substances cause toxic or sublethal effects to a range of organisms so that standards for the protection of the aquatic environment can be developed. There are, however, so many chemicals being produced and released to the environment that only a small proportion are tested on a range of organisms. A method of predicting their likely effect is to use the **quantitative structure-activity relationship** (QSAR). On the basis of their physico-chemical properties, the biological activities (bioaccumulation, toxicity, persistence, degradation) of molecules can be predicted. The most frequently used measure of chemical structure is the octanol–water partition coefficient, which is a measure of the equilibrium distribution of the chemical dissolved in a two-phase system of water and octanol:

$$\log_{10} P_{ow} = \log_{10} (P_o/P_w)$$

where P_o and P_w are the proportions of the total solute in the octanol and water phases respectively.

Compounds with high P_{ow}s are fat-soluble, toxic, carcinogenic, bioaccumulable and persistent (e.g. DDT, $P_{ow} = 5.7$). Those with low P_{ow}s are water-soluble and relatively innocuous (e.g. alcohol, $P_{ow} = -0.3$).

Acute toxicity

Examples of acute toxicity curves for fishes are illustrated in Figs 2.3 and 2.4. Note that both axes are on a logarithmic scale. Figure 2.3 illustrates a curvilinear relationship, which has been observed for many toxic pollutants. The incipient LC_{50} can be obtained approximately as the asymptote and is about 25 mg l^{-1} for ammonia. Figure 2.4 shows a linear relationship with metals and there is an abrupt asymptote, the incipient LC_{50} being estimated at a concentration of 50 $\mu g\, l^{-1}$ for copper and 600 $\mu g\, l^{-1}$ for zinc. The incipient LC_{50} can be obtained more precisely using log-probit methods (Abel, 1989a). The incipient LC_{50} is a useful value in that it enables the toxicities of different pollutants to be easily compared and it forms a basic measuring unit for predicting the joint toxicity of two or more pollutants, as well as for describing sublethal effects.

A number of factors influence the responses of organisms in toxicity tests (Fig. 2.5). Different species vary in their vulnerability to specific pollutants. A comparison of the median lethal concentrations in hard water of three metals to the stone loach (*Noemacheilus barbatulus*) and rainbow trout (*Oncorhynchus mykiss*) is shown in Table 2.2. Stone loach were more sensitive to zinc than rainbow trout, but they were more tolerant of cadmium. Both species showed similar sensitivities to copper. The mayfly nymph *Baetis rhodani* was found to be 50 times more sensitive to phenol than the oligochaete worm *Limnodrilus hoffmeisteri*, and the shrimp *Gammarus pulex* was more than twice as sensitive as another crustacean, the hoglouse *Asellus*

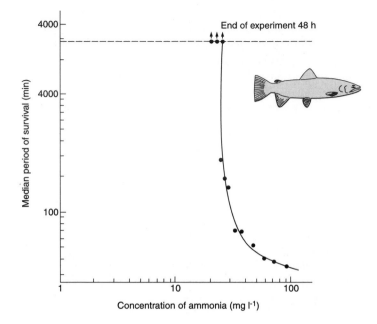

Fig. 2.3. A toxicity curve for trout in solutions of ammonia (NH_4Cl as $mg\,l^{-1}\,N$) at various concentrations (from Herbert, 1961).

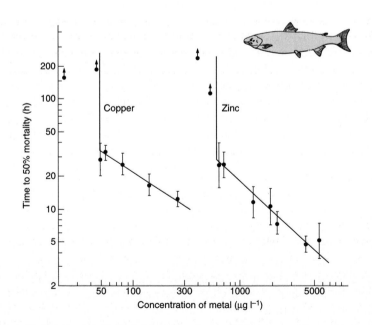

Fig. 2.4. Toxicity curves for salmon exposed to various concentrations of copper and zinc (from Sprague, 1964).

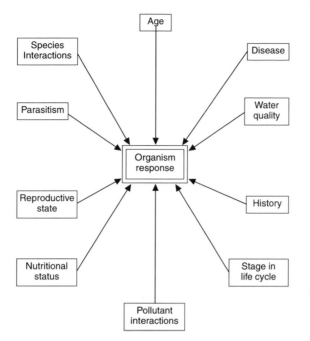

Fig. 2.5. Factors influencing the responses of organisms in standard toxicity tests (from McCahon and Pascoe, 1990).

Table 2.2 Median lethal concentrations (mg l^{-1}) of three metals to stone loach and rainbow trout (from Solbé and Cooper, 1976). Exposure time in days is in brackets

	Copper	*Cadmium*	*Zinc*
Stone loach	0.26 (63)	2.0 (54)	2.5 (5)
Rainbow trout	0.28 (119)	0.017 (5.5)	4.6 (5)

aquaticus (Green *et al.*, 1985). The pesticide lindane was the most toxic compound of four tested on the midge larva *Chironomus riparias*. By contrast copper was most toxic to *Gammarus pulex* after 96 h but lindane was more toxic over 240 h; two herbicides were less toxic to both species than either lindane or copper (Taylor *et al.*, 1991). Table 2.3 provides some comparative data on 96-hour LC$_{50}$ for cadmium, phenol and ammonia to several inverte-brate species. The sensitivity of individuals of a particular species to a pollutant may be influenced by internal factors, such as sex, age or size. For instance, the females of the crayfishes *Procambarus clarki* and *Faxonella clypeata* were much more tolerant of mercury than were the males. Females of both species survived throughout a 30- day exposure to 10^{-6} M mercuric

Table 2.3   Acute toxicity (96-h LC_{50}, mg l^{-1}) of cadmium, phenol and ammonia to several invertebrate species (from Williams *et al.*, 1984)

	Species	Cadmium	Phenol	Ammonia
Worm	*Limnodrilus hoffmeisteri*	2.9	780	1.92
Snail	*Physa fontinalis*	0.8	70	1.70
Shrimp	*Gammarus pulex*	0.03	69	2.05
Hoglouse	*Asellus aquaticus*	0.6	180	2.3
Mayfly	*Baetis rhodani*	0.7	15.5	1.7
Midge larva	*Chironomus riparius*	>200	240	1.65

chloride, whereas males suffered a 50 per cent mortality after only 3 days. Larger crayfish were better able to withstand high concentrations of mercury than small crayfish (Heit and Fingerman, 1977). Small *Asellus aquaticus* and *Gammarus pulex* were less tolerant of acidity than large individuals (Naylor *et al.*, 1990). In general the concentration of metals in invertebrates is inversely related to their body weight (van Hattum *et al.*, 1991).

These differences in the tolerance of poisons between individuals of a species make it dangerous to extrapolate from simple laboratory toxicity tests on a standardized organism to the field situation. The test organism may appear tolerant, but a particular stage in its life cycle may be especially sensitive, and this is the crucial stage in relation to the success of a population exposed to an environmental pollutant. The developmental or larval stages of an animal are generally more sensitive to toxic pollutants than adults (Mance, 1987; Kelly, 1988). Several studies have shown that eggs are more resistant to pollutants than larvae, probably because the cytoplasm contains all the necessary resources for cell division. In examining the acute toxicity of nine inorganic compounds to Arctic grayling (*Thymallus arcticus*), Coho salmon (*Oncorhynchus kisutch*) and rainbow trout it was found that alevins (larval stages) were less sensitive than juveniles (Buhl and Hamilton, 1991). In a review of toxicity studies, Woltering (1984) reported that the survival of larval fish was reduced in 57 per cent of studies, growth of larvae was inhibited in 36 per cent, reproduction in 30 per cent and the hatchability of eggs in 19 per cent. By contrast, adult survival was reduced in 13 per cent of studies and growth in only 5 per cent. In general therefore, early life stages are especially vulnerable.

Finally, the state of health may influence susceptibility to pollution, diseased or parasitized individuals succumbing more quickly. As an example, experiments were conducted in a Welsh stream to simulate acid events by adding sulphuric acid and aluminium sulphate. At a stretch downstream limestone was added to neutralize the acid, while an upstream stretch acted as a reference. Caged *Gammarus pulex*, half of which were parasitized with the intermediate cystacanth stage of the acanthocephalan (spiny-headed worm), *Pomphorynchus laevis* were added to each section. There was virtually no mortality in the reference section but animals quickly died in the experimental section, the time to death being significantly shorter in the parasitized animals. Mortality was reduced with the addition of lime but parasitized animals still suffered a significantly greater death rate (McCahon and Poulton, 1991).

Mixtures of poisons

Effluents are often complex mixtures of poisons. If two or more poisons are present in an effluent they may exert a combined effect on an organism which is *additive*. Alternatively, they may interfere with one another (*antagonism*), or their overall effect an an organism may be greater than when acting alone (*synergism*). A generalized scheme describing the combined effects of two pollutants is given in Fig. 2.6. In this scheme the concentration of one unit of pollutant A produces the response in the absence of B and one unit of B does the same in the absence of A. If, on combining the two pollutants, the response falls within the square, joint action is occurring, with the pollutants aiding one another. This joint action can be broken down into three special cases. If the response is produced by combinations represented by points on the diagonal (e.g. 0.5A + 0.5B), the effects are additive. If the response is produced by combinations falling in the lower triangle of the box (e.g. 0.5A + 0.2B) the effect is more than additive (synergistic), while if the response falls in the upper triangle (e.g. 0.8A + 0.7B), the effect is less than additive, though the pollutants are still working together in joint action. Antagonism occurs in the region outside the box, when, for example, more than one unit of A is required to produce the effect in the presence of B. Antagonistic effects make the comparison between laboratory studies and field conditions difficult because animals in the field which should theoretically be dead very often are not.

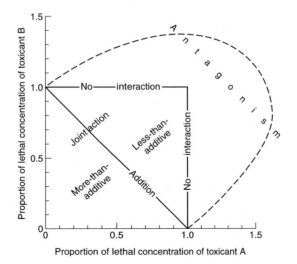

Fig. 2.6. Terms used to describe the combined effects of two pollutants (from Sprague, 1970).

An example of an additive interaction is the combined toxicity of zinc and cadmium to fish. Calcium is antagonistic to lead, zinc and aluminium. Copper is more than additive with chlorine, zinc, cadmium and mercury, whilst it decreases the toxicity of cyanide. Figure 2.7 illustrates the toxicity to rainbow

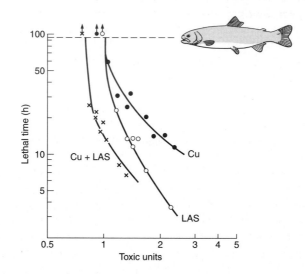

Fig. 2.7. Toxicity curves to rainbow trout of copper (Cu) alone, surfactant (LAS) alone and mixtures in various ratios of the two poisons (Cu + LAS). Lethal time to 50 per cent of animals in hours (h) and concentration in toxic units (from Calamari and Marchetti, 1973).

trout of copper and a surfactant (sodium laurylbenzenesulphonate, LAS) acting separately and in combination, and the toxicity can be seen to be markedly more than additive. It is generally considered that surfactants reduce the surface tension on the gill membranes, thus increasing the permeability of the gill to the surfactant and other poisons, but the physical effects are considered to be less important than the chemical effects of the surfactant.

Sublethal effects

Poisons are frequently present in freshwaters at concentrations too low to cause rapid death directly, but they may impair the functioning of organisms. These sublethal effects may be observed at the biochemical, physiological, behavioural or life cycle level. Many small changes have been related to pollution but it is essential to show that these changes have ecological meaning, that they reduce the fitness of an organism in its environment and are not merely within the organism's range of adaptation. The biochemical effects of pollution are basic, and these can then be related to the efficiency of tissues and organs, which can in turn be examined in relation to the performance of the organism and whether this has any adverse effect on the natural population (Sprague, 1971).

The early detection of specific molecular abnormalities in the tissues of organisms may provide an indication of exposure to pollutants long before any gross signs become apparent, and such biochemical indices may be valuable in signalling the development of sublethal abnormalities which could reduce the fitness of a population. The effects of salts of methyl mercury, cadmium and lead have been examined on six biochemical parameters in the brook trout (*Salvelinus fontinalis*), with exposure periods from six to eight weeks. Lead caused statistically significant increases in levels of sodium and chloride in blood plasma, with decreases in haemoglobin and glutamic dehydrogenase activity. Cadmium caused increases in plasma chloride and lactic dehydrogenase activity, while decreasing plasma glucose levels, whereas methyl mercury resulted in increases in haemoglobin and plasma sodium and chloride (Christensen *et al.*, 1977).

Among its other effects lead inhibits enzyme activity and acts at a large number of biochemical sites. Whitefish (*Coregonus* sp.), collected from a lead-polluted lake in northern Sweden, showed a marked inhibition of the activity of the enzyme alanine deaminase in erythrocytes. But despite a reduction in ALA-D activity of 87–88 per cent compared with whitefish from an unpolluted site, there was no consistent reduction in haemoglobin content or haematocrit value. Fish from the lead-polluted lakes also had higher blood glucose levels and lower blood plasma concentrations than fish from unpolluted sites (Larsson *et al.*, 1985).

Pollutants act at the biochemical level at a number of sites but an organism may be able to adapt by normal homeostatic mechanisms so that enzyme inhibition may not reduce its overall fitness. Enzyme bioassays remain, however, a useful technique in looking for sublethal effects of toxic pollution.

The effects of pollution on the genetic diversity of populations have been examined. Restriction fragment length polymorphisms were used to assess genetic variation in the mitochondrial genome of brown bullheads (*Aneiurus nebulosus*) from contaminated and clean sites in the Great Lakes region of North America. Genetic diversity estimates were always lower in bullhead populations from contaminated sites. Although these sites currently support good bullhead populations, it was considered that severe pollution events in the past had reduced numbers and eliminated much of the genetic diversity (Murdoch and Hebert, 1994). This reduced genetic diversity may compromise the long-term survival of populations in contaminated sites.

Figure 2.8 illustrates a generalized relationship between physiological impairment following increasing exposure to pollutants and the consequent disability of organisms. At low levels of pollution the organism is maintained in health by normal homeostatic mechanisms. As levels rise compensation occurs such that normal functioning is maintained without significant metabolic cost, but at higher levels still the organism becomes stressed and physiological breakdown occurs, the organism is unable to repair the damage and

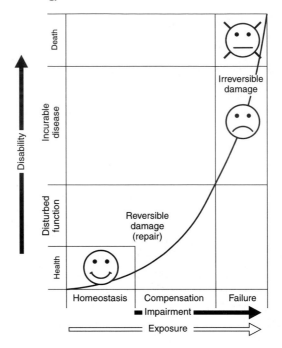

Fig. 2.8. Relationship between exposure to a pollutant and physiological impairment (after Fox, 1993).

becomes disabled. Still higher loadings of pollutants result in physiological failure and death.

The effects of toxic pollutants on the respiration of fishes and invertebrates have received widespread attention. By cannulating the blood system of fishes it is possible to measure the concentrations of oxygen, metabolites and pollutants and hence understand more fully the mode of action of toxic pollutants. Using cannulation techniques it was found that zinc reduced the oxygen level of blood leaving the gills, but zinc injected into the blood system had no effect. Zinc, therefore, reduces the efficiency of oxygen transport across the gill membrane so that the fish die of hypoxia (Skidmore, 1970). When the freshwater shrimp *Macrobrachium carcinus* was exposed to copper or zinc the respiration and ammonia excretion rates were reduced, resulting in a reduction in the ratio of oxygen to nitrogen in the tissues and increasing the animal's dependence on carbohydrate and fat reserves (Correa, 1987).

The effects of pollutants on the water balance of fishes were examined using an indirect method whereby the urine flow rate was measured in rainbow trout fitted with a urinary catheter. The results illustrating the effects of various concentrations of ammonia on urine production are shown in Fig. 2.9. Urine flow rate increased with an increase in the external ammonia concentration and all fish died at the highest concentration of $20\,mg\,N\,l^{-1}$. The urine flow rate decreased in the lower ammonia concentration after one day, suggesting that ac-

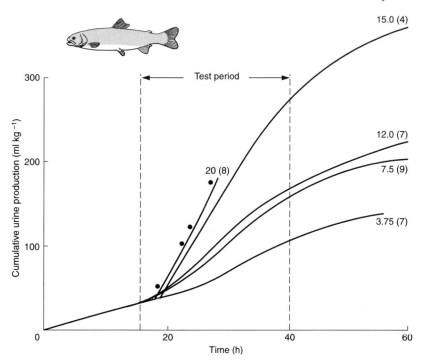

Fig. 2.9. Effect of concentrations of ammonia (as mg N l⁻¹) on urine production in rainbow trout. Ammonia concentrations are given against each curve, with the number of fish in brackets. The times for death at the highest concentration are shown (●). Temperature 10.5°C, pH 8.1 (from Lloyd and Orr, 1969).

climation was taking place, possibly by an increased rate of detoxification of ammonia. This might be dependent on an increased synthesis of glutamic acid, which is converted to glutamine in the presence of ammonia. The mechanism by which ammonia increases the water uptake by fish is unknown, though it could be related to increased activity, which may increase gill permeability.

Growth can be examined as an integrated measure of the sublethal effects of toxic pollutants on the biochemistry and physiology of organisms but, with homeostatic mechanisms operating, the effects on growth may be minimal. Bleached kraft mill effluent from the paper industry, for example, has been variously described as reducing growth, stimulating growth and having no effect on growth in different fish species. Nevertheless the *scope for growth* has been shown to be a good indicator of environmental stress in invertebrates. The scope for growth (or production) is the difference between energy intake and energy lost in metabolism and excretion. A pollutant may reduce the scope for growth by increasing energy expenditure or reducing energy intake, but the latter is more common. For example, the exposure of the freshwater shrimp *Gammarus pulex* to chlorinated ethers significantly reduced the rate of feeding, and hence the scope for growth, but had no effect on the rate of respiration

(Maltby, 1992). Growth was impaired by increasing copper concentrations in the same species (Maund *et al.*, 1992a). The pesticide lindane significantly reduced the growth of the midge larva *Chironomus riparius* and delayed the time to emergence (Maund *et al.*, 1992b). Because growth is easily measured and is an important indicator of the success of natural populations it should be recorded in laboratory studies of the effects of pollution.

Pollutants may cause changes in swimming performance, orientation, coordination of movement, aggression and comfort behaviour of fish. There was an increase, for example, in aggression in dominant members of a population of blue-gill sunfish (*Lepomis macrochirus*) when exposed to $0.034 \, \text{mg} \, l^{-1}$ copper (Henry and Atchison, 1986). A review of the effects of metals on the behaviour of fish is provided by Atchison *et al.* (1987) and Weber and Spieler (1994).

Reproductive behaviour may also be affected. *Gammarus pulex* engages in pre-copulatory pairing for up to ten days before mating, the behaviour allowing the male to protect the female from the attentions of other suitors. This pairing behaviour is disrupted by pollution (Poulton and Pascoe, 1990).

Animals may also avoid polluted water. This has been shown, for example, with Atlantic salmon (*Salmo salar*), where the avoidance of copper has been demonstrated in laboratory studies and verified by field observations (Atchison *et al.*, 1987).

Because the various life stages of an organism may be affected differentially by a toxic chemical it is necessary to study the species over its lifetime to find the weak link in its response to pollution. Such long-term experiments are essential in order to discover any carcinogenic or mutagenic effects of pollutants, or any teratogenic effects causing developmental abnormalities.

Environmental factors affecting toxicity

Environmental factors may modify the acute toxic effect of pollutants. Temperature is important because it not only influences the metabolic activity and behaviour of organisms, which may affect their exposure to a pollutant, but it may also alter the physical and chemical state of the pollutant. In general, toxicity increases with temperature, as is the case for metals (Felts and Heath, 1984; Khangarot and Ray, 1987). There are, however, many exceptions to the increase in toxicity with temperature. The time to death of rainbow trout exposed to phenol increased as temperature increased, but the LC_{50} decreased. Phenol causes paralysis and cardiovascular congestion, resulting in suffocation. The internal concentration of phenol is influenced by the relative rates of absorption and detoxification, both of which are directly proportional to temperature, but it is considered that temperature influences the rate of detoxification to a greater extent than the rate of absorption, at least at lower temperatures. Phenol therefore probably accumulates to higher levels at low temperatures, accounting for the greater toxicity in the cold (Brown *et al.*,

1967). Phenol was also less toxic at higher temperatures to *Asellus aquaticus* (Green *et al.*, 1988). It is considered that the rate of detoxification decreases more rapidly than the rate of absorption as the temperature is lowered. Similarly the accumulation of cadmium and copper, but not zinc, in *Asellus aquaticus* was dependent on temperature (van Hattum *et al.*, 1993).

The toxic effect of pollutants varies with the quality of water, pH and hardness being especially important. Hydrogen cyanide, for example, is especially toxic in the molecular form, so that any change in the pH that reduces the degree of dissociation will increase the toxicity of the solution without there being any change in the total concentration of cyanide. The toxicity of ammonia is also affected by pH (Smart, 1981).

The chemical speciation of some metals is markedly affected by pH. Metal 'species' can be grouped into three phases – an aqueous phase (free ions and dissolved complexes), a solid phase (particles and colloids) and a biological phase (incorporated into cells or adsorbed on to biological surfaces) (Gerhardt, 1993). Generally the ionic form of the metal is most toxic. Campbell and Stokes (1985) have described two contrasting responses of an organism to metal toxicity with a decrease in pH:

1. If there is little change in speciation and metal binding is weak at the biological surface, a decrease in pH will decrease toxicity due to competition for binding sites from hydrogen ions.
2. Where there is a marked effect on speciation and strong binding of the metal at the biological surface the dominant effect of a decrease in pH will be to increase metal availability.

Zinc and copper show the first response, lead the second. Reviews of the influence of pH on the uptake and toxicity of metals are provided by Spry and Wiener (1991) for fish and by Wren and Stephenson (1991) and Gerhardt (1993) for invertebrates.

An examination of the effect of hardness of water on toxicity is usually carried out after stabilizing the pH. Pollutants tend to be more toxic in soft waters (e.g. mercury, lead, copper, zinc). Figure 2.10 shows the relationship between the median survival time of rainbow trout and the concentration of zinc ions in waters of different total hardness. There was a linear relationship between survival and concentration of zinc in soft water, whereas the relationship was curvilinear at intermediate and high levels of hardness. Low calcium concentrations in water enhance the toxicity of metals because the permeability of the gill membrane to metals is inversely related to the aqueous calcium concentration. Calcium competes with other metal cations for binding sites on the gill surface. The fish is thus protected in hard waters because the direct uptake of metal ions is reduced. Metals (e.g. lead) may be precipitated in hard water conditions and soluble complexes may also be formed. The toxicity of zinc decreases with an increase in inorganic suspended solids as the metal is absorbed by or adsorbed on to the suspended particles.

A number of poisons become more toxic at low oxygen concentrations

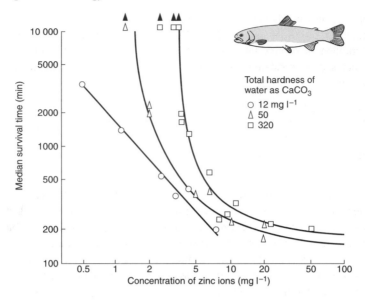

Fig. 2.10. Relationships between median survival time of rainbow trout and the concentration of zinc ions at three levels of hardness (after Lloyd, 1960).

because an increase in respiratory rate occurs, increasing the amount of poison the animal is exposed to. The toxicity of several poisons to rainbow trout increased in direct proportion to the decrease in oxygen concentration in the water (Lloyd, 1992).

The time of exposure to a pollutant also influences toxicity, and this may be important in episodic pollution events. *Asellus aquaticus*, immobilized by brief exposures to phenol, was able to recover if placed in clean water. The rate of recovery was influenced by exposure time, exposure concentration and temperature (Green *et al.*, 1988).

Mayer and Ellersieck (1988) have examined the effects of various factors on the toxicity of 410 chemicals (75 per cent of them pesticides) in nearly 5000 toxicity tests with various species of invertebrates and fish. Only 20 per cent of chemicals showed a change in toxicity with pH, but this nevertheless caused the greatest average change in toxicity of any factor examined. Hardness had little effect on the toxicity of organic chemicals. Temperature generally increased toxicity. Insects were usually the most sensitive group, followed by crustaceans, fishes and amphibians.

Neoplasms

A number of chemicals are known from laboratory experiments to be carcinogenic to fish, inducing neoplasms or cancers. These include aflatoxins, azo-compounds, nitroso-compounds, polynuclear aromatic hydrocarbons (PAHs), polychlorinated biphenyls (PCBs) and a number of pesticides. Neoplasms, especially of the skin and liver, are found in fish in the wild, and in some areas incidence may be quite high. Neoplasms have been divided into two broad categories in relation to exposure to toxic chemicals (Harshbarger and Clark, 1990):

a) not obviously associated with pollution – lesions in this category include haemic neural pigment cell, connective tissue and gonadal neoplasms;
b) associated with pollution – these include epithelial neoplasms of the liver, pancreas and gastro-intestinal tract.

Bottom-living fish are most likely to be affected by the second group, especially when they come into contact with contaminated sediments. Proving a link between contaminants and neoplasms in the environment is, however, extraordinarily difficult. Factors such as age, life history, feeding behaviour and interspecific responses act to confound relationships, as do the dose and duration of exposure to pollutants, which will be unique to a particular geographical location. Any polluted site is likely to have a number of compounds present that could induce neoplasms.

The most convincing studies relating neoplasms to environmental contaminants are from the marine environment, particularly Puget Sound in the western United States. Here PAHs, PCBs and DDT have been linked to liver lesions in English sole (*Parophrys vetulus*), white croaker (*Genyonemus lineata*) and starry flounder (*Platichthys stellatus*) (Myers *et al.*, 1994). Studies in freshwaters have been less comprehensive, but in the Great Lakes region of the United States PAHs appear to be responsible for neoplasms in brown bullhead (*Ictalurus nebulosus*) (Black and Baumann, 1991) and other species. Damage to the immune system by pollutants is likely to be a major cause of the susceptibility of fish to the development of neoplasms. A review of neoplasms in fish is provided by Metcalfe (1994) and by Moore and Myers (1994).

Transformations

Many of the compounds released into watercourses are subject to transformations within the environment, and this may render them more toxic. Mercury can be taken as an example, and some of its transformations are illustrated in Fig. 2.11. Inorganic mercury (Hg^{2+}) is converted to methyl (CH_3Hg^+) and dimethyl mercury in aquatic environments due to the activities of bacteria and fungi. The methylation process may occur under anaerobic (e.g. by *Clostridium*) or aerobic (e.g. by *Neurospora*, *Pseudomonas*) conditions. Several non-biological transformations of mercury also occur, depending on the environmental conditions.

Methyl mercury production in lake sediments is increased at low pH due to both increased methylation and decreased demethylation. Methylation is also

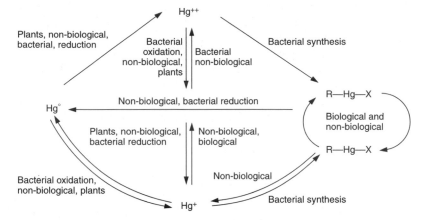

Fig. 2.11. Transformations of mercury (from Higgins and Burns, 1975).

stimulated by sulphate, for this increases the activity of sulphate-reducing bacteria which are important mediators of metal methylation (Gilmour and Henry, 1991). In lakes receiving acid rain the increased methyl mercury is accumulated in fish. Methyl mercury is exceptionally toxic to many animals.

Pesticides also undergo transformations in the environment, but these tend to be minor. For instance, aldrin is converted to dieldrin and DDT to DDE, but these products are still highly toxic. Chlorinated hydrocarbons (including PCBs) are very persistent in the environment and remain one of the great problems in environmental pollution (p. 177).

Tolerance

Populations can develop a tolerance to pollutants which enables them to survive in highly polluted environments. They may achieve this by functioning normally at high toxic loadings or by metabolizing and detoxifying pollutants. The mechanisms of tolerance to pollution are extremely complex, involving several metabolic systems, and species have solved the problem of tolerance to a particular pollutant in different ways.

In Welsh rivers polluted with heavy metals from old mine workings, the filamentous green alga *Hormidium rivulare* is highly tolerant of metals, and *Scapania undulata* is a highly tolerant moss. The moss *Fontinalis antipyretica* is less tolerant (McLean and Jones, 1975). In a wide-ranging survey of 47 sites with different levels of zinc, *H. rivulare* was found to be abundant at all sites, ranging from low zinc concentrations to as high as $30.2 \, \text{mg} \, \text{Zn} \, l^{-1}$. The closely related *H. fluitans* only occurred at sites with zinc up to a mean of $5.59 \, \text{mg} \, \text{Zn} \, l^{-1}$. *Hormidium* growing in streams high in zinc were more tolerant than populations taken from unpolluted waters, and this adaptation was genetically determined (Say *et al.*, 1977).

Non-tolerant isolates of the green algae *Scenedesmus* and *Chlorella* were subcultured into successively higher concentrations of nickel and copper and, after eight generations, the tolerance limits of both species had doubled (Stokes, 1975). Similarly the concentration of zinc,which was strongly inhibiting to growth of the cyanobacterium *Anacystis nidulans*, was raised from 2 to $15\,mg\,l^{-1}$ over 75 subcultures. Even after the tolerant strain was subcultured 20 times in the absence of zinc it quickly regained resistance, indicating that resistance was due to the selection of spontaneous mutants (Whitton and Shehata, 1982).

Some populations of the isopod *Asellus meridianus* are tolerant to copper and lead. Copper-tolerant animals were found to accumulate copper from solution and from food to levels which proved lethal to non-tolerant animals, while a lead-tolerant population accumulated concentrations up to $30\,\mu g\,Pb\,g^{-1}$, non-tolerant populations dying before tissue levels had reached $15–20\,\mu g\,g^{-1}$. The 48-hour LC_{50} for *Asellus* collected from a river contaminated with lead from mine-water was $3500\,\mu g\,l^{-1}$, whereas it was $280\,\mu g\,Pb\,l^{-1}$ for animals from a control river. Tolerance to copper in *A. meridianus* appears to confer tolerance to lead. The trace metals are stored in the hepato-pancreas of *Asellus* and copper and lead compete for sites, with lead being more readily bound (Brown, 1976; 1977; 1978).

After exposure for 24 hours to a copper concentration of $0.55\,mg\,l^{-1}$, rainbow trout showed a 55 per cent inhibition of sodium uptake and a 49 per cent reduction in affinity for sodium, which resulted in an overall decrease in total sodium concentration of 12.5 per cent. Within one week, however, the concentration had returned to normal. The rate of sodium uptake was still inhibited, but the rate of sodium loss was reduced. Copper accumulated in the liver and there was an increase in the concentration of a sulphydryl-rich protein (Laurén and McDonald, 1987a, b). The protein was thought to be a metallothionein; these low molecular weight proteins contain many sulphur-rich amino acids which bind and detoxify some heavy metals.

A previous exposure of an organism to low levels of a pollutant may make it more tolerant later. Pretreatment of rainbow trout eggs with cadmium gave protection to hatching larvae when they were exposed to cadmium (Beattie and Pascoe, 1978). Rainbow trout were pretreated at 0.001 and $0.01\,mg\,Cd\,l^{-1}$ and the 48-hour LC_{50} on later exposure to cadmium was 0.11 and $1.5\,mg\,Cd\,l^{-1}$ respectively. Control trout, not pretreated with cadmium, had a 48-hour LC_{50} of less than $0.1\,mg\,Cd\,l^{-1}$ when exposed later. It was suggested that pretreatment with low doses of metal stimulated the synthesis of metallothionein (Pascoe and Beattie, 1979).

Many bacteria quickly become tolerant to toxic pollutants, and the resistance gene is carried on plasmids. Plasmids are small, independent genetic elements which may be transferred from cell to cell by bacterial conjugation or by transduction (which involves transfer by bacterial viruses). The transferred plasmid replicates rapidly and is passed on to the progeny, so that resistance can spread very quickly. Resistant bacteria might enhance the transport of pollutants, especially metals, in the environment by mechanisms such as solubilization, concentration and conversion to organometallic species and their respective elemental forms, this constituting a potential environmental danger. Resistant bacteria can, however, also remove heavy metals from the environment and, with the application of modern techniques, genetically engineered microbes (GEMs) could be produced to play a valuable role in detoxifying industrial wastes. Reviews are provided by Gadd (1992) and Hardman *et al.* (1993).

Accumulation

It is necessary to distinguish clearly between the terms *biomagnification* and *bioconcentration* (or *bioaccumulation*). With biomagnification there are progressively greater amounts of contaminant along the food chain, carnivores containing greater concentrations than herbivores, which contain more than plants. Bioconcentration requires only uptake from the water and is independent of trophic level. Organochlorine pesticides have been shown to biomagnify along the food chain, but biomagnification is the exception for metals (Mance, 1987). Mercury is one such exception. Bioconcentration occurs with many toxic pollutants, very high levels being accumulated in organisms from very low levels in water. For example, in the moss *Fontinalis squammosa* the bioconcentration factor for lead is 20 000 and for zinc 22 000 (Say *et al.*, 1981). Bioconcentration factors for some organic compounds measured in the fathead minnow (*Pimephales promelas*) after 32 days of exposure are given in Table 2.4.

Table 2.4 Bioconcentration factors of organic compounds in the fathead minnow (*Pimephales promelas*) after 32 days exposure (from Veith *et al.*, 1979)

Compound	Bioconcentration factor
Lindane	180
Pentachlorophenol	770
Mirex	18 100
p,p′DDT	29 400
Aroclor 1260 (PCB)	194 000

The rate of accumulation of pollutants will depend on factors both external and internal to the organism. The concentration of pollutant in the water is clearly important, and many species carry higher loadings of pollutants when living in contaminated waters. For example, metal concentrations in algae and bryophytes are significantly correlated with concentrations in water. There appear to be no consistent correlations, however, between environmental levels of metals, other than mercury, and concentrations in invertebrates and fish (Kelly, 1988; Barak and Mason, 1989; van Hattum *et al.*, 1991).

Temperature influences the absorption, detoxification and excretion rates of pollutants (Green *et al.*, 1988), but not necessarily to the same extent, so that the overall bioconcentration may vary with temperature. An increase in the bioconcentration of cadmium, for example, was found with temperature in the stone loach (Douben, 1989).

Internal factors that influence bioconcentration include physiological condition. The concentration of lipophilic organochlorine compounds in different species of fish is directly related to the fat content of the fish (Sugiura *et al.*, 1978). Periods of fast growth are associated with a reduction in the level of contamination (Mance, 1987). Fish with higher metabolic rates also accumulate contaminants faster and, because feeding results in a higher metabolism, a greater uptake of pollutants across the gills may occur in feeding as opposed to starved fish, as has been shown for cadmium uptake in the stone loach (Douben, 1989). Age, sex and the presence of competing pollutants in the water may also influence accumulation rates.

The accumulation of a poison is a function of both uptake and elimination, and a generalized curve is shown in Fig. 2.12. Assuming uptake to be due solely to chemical diffusion, the process will continue until the internal level is equal to the level in the environment (point X in Fig. 2.12). Most pollutants can, however, be eliminated from the body, and this is an active biochemical and physiological process which cannot be described in simple diffusion terms. Pollutants may be oxidized in the liver by microsomal enzymes, the best known of which are the mixed function oxidases. These utilize NADPH (nicotinamide adenine dinucleotide phosphate) as a co-substrate. Two enzymes are involved in this type of oxidation: cytochrome P-450 and a flavoprotein, NADPH- cytochrome P-450 reductase, which are embedded in the phospholipid membranes of the endoplasmic reticulum. The flavoprotein transfers electrons from NADPH to P-450. This then inserts one atom of molecular oxygen into the toxic compound and reduces the second oxygen atom

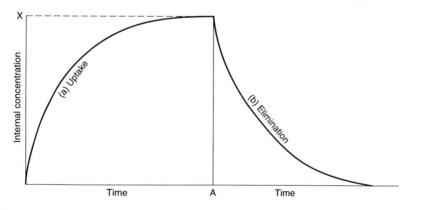

Fig. 2.12. A generalized curve for the uptake and elimination of a pollutant. The internal and external concentrations are equal at X, external concentrations are reduced to zero at time A (from Holdgate, 1979).

to produce water, using the hydrogen atoms from the reduced NADP. Multiple cytochrome P-450 isoenzymes occur in animals and these have different substrate specifications. The metabolites produced may be more or less toxic, but they are generally water-soluble and can be eliminated from the body. More details of the processes in fish can be found in Huckle and Millburn (1990) and Stegeman and Hahn (1994).

Conjugation or transformation may also occur to render compounds less toxic and more soluble. Both uptake and elimination of pollutants occur simultaneously and the plateau in Fig. 2.12 will depend on a balance between the factors determining the two processes.

A study of the bioconcentration of zinc by *Gammarus pulex* has led to the development of a kinetic model which may have more general significance (Xu and Pascoe, 1993):

$$BCF = Ca/Cw = K_1/K_2 + Co/Cw$$

where

Ca = concentration of the chemical in the animal's body at steady state following exposure $(mg\,kg^{-1})$

Co = basal level in the animal's body prior to exposure $(mg\,kg^{-1})$ (note that zinc is essential to life and hence naturally present)

Cw = concentration in water $(mg\,l^{-1})$

K_1 = rate constant of uptake (h^{-1})

K_2 = rate constant of elimination (h^{-1}).

Biomagnification can also be represented by a simple model (Spacie and Hamelink, 1985):

$$N_i = N_{iw} + f_i\,N_{i-1,w}$$
$$f_i = \alpha_i \cdot R_i/(K_{di} + G_i)$$

where

N_i = biomagnification factor for trophic level i

N_{iw} = bioconcentration factor for trophic level i

$N_{i-1,w}$ = bioconcentration factor for trophic level i−1

f_i = food chain transfer number for the consumer

α_i = assimilation efficiency of the contaminant from food R_i = weight-specific ration for feeding rate

K_{di} = rate of removal constant (d^{-1})

G_i = growth rate constant (d^{-1})

Contaminants may be taken up through the gut from the food or directly from the water. With invertebrates, uptake from water is generally the most significant (Gerhardt, 1993). In *Asellus aquaticus*, for example, the majority of cadmium was taken up directly from the water (van Hattum *et al.*, 1989), as was zinc in *Gammarus pulex* (Xu and Pascoe, 1993). There are exceptions, however. The burrowing mayfly *Hexagenia rigida* obtained most of its cadmium and zinc from sediment ingested as food (Hare *et al.*, 1991).

Field studies of toxic pollution

Toxic pollutants may affect ecosystems by killing out populations of organisms or by reducing their fitness, though sublethal effects have received little attention in field situations. The loss of one group of organisms can have serious repercussions on other groups. Toxic pollutants in sewage effluents, for example, may destroy those bacteria responsible for the biodegradation of organic matter, with the result that the oxygen sag curve may extend considerably further downstream than would otherwise be the case.

Willis (1985) studied populations of the leech *Erpobdella octoculata* above and below a discharge of mine waste on the Afon Crafnant, north Wales. Concentrations of zinc in the river below the mine were ten times greater than above the mine. Above the discharge the population density and life cycle of the leech were little different to those studied elsewhere. Below the discharge the density of leeches was much reduced and there was a delay in cocoon production and hence hatching of young leeches, while more mis-shapen and empty cocoons were recorded downstream. There was a lower proportion of juveniles at the downstream site. Toxicity tests in the laboratory showed that newly hatched leeches were more susceptible to zinc, being killed after a 50-day exposure to concentrations of $0.18\,\mathrm{mg\,l^{-1}}$. In higher concentrations of zinc fewer cocoons were produced and they took longer to hatch, with effects being recorded at concentrations of $0.1\,\mathrm{mg\,Zn\,l^{-1}}$ (Willis, 1989). The laboratory studies therefore strongly support the field observations that high zinc concentrations cause a reduction in the density of leeches, due largely to reduced reproductive capacity and poor survival of hatchlings. Without migration of leeches from above the mine it is likely that the population below the discharge would die out.

A study has been made of the fish community below a pulp and paper mill discharging effluent into the Mänttä watercourse in central Finland. The flow of the watercourse was small relative to the disharge of the effluent, resulting in a BOD (see p. 48) of $18\,\mathrm{mg\,l^{-1}}$ in the receiving water in 1977. In 1986 an activated sludge plant (p. 64) was built, and by 1990 BOD averaged $8\,\mathrm{mg\,l^{-1}}$. The effluent also raised phosphorus levels. Effluents from pulp mills can contain up to 200 compounds, including chlorophenolics, fatty acids and resin acids, which may be acutely toxic and bioaccumulative. Dehydroabietic acid (DHAA) has an LC_{50} to fish of $1\,\mathrm{mg\,l^{-1}}$.

The number of species of fish recorded at stations downstream of the discharge is shown in Fig. 2.13. In 1977 only roach (*Rutilus rutilus*) and ide (*Leuciscus idus*), two cyprinid species, were present in the first 5 km, and the population was very small (Fig. 2.14) due mainly to a lack of oxygen in the autumn. By 1990, following on-site effluent treatment, the number of fish species in the first 5 km had increased to nine, to include three species of bream (*Blicca bjoerkna*, *Abramis brama* and *A. ballerus*), pike (*Esox lucius*) and perch (*Perca fluviatilis*). There had also been a substantial increase in the biomass of the community. Nevertheless there was no significant recruitment

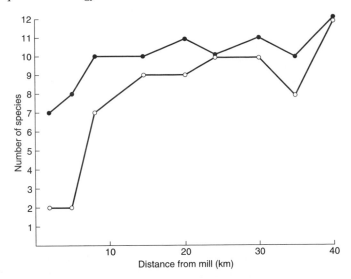

Fig. 2.13. Number of fish species at stations downstream of a pulp and paper mill before (○) and after (●) the installation of an effluent treatment plant (from data in Hakkari, 1992).

of young fish in the watercourse for 15 km below the mill, the population being maintained by immigration. Studies in Sweden of perch below bleached pulp mill effluents have revealed a 10 per cent incidence of malformed embryos, these having sharp bends in the posterior part of the spinal cord, and high mortality close to hatching, probably the result of acute toxicity (Karås *et al.*, 1991). In the Mänttä study area the fish caught in the first 15 km below the mill were strongly tainted (i.e., they tasted 'off' when cooked and eaten (p. 216)) in both 1977 and 1990. In the lower reaches, 60 per cent of fish were tainted in 1977 but none were in 1990. The installation of the effluent treatment plant at the mill had therefore led to some noticeable improvement in the fishery but it was still severely affected, with little improvement in recruitment for some distance below the mill, probably because of the presence of toxic wastes which were not removed by the treatment.

As well as observations on natural populations, caged fish (or indeed invertebrates) can be introduced into rivers to investigate toxic pollution. In Slate River, Colorado, *in situ* tests were conducted to explain the field distributions of fish in relation to a metal-contaminated tributary, Coal Creek (Davies and Woodling, 1980). Coal Creek had zinc and copper concentrations as high as $9.9\,\mathrm{mg\,l^{-1}}$ and $0.11\,\mathrm{mg\,l^{-1}}$ respectively and concentrations were still high, compared with an upstream control, at the lowest study site on Slate River, 15 km downstream of the confluence with Coal Creek. Upstream of Coal Creek, Slate River supported a thriving community of three salmonids, brown trout, rainbow trout and brook trout. Downstream, rainbow trout were recorded only occasionally, and only at the lowermost site, while populations

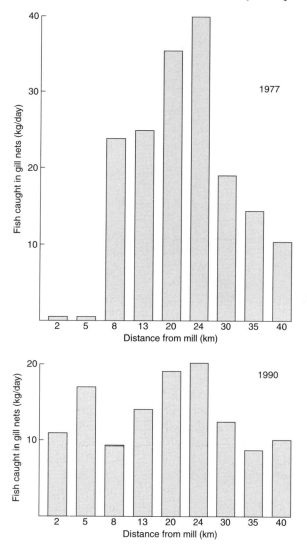

Fig. 2.14. Abundance of fish (kg gill-netted per day) at stations downstream of a pulp and paper mill before and after the installation of an effluent treatment plant (from data in Hakkari, 1992).

of brown trout were reduced ten-fold. Brook trout were the most abundant throughout, but did not occur in Coal Creek or at its confluence with Slate River. An over-abundance of gravid females 6.7 km below the confluence indicated that metal concentrations were preventing the upstream migration of brook trout to spawning sites. Tests with caged fish of all three species showed that rainbow trout were the most sensitive, mortality occuring at all sites except the control, explaining the virtual absence of this species below the confluence. All species suffered total mortality in Slate River 50 m below the

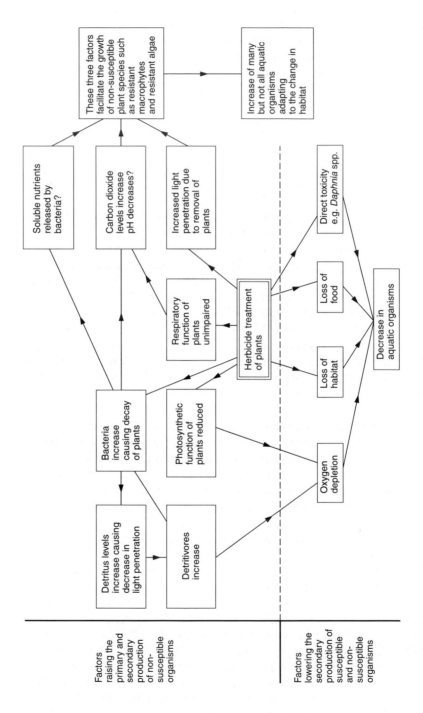

Fig. 2.15. Possible short-term ecological effects of herbicides in aquatic ecosystems (from Newbold, 1975).

confluence when river flow was low and metal concentrations high. Mortality did not occur in brown or brook trout when water levels were higher and metals were diluted. Metals were therefore likely to be only periodically toxic to wild populations of these two species.

The introduction to this chapter mentions that biocides are deliberately applied to watercourses to control undesirable animals and plants. Some of these poisons may be very specific, but the majority directly affect many more organisms than the target species, while resultant changes in the structure of food webs have still wider ramifications. Muirhead-Thomson (1987) provides a number of examples of the ecological effects of pest control programmes. Herbicides are especially widely used to control water plants that may impede the flow of water during the summer, when sudden heavy rain can cause flooding. The direct effect of herbicide addition is the loss of larger plants (macrophytes) and non-target organisms such as sensitive invertebrates and fish. Indirect effects include the death and decay of plants, with resultant changes in water chemistry and oxygen levels, the loss of habitat and the loss of food supply (Brooker and Edwards, 1975). Aquatic plants are often replaced by algae. The short-term ecological effects are illustrated in Fig. 2.15. Although herbicides may be acutely toxic to fish and invertebrates in laboratory studies, the concentrations in the field, when they are sensibly applied, are often too low to cause problems. The application of herbicides results in a rapid reduction in photosynthesis, with a consequent reduction in CO_2 uptake and oxygen output. The decomposition of large quantities of plant material may cause severe deoxygenation of the water, resulting in fish-kills, while the release of nutrients during decomposition may stimulate the growth of algae, causing blooms.

Bottom-living (benthic) invertebrates may increase in number following a herbicide application because of the increased detritus available as food, but those animals closely associated with macrophytes, such as molluscs, caddis and some chironomids, suffer a severe reduction in numbers, which may still be apparent in the following year after the plants have recovered. The change in invertebrates may affect the diet of predators and could eliminate specialists. The diet of eels changed after herbicide application to a reservoir, with benthic chironomids considerably increasing in importance (Brooker and Edwards, 1974).

It is clear that aquatic toxicology can be investigated from the subcellular to the ecosystem level. A review of molecular, biochemical and cellular perspectives is provided by Malins and Ostrander (1994). General reviews are provided by Abel (1989a, b) and Paasivirta (1991). Metal uptake, regulation and excretion in freshwater invertebrates are reviewed by Rainbow and Dallinger (1992) and accumulation by Timmermans (1992), while the papers in Newman and McIntosh (1992) discuss a range of issues in metal ecotoxicology. Maltby and Calow (1995) consider the methodology of ecotoxicology and the principles behind it, while Forbes and Forbes (1994) provide a critical review of ecotoxicological theory and practice.

Chapter 3

ORGANIC POLLUTION

Organic pollution occurs when large quantities of organic compounds, which can act as substrates for microorganisms, are released into watercourses. During the decomposition process the dissolved oxygen in the receiving water may be used up at a greater rate than it can be replenished, causing oxygen depletion, which has severe consequences for the stream biota. Organic effluents also frequently contain large quantities of suspended solids which reduce the light available to photosynthetic organisms and, on settling out, alter the characteristics of the river bed, rendering it an unsuitable habitat for many organisms. Ammonia is often present and this adds to toxicity.

A simple measure of the potential of biologically oxidizable matter for deoxygenating water is given by the *biochemical oxygen demand* (BOD). The BOD is obtained in the laboratory by incubating a sample of water for five days at 20°C and determining the oxygen used. It estimates the pollution potential of a waste water containing an available source of organic carbon by measuring the amount of oxygen used by the indigenous microorganisms in a standard sample. BOD provides a broad measure of the effects of organic pollution on a receiving water. Methods for determining BOD are given in American Public Health Association (1989), while the rationale, development and limitations of the test are discussed by Ellis (1989). Effluents with high BODs can cause severe problems in watercourses that receive them.

Organic pollutants consist of proteins, carbohydrates, fats and nucleic acids in a multiplicity of combinations. Organic wastes from humans and their animals may also be rich in disease-causing (pathogenic) organisms.

Origins of organic pollutants

Organic pollutants originate from domestic sewage (raw or treated), urban run-off, industrial (trade) effluents and farm wastes.

Sewage effluent is the greatest source of organic materials discharged to freshwaters. In England and Wales in 1990 there were over 4100 discharges releasing treated sewage to rivers and canals and several hundred more discharges of crude sewage, the great majority of them to the lower, tidal reaches of rivers or, via long outfalls, to the open sea (Department of the Environment,

1992). It is assumed that the sea has an almost unlimited capacity for purifying biodegradable matter. About 80 per cent of some 12 million cubic metres of sewage effluent produced each day in England and Wales is of domestic origin. Ninety-six per cent of this population is served by public sewers, and 80 per cent of the sewage produced receives secondary treatment. This removes 95 per cent of the polluting load before effluent is released to rivers, or 50 per cent before it is discharged to coastal waters (Department of the Environment, 1992). Similar levels of treatment are typical of much of the developed world. In the third world, however, where the capital for sewage treatment facilities is not available, the release of crude sewage into watercourses is still a major problem. I well remember, in the 1980s, in a restuarant in a small Algerian town, being presented with a carafe of water in which thick strands of sewage fungus were floating, and seeing women in Bombay washing clothes in an open sewer of black, stinking water. Almost one third of the disease in the world is the result of poor water supply and sanitation. The provision of wholesome water to populations of many millions of people remains an essential priority.

In urban areas, the run-off from houses, factories and roads can result in severe pollution, especially in storm conditions after periods of dry weather. This urban run-off may be routed through the sewage works (combined sewerage system), or it may be separately sewered and flow directly into rivers and streams. In the former case sewage treatment works may be severely overloaded during storm conditions and the effluent discharged may be of a much lower quality than under normal operating conditions, though the dilution is greater when the receiving river is in flood. Where urban run-off drains directly into rivers, pollution can be severe because drainage from the hard surfaces is so rapid that it may reach a river which has not yet increased above dry weather flow, so that the dilution will be minimal. Even in rural areas the volume of traffic on many highways is so great that run-off can cause significant local pollution in watercourses (Hamilton and Harrison, 1991).

With urban run-off there is often a first flush of highly polluting water, and much of this may come from the catchpits of roadside drains, which have thick bacterial scums and are usually anoxic. For example, in Saginaw, Michigan, the average first flush of storm sewage had a BOD of $370 \, \text{mg} \, l^{-1}$, which fell to $120 \, \text{mg} \, l^{-1}$ after 30 minutes and to $90 \, \text{mg} \, l^{-1}$ after one hour (Finnemore and Lynard, 1982). The constituents of urban run-off are obviously very variable. The effluent from urban run-off will be rich in dog faeces, suspended material, heavy metals and, seasonally, chloride from road-salting operations. The quality of the run-off is also highly variable, but BODs as high as $7700 \, \text{mg} \, l^{-1}$ have been recorded. Ellis (1989), Walesh (1989) and Viessman and Hammer (1993) provide more details on surface water run-off from towns.

Industrial effluents are a further source of organic pollution. These may be routed via the sewage treatment works or they may be released, with or without treatment, directly into a waterway. Amongst the industries producing effluents containing substantial amounts of organic wastes are the food processing and brewing industries, dairies, abattoirs and tanneries and textile and paper-making factories. Many trade effluents can be effectively treated by mixing with domestic sewage, but some can inhibit the microbiological activity in the treatment works, resulting in a poor quality effluent. These must be treated on the industrial site.

Farm effluents have become an increasing pollution problem, particularly with the intensification of livestock production in recent years. Over the decade from 1979 the average size of dairy herds in England and Wales increased from 46 to 320, and of pig herds from 200 to 345. The catchment of the River Torridge, a dairying area in southwest England, supports some 84 000 cattle, producing as much waste as 600 000 people. The human population of the catchment is little more than 15 000, most of them, in contrast to the cattle, connected to the sewerage system. Within the southwest region less than 25 per cent of a total of 3000 km of river had water of first class quality in 1985, compared with over 60 per cent in 1980. Most of this decline was due to farm pollution.

The increase in reported farm pollution incidents in England and Wales since 1979 is shown in Fig. 3.1. Over the period 1988–92, organic farm wastes accounted for 20 per cent of all pollution incidents in the United Kingdom. Data for Scotland for the period 1982–90 showed that 52 per cent of farm pollution incidents were caused by silage effluent and 28 per cent by livestock slurries. Chemical fertilizers, sheep dip, fuel oil and farm tips accounted together for about 10 per cent, the remainder being due to miscellaneous fac-

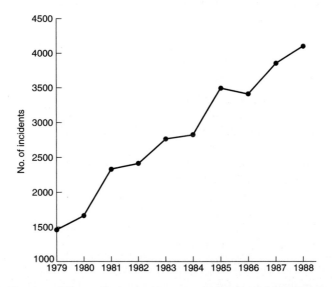

Fig. 3.1. Number of farm pollution incidents reported in England and Wales, 1979–88 (from data in Water Authorities Association, 1989). New regulations controlling the storage of silage, slurry and fuel oil on farms have led to a decline in incidents since 1988.

tors (Cole *et al.*, 1994). Undiluted farm slurry is 80–100 times more polluting, and silage liquor 200 times more polluting, than untreated sewage. In 1985 a tank burst, allowing some 250 000 l of pig slurry to flow into the River Perry in the West Midlands of England. The river completely deoxygenated, killing 10 000 brown trout (*Salmo trutta*) and 100 000 other fish. In August 1994, slurry from a dairy farm killed 100 000 mature trout in a fish farm taking its water from the River Camel in Cornwall. The more intensive and mechanized ways of dealing with vegetable crops may also cause pollution. For example, the mechanical vining of peas and especially the drainage from pea haulm silage can cause severe, though seasonal, problems. One such effluent had a BOD of 15 000 mg l^{-1}.

Farms have, therefore, led to an increase in the chronic pollution of rivers and streams over the last two decades, often in areas where water was previously of a high quality. There has also been an increase in accidents (episodic events) having major effects on fish and other river life. As farms are distributed over wide areas and are often remote, pollution incidents are difficult to detect and police.

Fish farming may also result in a deterioration of water quality downstream. Farms mostly rear trout, which require clean water, but fish are kept at high densities. Effluents leaving the farm are high in suspended solids, BOD, ammonia and nutrients from faeces and excess food. The effluent may also contain compounds used in fish husbandry, such as antibiotics, algicides and fungicides.

In general the origins of effluents and their composition are extremely diverse, and we can expect a similar diversity in the effects they have on receiving waters.

Pathogens

Diseases contracted from water kill some 25 million people, most of them children, each year, while many millions more are debilitated by waterborne diseases. Faecal contamination of water can introduce a variety of pathogens into waterways, including bacteria, viruses, protozoans and parasitic worms (Table 3.1). Waterborne diseases remain a major hazard in many parts of the world. Four classes of water-related diseases have been recognized (Cairncross and Feacham, 1983). Class 1, the true waterborne diseases, are contracted by drinking water that contains the pathogenic organisms, usually because of faecal contamination. Examples include cholera, typhoid and hepatitis A. Class 2 diseases are indirect infections associated with a lack of personal hygiene (e.g. hand washing) which can be reduced by providing adequate amounts of water for bathing and washing. To control such diseases it is necessary to provide people with *sufficient* water of *reasonable* quality; achieving a *high* bacteriological quality is a secondary consideration (Ellis, 1989). This second class includes all the diseases in class 1, together with

Table 3.1   Some water-related diseases and their causative organisms

Causative organisms		Disease or symptoms
Bacteria	*Salmonella typhi*	typhoid fever
	S. paratyphi	paratyphoid fever
	Salmonella spp.	gastroenteritis
	Shigella spp.	bacterial dysentery
	Vibrio cholerae	cholera
	Escherichia coli	gastroenteritis
	Leptospira icterohaemorrhagiae	Weil's disease
	Campylobacter spp.	intestinal infections
	Francisella tularensis	chills, fever, weakness
	Mycobacterium	tuberculosis
Viruses	Enteroviruses	many diseases, including poliomyelitis, respiratory diseases, meningitis, and infectious hepatitis
	Rotavirus	diarrhoea and enteritis
Protozoa	*Entamoeba histolytica*	amoebic dysentery
	Giardia lamblia	diarrhoea, malabsorption
	Naegleria fowleri	amoebic meningoencephalitis
	Cryptosporidium sp.	diarrhoea
Helminths	*Diphyllobothrium latum*	tapeworm infection
	Taenia saginata	tapeworm infection
	Schistosoma spp.	bilharzia
	Clonorchis sinensis	trematode infection
	Dracunculus medinensis	Guinea-worm

various other diarrhoeal diseases, some infections of eyes (e.g. trachoma) and skin (e.g. ringworm) and infections carried by lice and mites. Class 3 are diseases caused by helminths that spend part of their life cycle in water, while class 4 are diseases that require a water-related insect vector (e.g. yellow fever, malaria, river-blindness, filariasis), though these are not associated with polluted waters.

Bacteria

Bacteria of the genus *Salmonella* can cause acute gastroenteritis (i.e. food poisoning), with diarrhoea, fever and vomiting, and *Salmonella typhi*, the most virulent of the genus, causes typhoid fever. Some 200 immunologically distinguishable types (serotypes) of *Salmonella* are known to be pathogenic to humans and there are many more which infect animals, including livestock. Cross-infection between humans and animals can occur via water pollution. Currently there appears to be an increase in the spread of *Salmonella*, and this has been related to modern living conditions, such as mass food production (e.g. of poultry) and communal feeding. Salmonellae in rivers can survive for some distance downstream of the source. Large concentrations of gulls, feeding at rubbish tips and roosting on water-supply reservoirs, may excrete high numbers of *Salmonella* and faecal streptococci into the water.

Shigella species, especially *S. dysenteriae* and *S. sonnei*, are among the most commonly identified causes of acute diarrhoea in the United States. *Shigella* appears to be almost exclusively associated with humans and it is transmitted by the consumption of faecally contaminated water or food. If untreated, up to 15 per cent of the poorest populations may die of shigellosis.

Leptospirosis is an acute infection of the liver, kidneys and central nervous system and is caused by a group of spiral-shaped, motile bacteria, of which more than 100 serotypes are known. The primary hosts are rodents, which carry the organisms in their kidneys, and humans may become infected by wading or swimming in water contaminated with the rodents' urine. Most cases occur from infection through skin abrasions or from accidentally falling into contaminated water. Weil's disease is a particularly serious form of leptospirosis, and all workers likely to come into contact with sewage in the United Kingdom now carry special cards warning doctors of their potential exposure to the disease should they become ill.

Gastroenteritis and diarrhoea, especially in young children, may be caused by four different groups and various serotypes of *Escherichia coli*. Most urinary infections of adults are also caused by pathogenic *E. coli*, though urinary infection is usually by spread of *E. coli* from the person's own intestinal flora, rather than from water supplies. Nowadays people holidaying abroad often suffer from traveller's diarrhoea (Montezuma's revenge or Delhi belly), a common cause of which is enterotoxogenic strains of *E. coli*, or serotypes not previously encountered by the victims.

Since 1977 *Campylobacter* species have been increasingly recognized as one of the commonest causes of outbreaks of an intestinal infection which mimics *Shigella* dysentery. In young children it causes diarrhoea with obviously bloody faeces. Isolation of the organism, *C. jejuni*, requires special medium, which is why it was not discovered until relatively recently. The natural hosts are birds, particularly poultry, and the disease can occur in young dogs and cats. During the 1980s in the United Kingdom, notified cases of food poisoning caused by campylobacters increased from 10 000 to over 32 000, and the problem may be 100 times worse (Jones and Telford, 1991). Waterborne outbreaks have also been reported in the United States.

Vibrio cholerae causes cholera, an acute intestinal disease which can result in death within a few hours of onset. The El Tor vibrio lives on the gut wall and produces a protein exotoxin consisting of two subunits, A and B. The B subunit binds to receptors on the surface of cells of the gut wall, while the A subunit acts as a hormonal mimic, causing the cells greatly to increase their output of water and of sodium, bicarbonate and potassium ions. This fluid then leaves the body as 'rice-water' diarrhoea, which helps to disseminate the vibrios and also results in severe dehydration which can be fatal if untreated. Cholera is largely under control in countries where widespread sewage treatment is practised, but it is still rife in many parts of the world and outbreaks occur when disasters happen, e.g. famines, earthquakes and floods. A recent outbreak occurred in 1991 in Lima, Peru where provision of clean water and disposal of sewage have not kept pace with the population increase, from 3 million to 7 million since 1970. A cholera epidemic also broke out in 1994 in camps housing refugees from the war in Rwanda.

Mycobacterium tuberculosis is responsible for tuberculosis. Transmission by water appears to be uncommon, but this may in part be due to the long time between infection and the appearance of symptoms, which makes the origins of the disease difficult to trace in any particular incident. Tubercle bacilli are able to survive in water for several weeks.

Legionnaire's disease is so named after an outbreak in Philadelphia in 1976 when 183 delegates to a large military convention developed pneumonia and 29 of them died. Six months of intensive epidemiological and microbiological investigation was needed before the causitive organism, *Legionella pneumophila*, a previously unknown bacterium, was grown and the source of infection identified. The Legionellaceae occur naturally in pond water but the pathogenic variety colonizes the water in air-conditioning cooling towers, condensers, bathroom shower-heads and other man-made appliances that allow it to flourish. It becomes an opportunistic pathogen when minute infective water droplets from these sources are inhaled by unsuspecting individuals. It can also survive and multiply within amoeboid protozoa, which may then act as vectors for the disease. The disease can be prevented if water systems in large buildings are regularly cleaned, the water continuously chlorinated and hot water supplies maintained above 55°C. About one third of the 200 cases diagnosed annually in Britain are contracted abroad.

Viruses

Viruses seem to survive in water for much longer than faecal bacteria and there is increasing concern over their occurrence and control in sewage effluents and in water supplies. Most of the hundred or so species of enteroviruses (viruses which infect the digestive tract) are not destroyed by chlorination. In some large cities the recycled drinking water may come from rivers that have been contaminated by treated sewage. Such drinking water is bacteriologically safe, but virus contamination may persist. Up to 10 million viruses per gram of faeces have been recorded, and they may be shed over long periods. Polio virus, for example, may be excreted for 50 days (Gray, 1994). Animals such as dogs also carry viruses that are pathogenic to humans so that storm water flows, especially from paved areas, may be another source of contamination of freshwaters.

Infectious hepatitis (jaundice) is endemic in many countries and, of the two main kinds, hepatitis A is transmitted by contaminated water. Like cholera, it thrives during disasters, such as floods, or in refugee camps. Hepatitis A virus is relatively resistant to the chlorine used in the water treatment process. Some 87 per cent of all waterborne viral outbreaks in the United States are caused by hepatitis A.

Poliomyelitis virus may occasionally be carried by water but most transmission is by personal contact. Enteritis can be caused by Coxsackie and ECHO viruses and outbreaks may develop very rapidly if water supplies become contaminated with untreated sewage.

Contaminated water can be involved in the transmission of rotaviruses, which are a major source of diarrhoea in both adults and children. It is thought that up to 50 per cent of cases of acute diarrhoea in children under two years old are caused by rotaviruses, resulting in some 6 million deaths annually in developing countries. They may also have a role in traveller's diarrhoea in adults.

Reviews of viruses in water have been provided by Gerba and Rose (1990) and Gerba *et al.* (1991).

Protozoa

Amoebic dysentery, caused by the parasitic protozoan *Entamoeba histolytica*, is endemic in the tropics and subtropics and the contamination of drinking water by amoebic cysts from faeces is a major transmission route. The trophozoite, a stage in the life cycle that causes the disease, may also get into the lymph and blood vessels to be carried to the liver, lungs and brain. It may cause bowel ulcers and liver abscesses. Chronic infection results in recurrent episodes of diarrhoea, accompanied by blood and mucus, alternating with constipation. Other protozoans causing waterborne disease include *Giardia lamblia*, which results in diarrhoea and a reduced absorption of nutrients and vitamins across the gut wall, and a pathogenic strain of *Naegleria fowleri*, which is contracted while swimming in the warm waters of small lakes, swimming pools or polluted estuaries and causes a rare, but fatal, amoebic meningoencephalitis.

A pathogenic protozoan which appears to be on the increase is *Cryptosporidium*. The organism multiplies in the intestine, causing stomach pains, vomiting, diarrhoea and fever. The illness usually abates within three weeks but certain groups, including children, the elderly and immuno-compromised individuals suffering from diseases such as leukaemia and AIDS may be affected much more severely. At present there is no known cure. The parasite lives in the gut of livestock and passes with the faeces into the environment where the spores (oocysts) can survive for up to 18 months. The intensification of livestock rearing, with attendant problems of waste disposal, may be one cause of the increase of *Cryptosporidium*. A recent survey in the United States revealed that up to 77 per cent of surface waters examined contained the pathogen. An outbreak of the disease in Georgia affected 13 000 people and in Britain the number of cases increased sixty-fold between 1983 and 1986, with two major outbreaks since 1988. A single oocyst may be sufficient to initiate an infection (Blewett *et al.*, 1993). *Cryptosporidium* is resistant to chlorine and standard tests used by water authorities to monitor drinking water cannot detect it. To provide a sample, 1000 litres of water need to be centrifuged, and the organism can then be identified using genetically engineered monoclonal antibodies.

Parasitic worms

The parasitic worms include the phyla Nematoda (roundworms) and Platyhelminthes (tapeworms and flukes). The ova of these parasites are passed out in the faeces and urine, and are often resistant to sewage treatment processes. A tapeworm can produce up to 1 million eggs in a day. The majority of life cycles involve an intermediate host. In the case of the beef tapeworm, *Taenia saginata*, cattle, the intermediate host, may become infected by grazing on pastures sprayed with sewage sludge or by drinking water contaminated with sewage.

Dracunculus medinensis, the Guinea-worm, is prevalent in tropical regions. The female worm can be up to 1.3 m long and lives under the skin of the legs. The worm produces a blister near the ankle and when immersed in water this bursts to release large numbers of larvae. These infect copepods and the cycle is completed when people drink unfiltered water containing the copepods.

Schistosomiasis (bilharzia) in humans is mainly caused by one of three species of blood fluke: *Schistosoma mansoni*, particularly in Africa and South America, *S. haematobium* in much of Africa, and *S. japonicum* in the Far East. The last species also occurs in animals. The disease is of particular importance in that 300 million people, mainly in the tropics, are infected. It is a debilitating illness which, though not often fatal, causes general weakness in the sufferer and results in an annual economic loss of hundreds of millions of dollars. Eggs of the flukes pass out with human faeces or urine and, if they reach fresh water, develop into miracidia larvae which infect snails. Cercariae develop in the snails which, on leaving those hosts, penetrate the skin of humans wading in the water. After extensive migration in the body each parasite settles and produces ova in the venous system that the particular species frequents (e.g. *S. haematobium* in the bladder and urinary tract, *S. mansoni* in the portal blood system of the gut and liver). The late clinical features of these three pathogens depend on the site and injurious effects of the different types of ova. A single infected person who is passing parasite eggs daily in faeces or urine could contaminate a whole river if there are plenty of appropriate snails to perpetuate the disease.

The incidence of bilharzia appears to be influenced by changes in land use. With the filling of Lake Nasser behind the Aswan Dam in Egypt the surrounding land is now irrigated by canals and ditches rather than by seasonal flooding. This has led to the build-up of large populations of snail hosts, which now have a permanent rather than a seasonal habitat, and the incidence of bilharzia has greatly increased. Forest clearance and agricultural development in Africa lead to the appearance of *Schistosoma haematobium*, because the new conditions allow its snail host *Bulinus rohlfsi* to spread.

Bitton (1994) provides a review of the health risks associated with sewage discharges. To avoid waterborne diseases it is obviously essential that as many pathogens as possible are removed from sewage before it enters watercourses.

Sewage treatment

Historical

Until the last 200 years or so, the deterioration of watercourses due to organic pollution was not a serious problem because a relatively small human population lived in scattered communities. The natural self-purification properties of rivers could cope with the waste dumped into them. Introduced pathogenic organisms, by contrast, were a severe problem, though the link between disease and faecal contamination had of course not yet been made.

Water pollution became a severe problem with industrialization, coupled with the rapid acceleration in population growth. Industrialization led to urbanization, with people leaving the land to work in the new factories. Domestic wastes from the rapidly expanding towns and wastes from industrial processes were all poured untreated into the nearest river.

The historical pollution of the River Thames has been described by Wood (1982). Though pollution in the Thames at London was recorded as early as the thirteenth century, it was not until the eighteenth century that problems became acute. The population of London doubled between 1700 and 1820 to 1 250 000. Sewage was collected in individual or communal cesspits, which were periodically emptied to fertilize the surrounding land. In 1843 main sewers were laid and 200 000 cesspits were abolished. The sewage drained directly into the Thames, causing gross pollution and epidemics of cholera, whilst noxious odours rising from the river periodically disrupted the work of Parliament and the Law Courts, 1858 being known as the 'Year of the Great Stink' (Fig. 3.2).

In 1865 a scheme designed by Sir Joseph Bazalgette diverted the sewage, via three main sewers, to outfalls situated 10 miles below London Bridge, where it was discharged untreated on the ebb tide. The river through London showed some improvement, though sewage and industrial effluents from many other discharges continued to add to the pollution. The situation around the outfalls downstream of the capital was atrocious.

During the latter part of the nineteenth century a Rivers Pollution Committee recommended various treatment processes, and in 1882 a Royal

FARADAY GIVING HIS CARD TO FATHER THAMES;

And we hope the Dirty Fellow will consult the learned Professor.

Fig. 3.2. The 'Year of the Great Stink', 1858 (from the magazine *Punch*).

Commission on Metropolitan Sewage Disposal was set up, eventually concluding that the suspended solids in sewage should be separated from the liquid before it was discharged into the river. At this time Mr W.J. Dibdin put before the Commission the idea that sewage could be treated biologically, though this was not taken further at the time. The solids in sewage were separated at the outfalls from 1889 and the resulting sludge was dumped at sea.

Dibdin's ideas on bacteriological treatment of sewage were researched further throughout the 1890s, and the first treatment plant was put into commission in 1914 – though in Manchester, not London. Three experimental plants for biological treatment were installed downstream of London in 1920 and a major plant was erected in 1928, with greatly expanded facilities provided since then. A marked improvement in the quality of the water in the River Thames became apparent during the 1960s and a diverse flora and fauna, absent for many years, began to colonize the river (see p. 89).

Similar stories could be told of cities throughout the world.

The basic treatment process

There are three objectives in the treatment of sewage: to convert sewage into suitable end products, that is an effluent that can be satisfactorily discharged into the local watercourse and a sludge that can be readily disposed of; to carry this out without nuisance or offence; and to do so economically and efficiently.

There are four stages in the treatment of sewage but, depending on the quality of the effluent required, not all the stages may be used:

Preliminary treatment, which involves screening for large objects, maceration and removal of grit, together with the separation of storm flows;

Primary treatment (sedimentation), where the suspended solids are separated out as sludge;

Secondary (biological) treatment, where dissolved and colloidal organics are oxidized in the presence of microorganisms;

Tertiary treatment, which is used when a very high quality effluent is required. It may involve the removal of further BOD, bacteria, suspended solids, specific toxic compounds or nutrients.

The amount of treatment that is provided will depend to some extent on the amount of dilution that is available in the receiving water and on the quality objectives for that water. Preliminary treatment only may be given for effluents being discharged out to sea, whereas tertiary treatment may be required if water is abstracted for potable supply downstream of a discharge. In Britain the minimum requirement for an effluent, which can be achieved with secondary treatment, is the Royal Commission Standard, allowing no more than $30 \, mg \, l^{-1}$ of suspended solids and $20 \, mg \, l^{-1}$ BOD (a 30:20 effluent). To achieve these standards, the Royal Commission envisaged that the effluent would be diluted with eight volumes of clean river water having a BOD of $2 \, mg \, l^{-1}$. Such a dilution may not always be available, so that a more stringent standard than $30 : 20$ may be required for the effluent.

A modern sewage treatment works is illustrated in Fig. 3.3, though the majority of works serve populations much smaller than that of London. The basic organization of a sewage works is shown in Fig. 3.4.

Preliminary treatment

The sewage is initially passed through screens (rows of iron bars with a spacing of 5–10 cm), which removes large debris such as wood, paper and bottles. The screens are operated automatically and the screenings are either burnt or macerated and returned upstream of the screens. Grit and small stones are removed either by passing the sewage along a constant velocity channel or through a grit chamber.

Fig. 3.3. An aerial photograph of the Crossness sewage treatment works serving 1.7 million people in 240 km² of London (photograph courtesy of Thames Water). Primary sedimentation tanks can be seen in the foreground, followed by activated sludge tanks, final sedimentation tanks, and at the back secondary digestion tanks, with primary digestion tanks central. The building in the centre of the photograph is the power house. The River Thames, with a sludge barge, is in the background.

Primary treatment

In the primary treatment or sedimentation process, sewage is passed slowly and continuously through tanks to remove as much solid matter as possible by sedimentation. The raw sludge can then be passed to the sludge digestion tank, the supernatant liquid (primary effluent or settled sewage) being given secondary treatment. There are a variety of designs of sedimentation tank but the ones most frequently installed are of a shallow, radial design, equipped with mechanical gear to remove the sludge. The sewage is retained for several hours and about 50 per cent of the suspended solids settle out as primary sludge. Sedimentation is cheaper than biological treatment in terms of unit

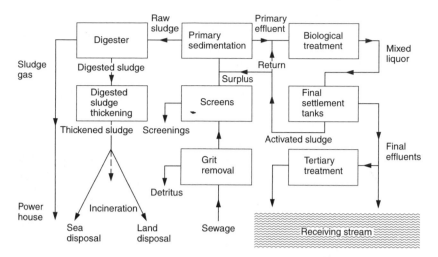

Fig. 3.4. Flow diagram of a typical sewage treatment works (from Wood, 1982).

removal of pollution, so that the tanks need to be operated at their maximum efficiency.

Secondary (biological) treatment

Secondary treatment involves the oxidation of dissolved and colloidal organic compounds in the presence of microorganisms and other decomposer organisms. The aerated conditions are obtained usually either by trickling filters or activated sludge tanks, while in warmer climates oxidation ponds may be used. The secondary sludge which results from biological treatment is combined with the primary sludge in sludge digestion tanks, where anaerobic breakdown by microorganisms occurs.

There are advantages and disadvantages in the methods of biological treatment. Filters are generally installed in small sewage works serving populations of less than 50 000. They tend to be higher in capital cost but lower in running costs than activated sludge plants. Filters take up a greater amount of land than activated sludge plants, but they require less skill and active control in functioning. Trickling filters may breed flies, which can cause a nuisance to local residents, but activated sludge plants are noisy. Filter beds oxidize more nitrogen than activated sludge plants but the final effluent carries more suspended solids. Oxidation ponds are much cheaper than other methods of treatment to construct and maintain and they produce a good quality effluent. They only function in warm, sunny climates, however, and require large areas of land. The effluent is turbid due to large populations of algae, and the ponds can be the breeding ground for noxious insects, especially mosquitoes.

Trickling (percolating) filters A cross-section of a typical trickling filter is illustrated in Fig. 3.5. Trickling filters are circular or rectangular tanks, some 1–3 m high and filled with a packed bed of mineral or plastic. The mineral may be broken rock, gravel, clinker or slag, but it must be uniformly graded so as to give a large proportion of spaces (voidage). The size range is usually 3.8–5.0 cm, with a specific surface area of 80–110 m^2 m^{-3} of volume and a proportion of spaces of 45–55 per cent of the total volume. Plastics are nowadays widely used instead of minerals because a high voidage can be obtained. The sewage from the primary settling tank is applied to the bed from above, either from rotating arms (circular tanks) or from pipes which travel backwards and forwards over the bed (rectangular tanks). The effluent flowing from the base of the bed contains suspended matter (humus) which is settled out and may be added to the primary settling tank. The clarified effluent may be recycled through the filter so as to dilute the incoming waste water. The main factors influencing the rate of removal of BOD are the specific surface area of the filtration medium, the hydraulic loading and the temperature of the sewage.

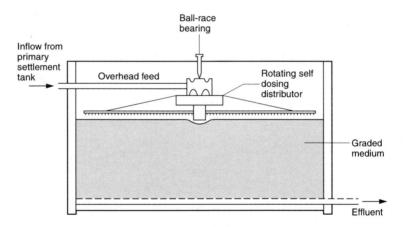

Fig. 3.5. Section through a trickling (percolating) filter.

With the application of settled sewage to the surface of the filter a biological community is gradually established as a slime on the surface of the material, and its constituent organisms oxidize the pollutants in the waste water. Bacteria are the most numerous organisms and form the base of the food web. A very wide range of bacteria has been recorded but the dominant aerobic genera appear to be the Gram-negative rods *Zoogloea*, *Pseudomonas*, *Achromobacter*, *Alcaligenes* and *Flavobacterium*. Fungi are normally outnumbered 8:1 by bacteria and are most abundant in the top 15 cm. The dominant genera are often *Sepedonium*, *Subbaromyces*, *Ascoidea*, *Fusarium*, *Geotrichium* and *Trichosporon*. These heterotrophic bacteria and fungi are responsible for the primary oxidation of the effluent. Autotrophic bacteria tend to be more predominant in the lower layers of the filter, with *Nitrosomonas* oxidizing ammonium to nitrite and *Nitrobacter* oxidizing nitrite to nitrate.

Algae (e.g. *Chlorella, Oscillatoria, Ulothrix*) are commonly found in percolating filters but they seem only to play a minor role in the purification process, and in large numbers they may reduce the efficiency of the filter.

Protozoa are present in proportions similar to fungi and 218 species have been identified, of which 116 were ciliates. Ciliates are also usually the most numerous, including species of *Carchesium, Chilodonella* and *Colpoda*. The major role of protozoa is to remove bacteria that are freely suspended in the liquid film, so that the effluent is clarified.

A diverse grazing fauna is present in percolating filters. It consists of rotifers, nematodes, enchytraeid and lumbricid worms (e.g. *Eiseniella, Dendrobaena*), larval and adult flies (Diptera such as *Anisopus, Psychoda, Metriocnemus*) as well as beetles (Coleoptera) and springtails (Collembola). Filters with macroinvertebrates produce a much better effluent than those without. This is partly due to the grazing activity of the animals, which prevents the accumulation of too much film and increases oxygen diffusion. The film material is made more settleable by passing through the animals' guts. Material which has been converted to the chitin of the moulted exuviae (skins) of larval and pupal *Psychoda* settles quickly, and a considerable proportion of the humus solids in percolating filters may consist of chitin. The respiration of the macroinvertebrate community may be responsible for between 3 per cent (in winter) and 10 per cent (in summer) of the total carbon dioxide dissipated, and this is a valuable contribution to the purification of sewage in biological filters (Solbé, 1971).

To function efficiently, the percolating filter requires a continuous inoculation of microorganisms in the sewage. To assist in the establishment of an active film, a new filter bed may be seeded with humus sludge or activated sludge solids. Establishment is most rapid in the summer but it may be up to two years before the community is fully developed.

The community developing will depend on the depth within the filter bed, the time of the year and the composition of the waste to be treated. The rate of accumulation of the film is determined by the difference between the rate of growth and the rate of removal, and its control is important in the efficient functioning of a filter bed. A thick film impedes the flow of water through the bed and the growth rate of bacteria may be reduced because the rate of diffusion of nutrients through the slime decreases. Recirculation of effluent after it has passed through the filter dilutes the incoming effluent, thus reducing the strength of feed. It also increases the hydraulic load and together these reduce the rate of growth of the film. Alternate double filtration is frequently used, in which two filters are operated in series. The waste water is applied at a relatively high rate to the primary filter and its effluent, after settlement, is passed to the second filter. The filters are switched at daily or weekly intervals. The primary filter quickly grows a thick film, but when it is switched to the secondary position the film rapidly shrinks. The overall costs are generally lower than for a single filtration. Horan (1990), Gray (1992) and Viessman and Hammer (1993) provide detailed accounts of these fixed film reactors.

Activated sludge process   Activated sludge tanks can be seen in Fig. 3.3. The settled sewage is mixed with a flocculent suspension of microorganisms and aerated in a tank for from 1 to 30 hours, depending on the treatment required. The medium is rich in dissolved and suspended nutrients, rich in oxygen and is violently agitated. The suspended and colloidal matter adsorbs on to the microbial flocs. Microbial metabolism then breaks down the flocs and the dissolved nutrients, a process known as stabilization. The sludge, which increases by 5–10 per cent during the process, is removed from the purified effluent in a sedimentation tank and returned to the inlet of the aeration tank. Any excess is returned to the inlet of the primary sedimentation tank.

By contrast to the percolating filter, the activated sludge is a truly aquatic environment. The turbulent conditions within the tank are unsuitable for macroinvertebrates so that the community lacks the higher links in the food web. The amount of microbial mass in the system is controlled by withdrawing excess sludge, whereas in the filter excess film is removed chiefly by biological agencies. In the activated sludge tank the microbial community is initially associated with the untreated waste and finally with the purified effluent, whereas in the filter bed a succession of communities is established at different depths in the bed and is associated with different degrees of effluent purification.

Most bacteria present in the activated sludge belong to Gram-negative genera such as *Pseudomonas, Zoogloea* and *Sphaerotilus*. Fungi are not usually dominant, though they may grow profusely if bacteria are inhibited, often when industrial effluent is present in the sewage. Fungi may cause 'bulking' in the sludge, in which loose-flocculent growths of microorganisms impede settling.

Over 200 species of protozoan are associated with activated sludge, some 70 per cent of them being ciliates. Activated sludge is therefore more species-rich than percolating filters, but a much greater proportion of species are ciliates. Densities of protozoa are of the order of 50 000 cells ml^{-1} and a generalized succession can be observed. Rhizopods and flagellates, able to use dissolved and particulate wastes, initially predominate but later, as bacteria increase, predatory flagellates and free-swimming ciliates are dominant. Crawling and attached ciliates predominate in the later stages of floc formation. The role of protozoa is, as in the filter, to clarify the effluent. Small numbers of nematodes and rotifers are also present in activated sludge.

For efficient operation the activated sludge process requires that the concentrations of substrate and microorganisms should be low. The basic activated sludge system has been developed in a number of ways, giving the process versatility and enabling it to be adapted to a wide range of operational circumstances. It is nowadays the most widely used biological process for the treatment of organic and industrial waste waters. Detailed reviews of the activated sludge process are provided by Hanel (1988), Horan (1990), Gray (1992) and Viessman and Hammer (1993), while Sell (1992) describes its use in dealing specifically with industrial wastes.

Sludge digestion and disposal

The sludge produced by primary and secondary treatment processes is passed to sludge digestion tanks, where it is decomposed anaerobically. Alternatively, it may be stored for later disposal. The sludge amounts to some 50 per cent of the initial organic matter entering the sewage works. The role of anaerobic bacteria is to convert the raw sludge into a stable and disposable product, which neither gives rise to offensive smells nor attracts harmful insects or rodents. The major constituents of raw sludge are proteins, fats and poly-saccharides and they are degraded in three processes. Hydrolysis involves the formation of long-chain fatty acids, amino acids, monosaccharides and disac-charides. The second process, acid formation, results in the production of a range of fatty acids, alcohols, aldehydes and ketones, together with ammonia, carbon dioxide, hydrogen and water. The third process, methanogenesis, results in methane, carbon dioxide and water. There is also, of course, an increase in bacterial biomass.

Sludge digestion often takes place in two stages. The first stage involves digestion in closed tanks, heated to a temperature of 27–35°C for 7–30 days; most of the gas evolution occurs here. Some of the gas evolved is used to heat the tanks. In the second stage, further digestion occurs for 20–60 days in open tanks at ordinary temperatures and the sludge consolidates and dries with the separation of a supernatant liquor, which is then returned to the sewage inlet for treatment with the sewage.

The sludge, after digestion, has been reduced in volume by two-thirds and its disposal presents some problems. It may be dumped at sea (though this will become illegal in the United Kingdom in 1998), it may be incinerated or it may be used as land-fill. It is also a potentially valuable fertilizer. In 1991 in England and Wales, 945 000 t of dried sewage sludge were produced, of which 22 per cent was disposed of at sea, 7 per cent incinerated, 11 per cent land-filled and 51 per cent used as fertilizer (Kinnersley, 1994). Sewage sludge may, however, contain toxic materials such as metals or fluoride, which will gradually build up in cultivated soils. Metal availability to crops increases at lower pH and may affect productivity or accumulate in those parts eaten by humans (Smith, 1994a, b). Liming of soils will reduce the uptake of metals and, at pH 6 or more, current regulations provide high protection of the human food chain from this potential source of contamination. Whether natural food chains, for example via metal uptake by earthworms, are so protected is less certain. Sewage sludge is a good fertilizer, however, and improves soil aggre-gation. With the soaring cost of fertilizers, it is economically attractive to farmers, particularly in situations where transport costs are low.

Oxidation ponds

Oxidation (or stabilization) ponds are used in warm climates to purify sewage, and the process involves an interaction between bacteria and algae (Fig. 3.6). The ponds are shallow lagoons, with an average depth of 1 m. Settled sewage passes through the pond in 2–3 weeks, but raw sewage may be retained for up to 6 months. The bacteria in the pond decompose the biodegradable organic matter to release carbon dioxide, ammonia and nitrates. These are used by the algae, together with sunlight, and the photosynthetic process releases oxygen, enabling bacteria to break down more waste. A layer of organic sludge settles on the bottom of the pond and anaerobic decomposition results in the release of methane. Oxidation ponds are used, not only for treating sewage, but also for treating wastes from food processing, paper production and other industries.

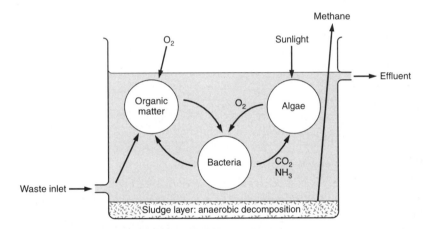

Fig. 3.6. Summary of processes in an oxidation pond.

Oxidation ponds not only purify wastes but they can also be used to provide energy. Algae have a solar conversion efficiency of 3–5 per cent and average production is 70 t ha^{-1} yr^{-1} (35 t ha^{-1} yr^{-1} algal protein). When harvested they may be fed directly to animals, used as fertilizer, as a source of chemicals, fermented to produce methane or burnt to produce electricity (Polprasert, 1989; Laliberté *et al.*, 1994). Fish also grow extremely quickly in stabilization ponds, yields as high as 14 600 kg ha^{-1} yr^{-1} being claimed, though in tropical lands yields of 3000–8000 kg ha^{-1} yr^{-1} seem more typical, with lower yields (generally less than 1000 kg ha^{-1} yr^{-1}) in temperate waste ponds (Polprasert, 1989). If fish are used for human consumption, care must be taken to avoid transferring pathogens and helminths.

Duckweed (*Lemna*) has been used in oxidation lagoons instead of algae (Hancock and Buddhavarapu, 1993). This simple, higher plant floats on the surface of the water, doubles its weight in 24 hours and is easily harvested and dried. It can be used directly to feed livestock or transferred to adjacent fish-ponds which are producing *Tilapia* (Cave, 1991).

Tertiary treatment

In many situations the dilution available to an effluent in the water that receives it is insufficient to prevent a deterioration of water quality. To prevent this a higher quality effluent must be produced and this effluent 'polishing' is known as tertiary treatment. The removal of phosphates and nitrates may also be necessary for environmental or public health reasons.

Nitrogen is usually removed from waste water by biological processes involving nitrification and denitrification, and when this is preceded by secondary treatment, over 90 per cent removal of total nitrogen can be achieved. Nitrification involves the oxidation of ammonia to nitrate, with nitrite as an intermediate:

$$2NH_4^+ + 3O_2 \rightarrow 2NO_2^- + 2H_2O + 4H^+$$
$$2NO_2 + O_2 \rightarrow 2NO_3^-$$

The reactions are carried out by the bacteria *Nitrosomonas* and *Nitrobacter* respectively. Denitrification involves the conversion of nitrate to nitrogen gas, and a number of facultative heterotrophs use nitrate instead of oxygen as the final electron acceptor during the breakdown of organic matter under anoxic conditions. With methanol as the organic carbon source the reaction is:

$$6NO_3^- + 5CH_3OH \rightarrow 3N_2 + 5CO_2 + 7H_2O + 6OH^-$$

Because nitrified effluent contains little carbon, a carbon source is normally added. This is frequently methanol because it is almost completely oxidized, thus producing less sludge for disposal, and it is relatively inexpensive. Reviews are provided by Robertson and Kuenen (1992) and Hardman *et al.* (1993).

The oxidation ponds described above may be used in conjunction with a standard sewage treatment works, secondary effluent being run into shallow lagoons to mature, resulting in a high quality final effluent. Effluents can also be passed through rapid sand filters, or micro-strained, resulting in a final effluent with a BOD of less than $10 \, \text{mg} \, l^{-1}$ and a similar level of suspended solids.

Fluidized bed systems treat wastes to a high quality and they are also efficient at removing nitrates. The fluidized bed combines features of both the trickling filter and activated sludge processes. Waste water is passed upward through a reaction vessel which is partially filled with a fine-grained medium. The velocity is sufficient to fluidize the bed (i.e. impart motion to it) and a microbial community develops on the surface of the medium. Activated carbon and sand are typical media and offer large surface areas per volume of reactor. As the particles are in fluid motion there is no contact between them, allowing the entire surface to be in contact with the waste water, the microbial biomass attained being several times that of the trickling filter or activated sludge process. The required degree of treatment can therefore be achieved in a much smaller reactor volume, saving considerable amounts of land. Operating costs are, however, greater because pure oxygen must be injected into the incoming waste water stream to support the increased microbial biomass. The sludge produced is more concentrated (10 per cent solids compared with 2 per cent for activated sludge). If the main aim is the removal of nitrates by denitrification then the expensive oxygen, of course, is not required.

Secondary effluent may also be passed through beds of aquatic macrophytes to remove wastes, the pollutants being immobilized and degraded by the normal physical and biological processes that operate in a wetland ecosystem (Hardman *et al.*, 1993). Macrophytes may be submerged (e.g. Canadian pondweed, *Elodea canadensis*), floating (e.g. water hyacinth, *Eichhornia crassipes*), or emergent (e.g. reed, *Phragmites australis*). Water hyacinth has a very high productivity in the tropics and is being used in countries such as Brazil as a tertiary treatment system, removing nitrogen and phosphorus. The biomass is harvested frequently to sustain maximum productivity and to remove incorporated nutrients. Water hyacinths can also be used as integrated secondary and tertiary treatment systems, removing BOD as well as nutrients, both decomposition of organic matter and the microbial transformation of nitrogen proceeding simultaneously. In this system harvesting is only carried out for maintenance purposes and performance with respect to phosphorus removal is poor (Brix and Schierup, 1989). Polprasert (1989) provides a review.

There is much current interest also in reedbed systems, especially for the cost-effective secondary treatment of wastes from small communities and for tertiary treatment. They may also be useful in treating the run-off from major roads prior to its discharge to watercourses (Ellis *et al.*, 1994). A typical reedbed system is illustrated in Fig. 3.7. Reedbed designs are of two types, those which have a subsurface flow and those in which water flows over the surface. A bed is dug to a depth of 60 cm, lined with an impervious liner (clay or butyl rubber), filled with gravel or sand (subsurface system) or back-filled with soil (surface system), and planted with reeds or other emergent macro-

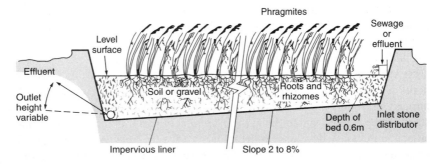

Fig. 3.7. Section through a typical reedbed treatment system (from Cooper *et al.*, 1989).

phytes. In the subsurface system the rhizomes grow deep into the substrate, providing channels along which the water can run. Oxygen is transported to the roots, some of which diffuses out to allow aerobic treatment of wastes by bacteria in the rhizosphere. Anaerobic treatment takes place in the surrounding substrate. In the surface flow system water flows to a depth of about 10 cm and waste is decomposed aerobically in the above-ground layer of litter formed from dead leaves and stems. The reedbed system works on the same basic principle of trickling filters but the reedbed has a greater structural diversity and provides a large, vertical gradient in redox potential, which it is believed allows a greater range of chemical and biochemical processes to take place. Reedbed systems appear to be very efficient at removing BOD and suspended solids. Nitrates are removed mainly by microbial action, while phosphates and metals are largely precipitated in complexes, a pH-dependent process. The success in nutrient removal is very variable and reedbed systems appear to be rather poor at removing phosphorus in the long term (Richardson and Craft, 1993; Merritt, 1994).

Potential advantages of the reedbed system include low capital costs, simple construction, low maintenance costs, robust operation, a consistent effluent quality and potential to provide wildlife habitat. They require, however, relatively large areas of flat land for construction and may take up to three years to reach maximum efficiency. Over 500 reedbed systems have been constructed in Western Europe since 1984 and many more in the United States, though they should still be considered experimental as their performance over long time scales is unknown. It must be stressed that systems for purifying waste water using macrophytes should be custom-built; natural wetlands are too scarce and vulnerable a resource to be used for waste disposal. Reviews of reedbed treatment systems are provided by Hammer (1989), Cooper and Findlater (1991), Olson and Marshall (1992), Moshiri (1993), Hardman *et al.* (1993), Bavor and Mitchell (1994) and Merritt (1994).

Removal of pathogens

The reduction in the numbers of pathogens during the sewage treatment process is governed by the length of retention time during treatment, the chemical composition of the wastes and their state of degradation, antagonistic forces in the biological flora, pH and operational temperature, together with other less well understood factors. Trickling filters have been found to reduce densities of *Salmonella paratyphi* by 84–99 per cent, enteric virus by 40–60 per cent and cysts of *Entamoeba histolytica* by 88–99 per cent, while waste stabilization lagoons result in more losses of pathogens. With further considerable dilution in receiving waters the sewage treatment process can be seen as an efficient way of reducing the incidence of pathogens. Greater potential problems are associated with home septic tanks, which are often overloaded and inefficiently maintained, and also with storm drainage.

Treatment of farm wastes.

Wastes produced by intensive livestock units present considerable problems of disposal, and pollution incidents caused by farm wastes are now of great concern. The wastes can however be treated in small, on-farm units which can be designed to suit the needs of the individual farming operation. Anaerobic digestion is the basis of many such systems. A typical example is illustrated in Fig. 3.8. Digesters vary in capacity from $10\,m^3$ to $1000\,m^3$. The anaerobic digester tank is heated, using methane produced in the digestion process, and stirred. Material progresses through the digester in 10–20 days and sufficient excess methane is produced to provide energy for heating and power generation on the farm. The polluting power of the waste is reduced by as much as 80 per cent. A separation unit can split the digested waste into liquid and solid components. The liquid can be pumped for long distances through small-bore pipe to be used as a fertilizer when and where it is required, while the solid waste is easy to handle and can be used on the farm or even sold as a high-quality compost. A review of biogas production and the operation of digesters is provided by Polprasert (1989), while Fallowfield *et al.* (1992) consider the merits of both aerobic and photosynthetic systems in treating animal wastes.

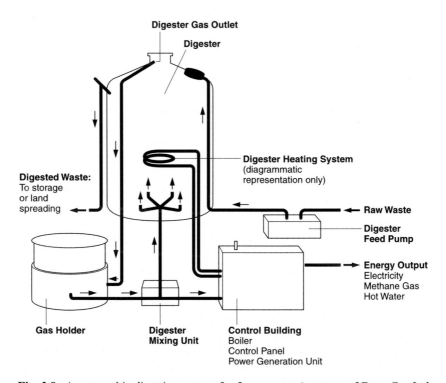

Fig. 3.8. An anaerobic digestion system for farm wastes (courtesy of Farm Gas Ltd, Bishop's Castle).

Conclusions

This section has largely been concerned with the treatment of sewage, but the principles apply equally well to the treatment of industrial organic wastes. Many factories have small-scale effluent treatment plants to deal with their organic wastes before these are discharged into rivers.

Sewage treatment has been described in detail because it demonstrates how biological systems can be applied effectively to industrial processes. A study of these processes can in turn help in the development of new fundamental biological theory, for instance in understanding the dynamics of ecological communities. The discussion has also shown that there are many potential by-products of economic importance to be gained from the treatment process, such as energy, organic compounds and food, either directly as fish or indirectly as animal feed. The exploitation of this potential will involve the skilled manipulation of biological systems. There has been considerable recent interest in methods to recover or produce compounds from organic and industrial wastes. These include the direct extraction of proteins, fats and other organic compounds, the microbial conversion of carbohydrate to protein and the recovery of metals (Wheatley, 1987; Brown, 1991). The development of genetically engineered microorganisms (GEMs) may enable many valuable materials to be recovered or synthesized from domestic or industrial wastes (Berry and Senior, 1987).

Microorganisms can also be used to break down recalcitrant compounds, which are not broken down when released to the environment (Bewley *et al.*, 1991; Müller, 1992). The majority of these are synthetic organic molecules, including detergents and pesticides. Some microorganisms can metabolize particular compounds. The bacteria *Alcaligenes*, *Azotobacter* and *Flavobacterium* use aromatics as a substrate, for example, and the fungus *Trichosporon cutaneum* uses phenols.

Microbial cultures to degrade these compounds can be made by enrichment culture or gene manipulation. In enrichment culture an inoculant of microorganisms is taken from a habitat already exposed to the compound of concern and is cultured and exposed to increasing concentrations of the compound until it becomes a source of essential nutrient. Gene cloning techniques have recently been used to produce bacteria that can break down a wide range of recalcitrant organics, and some of these microorganisms have been patented. Packaged microbes are also now available for dealing with wastes that are difficult to treat. Packaged microbes can, for example, be used to improve BOD removal in treatment plants, or to improve methane generation in digesters, or for degrading spilled oil (p. 218). Hardman *et al.* (1993) provide a valuable overview of the use of microorganisms in treating wastes.

In an ideal world, all the materials in sewage would be economically recovered and water alone would be returned to the hydrological cycle. The world, of course, is not ideal. Even where sewage is receiving secondary treatment, the effluent produced may be sub-standard because the design capacity of the works may not be sufficient to cope with the population it is now having to serve, and resources may not be available to expand the plant (Fig. 3.9). Similarly, many plants are old and inefficient or become overloaded during unusual conditions such as storms. Sewage is also changing with time (e.g. increased amounts of oils and plastics) and these may present problems with which old plants are not designed to cope. In England and Wales effluents from sewage works are one of the main causes of deterioration of river water quality, and 28 per cent of pollution incidents in 1991 were caused by sewage effluents (Department of the Environment, 1992). In addition to sewage, other forms of organic pollution may be affecting the receiving water and many parts of the world lack effective waste treatment facilities at all. In India, for example, 600 of the 2525 km of the Ganges are dangerously polluted with human and animal wastes and only 12 of the 132 industrial plants releasing effluents to the river have functioning waste treatment plants. Varanasi, and many other cities, are without sewage treatment facilities (Lean *et al.*, 1990). The next section will outline the effects of organic pollution on the receiving stream.

Fig. 3.9. A poor-quality effluent discharged from a sewage treatment works (photograph by Dr S. M. Macdonald).

Effects of organic pollutants on waters that receive them

The oxygen sag curve

When an organic polluting load is discharged into a river it is gradually eliminated by the activities of microorganisms in a very similar way to in the sewage treatment process. This *self-purification* requires sufficient concentrations of oxygen, and involves the breakdown of complex organic molecules into simple inorganic molecules. Dilution, sedimentation and sunlight also play a part in the process. Attached microorganisms in streams have a greater role than suspended organisms in self-purification (Esser, 1978). Their importance increases as the quality of the effluent increases, since attached microorganisms are already present in the stream, whereas suspended ones are mainly supplied with the discharge.

The deoxygenation of a river as a result of organic wastes is generally slow so that the point of maximum deoxygenation may occur considerably downstream of a discharge. The degree of deoxygenation depends on a number of factors such as the dilution that occurs when the effluent mixes with the stream, the BOD of the discharge and of the receiving water, the nature of the organic material, the total organic load in the river, temperature, the extent to which re-aeration occurs from the atmosphere, the dissolved oxygen in the stream and the numbers and types of bacteria in the effluent.

Figure 3.10 illustrates generalized curves, known as *oxygen sag curves*, which are obtained when dissolved oxygen is plotted against the time of flow downstream. The curves show slight, moderate and gross pollution. A poor quality effluent contains a high concentration of ammonia and the eventual oxidation of this to nitrite and nitrate may cause a further sag in the oxygen curve.

Figure 3.11 illustrates a real example of an oxygen sag curve, that measured during the four quarters of the year 1967 in the River Thames below

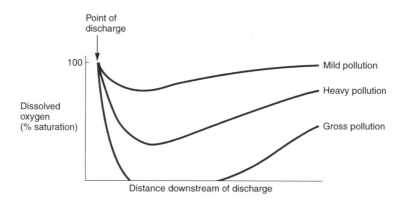

Fig. 3.10. Effect of an organic discharge on the oxygen content of river water.

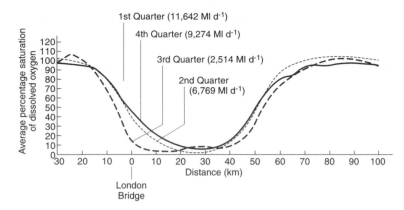

Fig. 3.11. Oxygen sag curve in the tidal stretch of the River Thames in the four quarters of 1967. Average daily flow over Teddington Weir in brackets (adapted from Arthur, 1972).

Teddington Weir. Note that the shape of the sag curve (the downstream position of the minimum oxygen and the length of river over which low oxygen conditions prevail) was largely determined by the amount of water flowing over Teddington Weir. With low flows in the autumn, minimum oxygen conditions prevailed over a distance of some 40 km downstream from London Bridge, whereas with high flows in the first quarter of the year only 12 km had minimum oxygen and this situation was not reached until 22 km below London Bridge.

If the oxygen demand of a sewage effluent is studied over a number of days, it is found that oxidation proceeds quite rapidly to begin with, but then slows down over a period of 15–20 days. There are often, however, two further stages in oxidation, which may account for a large proportion of the total oxygen consumption. The oxygen demand in the first 20 days is caused by the oxidation of organic matter (carbonaceous BOD), while later demand involves the oxidation of ammonia to nitrite and then nitrate.

To increase the rate of aeration and speed up self-purification below discharges, weirs are often built into rivers to considerable effect. To combat the disastrous effects of accidental pollution incidents, techniques have been developed to pump pure oxygen directly into threatened rivers. It is essential that the pollution be spotted in time, but the technique can be especially valuable where prime fisheries are at risk.

Effects on the biota

Organic pollution affects the organisms living in a stream by lowering the available oxygen in the water. This causes reduced fitness, or, when severe, asphyxiation. The increased turbidity of the water reduces the light available to photosynthetic organisms. Organic wastes also settle out on the bottom of the stream, altering the characteristics of the substratum. The general effects of fairly severe organic pollution are illustrated in Fig. 3.12. The top part of the figure (A) shows the oxygen sag curve, together with a massive increase in BOD, salts and suspended solids at the point of discharge, followed by a

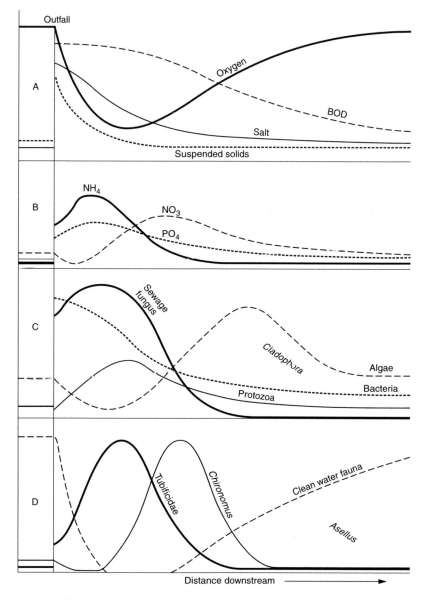

Fig. 3.12. Schematic representation of the changes in water quality and the populations of organisms in a river below a discharge of an organic effluent (from Hynes, 1960). A, physical changes; B, chemical changes; C, changes in microorganisms; D, changes in macroinvertebrates.

gradual decline in these parameters as self-purification occurs. The peak of ammonia (B) is replaced by a peak of nitrate as nitrification proceeds, and both are gradually diluted as the polluting load travels downstream.

There is a large increase in the number of bacteria (C) immediately below the outfall and these gradually decline as their substrate is depleted. There is also a large increase in sewage fungus (see later), which disappears as the stream re-oxygenates. The protozoa are chiefly predators on bacteria and increase in response to bacterial increases, subsequently to decline as bacterial numbers fall. Algae, especially *Cladophora* (blanket weed, a large filamentous green alga) increase in numbers as recovery begins, light conditions improve and nutrients are released from the oxidizing organic matter. They decrease as nutrients are used up or diluted.

The stream animals are shown at the bottom of Fig. 3.12. The clean water fauna is eliminated at the point of discharge of the pollutants, unable to tolerate the lowered oxygen tension. Sludge worms (Tubificidae) may be the only macroinvertebrates present immediately below the discharge and, in some cases of very severe pollution, even these may be absent. As conditions improve, blood worms, the larvae of the midge *Chironomus*, become abundant, and as further amelioration occurs, large populations of the isopod *Asellus* build up. As the stream gradually re-oxygenates, the clean water fauna increases in numbers and diversifies.

The absence of a particular species or group from a river may not be indicative of pollution because not all reaches are suitable for them. Stoneflies, for example, are largely confined to eroding substrata with fast currents and do not occur in the slow-moving, silty, lowland reaches of rivers, even where these are free of pollution. Various zones can be recognized along a river and these have characteristic animals and plants associated with them (Hawkes, 1975).

Microorganisms The primary effect of organic pollution is to increase the numbers of bacteria which use the waste as a substrate. Bacteria increase markedly below the sewage outfalls of large cities, with direct counts of up to $36 \times 10^6 \, \text{ml}^{-1}$. Most of the bacteria below an outfall are suspended, about 10 per cent being attached to plant surfaces. Viruses may also be present in significant numbers, the majority of them deriving from sewage effluents (Walter *et al.*, 1989).

Under fairly heavily polluted conditions a benthic community of microorganisms, known as *sewage fungus*, develops. Sewage fungus (Fig. 3.13) is an attached macroscopic growth, which may form a white or light-brown slime over the surface of the substratum, or may exist as a fluffy, fungoid growth with long streamers. It is usually bacteria, not fungi, that dominate the sewage fungus community. The organisms making up sewage fungus are listed in Table 3.2, and some organisms are illustrated in Fig. 3.14. The dominant species are usually *Sphaerotilus natans* and zoogloeal bacteria.

Sphaerotilus natans consists of an unbranched filament of cells, enclosed in

Fig. 3.13. Growth of sewage fungus in a polluted river (photograph by the author).

a sheath of mucilage. The sheath enables the bacterium to attach to solid surfaces, as well as protecting the organism from parasites and predators. These bacteria use a wide variety of organic compounds as substrates and they can also use inorganic nitrogen sources, though growth is less luxuriant than with organic nitrogen. When glucose and acetate were added as carbon sources to artificial channels, slime did not form below a concentration of 1 mg substrate l^{-1}, while above this concentration slime formation was proportional to oxygen concentration (Curtis *et al.*, 1971). The bacteria are aerobes, but they can survive in oxygen concentrations down to $2 \, mg \, l^{-1}$. For typical growth, a certain amount of flow is also necessary. The sewage fungus community extends further downstream during the winter than in summer, because the oxidation of organic matter in the effluent proceeds more slowly, such that the pollution plug extends further downstream. Also the sewage fungus bacteria are able to compete more effectively with other heterotrophic bacteria at lower temperatures.

Eleven protozoan species occur in very large numbers in the sewage fungus community (Table 3.2). *Carchesium*, a stalked, ciliated protozoan, tends to be most abundant at the lower end of the sewage fungus zone, where oxygen conditions are somewhat improved. It is a predator, feeding mainly on the large

Table 3.2   Typical organisms of the sewage fungus community (from Curtis and Curds, 1971)

Bacteria	*Sphaerotilus natans* *Zoogloeal* bacteria *Beggiatoa alba* *Flavobacterium* sp.
Fungi	*Geotrichum candidum* *Leptomitus lacteus*
Protozoa	*Colpidium colpoda* *Colpidium campylum* *Chilodonella cucullulus* *Chilodonella uncinata* *Cinetochilum margaritaceum* *Trachellophyllum pusillum* *Paramecium caudatum* *Paramecium trichium* *Uronema nigricans* *Hemiophrys fusidens* *Glaucoma scintillans* *Carchesium polypinum*
Algae	*Stigeoclonium tenue* *Navicula* spp. *Fragilaria* spp. *Synedra* spp.

populations of bacteria present in sewage effluent. *Carchesium*, when present in large numbers, can cause silting of the river bed because it flocculates suspended matter. *Colpidium colpoda* and *Chilodonella cucullulus*, both holotrichs, are very abundant in the sewage fungus. *Colpidium* feeds mainly on bacteria, while *Chilodonella* is a more generalized feeder on bacteria, diatoms, filamentous growths and flagellates. Gray (1985) provides a review of sewage fungus.

Algae   With heavy organic pollution, deoxygenation and low light, algae are eliminated from rivers, but there is a gradual reappearance as conditions improve and populations and growth are stimulated by the large concentrations of nutrients present. Algae have been investigated by immersing glass slides in organically enriched rivers. The frequency of species in three rivers is shown in Fig. 3.15. The filamentous *Stigeoclonium tenue* became common in the zone immediately below the region of gross pollution. *Stigeoclonium* also occurs in the sewage fungus community (Table 3.2). The characteristic species of the recovery zone were the diatoms *Nitzschia palea* and *Gomphonema parvulum*. The cyanobacterium *Chamaesiphon* sp., the green *Ulvella frequens* and the diatom *Cocconeis placentula* appeared when the pollution had dispersed. A long distance is required for the complete recovery of the algal com-

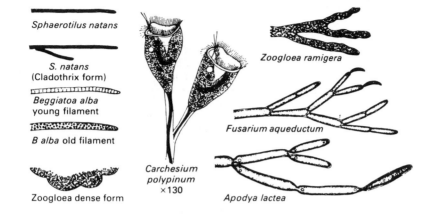

Fig. 3.14. Some of the characteristic species in the sewage fungus community (×330, except *Carchesium*) (from Hynes, 1960).

munity in grossly polluted rivers. For example, this amounted to 70 km downstream of the discharge on the River Tame and 56 km on the River Trent.

The diversity of algal species in clean waters can be very variable. Heavily polluted environments always have communities low in numbers of species because sensitive species are gradually eliminated as the pollution load increases. At low levels of organic pollution, however, tolerant species tend to increase, but conditions are not sufficiently severe to cause a great loss of sensitive species. Mildly polluted rivers can thus have high diversity.

The filamentous green alga, *Cladophora*, becomes abundant in the recovery zone and forms dense blankets over the substratum, where it provides both cover and food for invertebrates. The ecology of the plant has been reviewed by Dodds and Gudder (1992). Its dense growth in the recovery zone is linked to an increase in nutrients, and especially to phosphate, but other factors are probably also involved. *Cladophora* may be so prolific in organically enriched streams that it deoxygenates the water at night, causing fish to die. It also interferes with river recreation, such as angling and boating.

Higher plants Macrophytes are also adversely affected by organic pollution and may be eliminated by severe pollution events. The load of suspended solids reduces light and, by settling out, may render the bed of the river unstable. When conditions improve downstream, the rich supply of nutrients can result in a very dense growth of some species. Haslam (1987) has described the effects of organic pollution on plants in rivers. She recorded a decrease and loss of those species most sensitive to pollution, with a decrease in overall species richness, and an increase in any species favoured by pollution. The only species whose range appears to be increased by organic pollution is the pond weed *Potamogeton pectinatus*, which is very tolerant. Five species of higher plant were described as tolerant, being common in both clean and polluted streams, namely *Mimulus guttatus* (monkey flower), *Potamogeton*

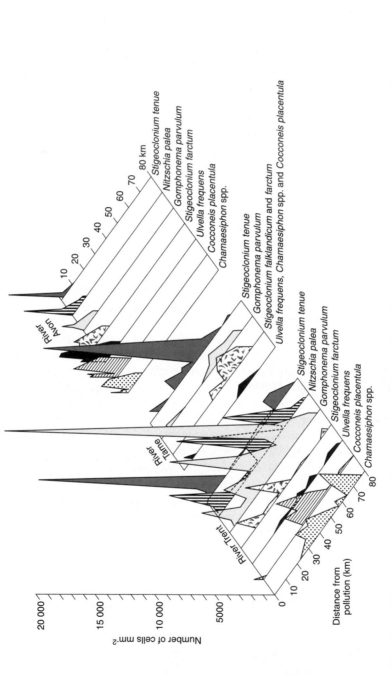

Fig. 3.15. Succession of the principal members of the algal flora growing on glass slides placed at sites downstream of an organic discharge in three English Midland rivers (from Hellawell 1986, after Butcher, 1947).

crispus (curled pond weed), *Schoenoplectus lacustris* (club-rush) and the bur-reeds, *Sparganium emersum* and *S. erectum*. A further seven species were fairly tolerant.

Figure 3.16 illustrates the distribution of some macrophyte species in the River Mersey catchment in northwest England. The Mersey has three main tributaries, the Goyt, the Tame and the Etherow. The River Goyt receives the greatest number of industrial discharges in the catchment, including four textile firms and two sewage works. Heavy pollution begins at the head of the river and, although there are some improvements downstream, water quality is still poor at its confluence with the Mersey. The River Tame is bounded on both sides by industrial and urban development throughout its length and receives effluents from six sewage works, all of which also treat industrial effluent, and several sewer overflows. The lower stretches of the river have continuous growths of sewage fungus. The River Etherow has a higher water quality than its tributaries but it deteriorates at the confluence with the Goyt, and the Mersey remains of poor quality until it discharges into the estuary.

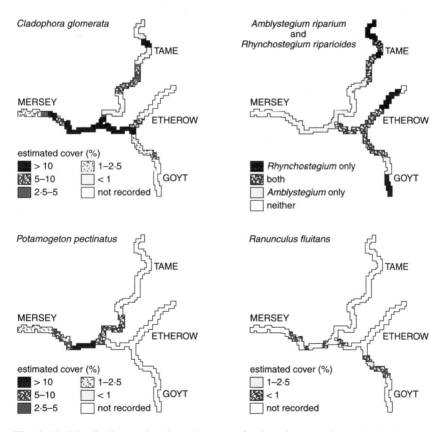

Fig. 3.16. Distribution and estimated cover of selected macrophytes in the Mersey catchment (from Holland and Harding, 1984).

The aquatic vegetation in the headwaters of the catchment is generally rich and dominated by bryophytes (mosses and liverworts). For example, the top of the Goyt supports a diverse and dense macroflora but in its middle reaches the river receives a severely polluting effluent from a textile works and all aquatic plants are absent. There is then a progressive increase in diversity and cover downstream. Plant communities undergo rapid changes in distribution, as evidenced by the discontinuous pattern of distribution and abundance shown in Fig. 3.16. The relative abundances of the mosses *Rhynchostegium ripariodes* and *Amblystegium riparium* appear to be related to organic pollution. In waters free of pollution neither species occurs. As enrichment increases *Rhynchostegium* appears, to be joined by *Amblystegium* with further enrichment. As pollution becomes more severe, *Rhynchostegium* declines, followed by *Amblystegium*, and neither species is present in conditions of gross pollution. The patchy distribution of the alga *Cladophora* is related to heavy metal contamination rather than organic pollution.

Overlying the effects of pollution on macrophytes are many natural characteristics of rivers, such as substratum type, flow regime, current speed, water chemistry and shading. These factors make it difficult to relate a particular plant community to pollution conditions, unless neighbouring streams with similar characteristics, but different loads of pollutants, are compared.

Invertebrates Protozoa dominate the animal community in polluted rivers where oxygen is severely limiting, and some species have already been mentioned in the discussion of sewage fungus. The response of protozoan communities to pollution is difficult to evaluate because large numbers are washed into streams from the surrounding land and in the effluents from the secondary sewage treatment process, making it difficult to distinguish between natural and intrusive members of the fauna.

There has been a great deal of work on the effects of organic pollution on macroinvertebrate communities. Heavy pollution affects whole taxonomic groups of macroinvertebrates, rather than individual species. Specific differences only become important in cases of mild pollution. In general those organisms associated with the silted regions of rivers are the most tolerant of organic pollution, while species associated with eroding substrata and swiftly flowing water are the most sensitive. Siltation of the river bed as solids settle out may clog the gills of species associated with eroding substrata. For example, the mayflies *Rhithrogena semicolorata* and *Ephemerella ignita* are particularly sensitive to siltation (Scullion and Edwards, 1980). Invertebrates from swiftly flowing waters also generally have higher metabolic rates than those from slow flowing waters and they would hence tend to be more sensitive to decreases in oxygen content in the water. Respiratory adaptations in invertebrates are discussed by Jeffries and Mills (1990) and Williams and Feltmate (1992). Some species tolerant and intolerant of organic pollution are illustrated in Fig. 3.17 and a summary of the invertebrate taxa associated with different degrees of organic enrichment has been given by Hawkes (1979).

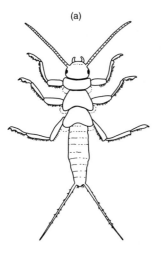

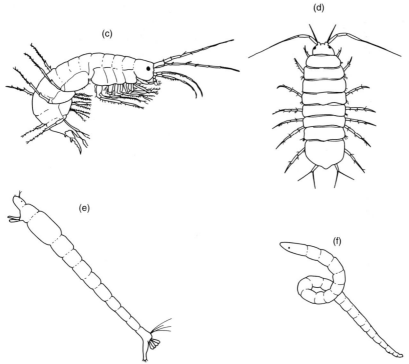

Fig. 3.17. Some species of freshwater invertebrates, in order of increasing tolerance to organic pollution. (a) *Dinocras cephalotes* (Plecoptera); (b) *Ecdyonurus venosus* (Ephemeroptera); (c) *Gammarus pulex* (Amphipoda); (d) *Asellus aquaticus* (Isopoda); (e) *Chironomus riparius* (Diptera); (f) *Tubifex tubifex* (Oligochaeta) (adapted from Macan, 1959).

In heavily polluted waters tubificid worms are frequently very abundant, often forming monocultures with densities of over $10^6 \, m^{-2}$. Successive species of tubificids are eliminated as conditions become more severe, and in very severe conditions only *Limnodrilus hoffmeisteri* and/or *Tubifex tubifex* remain. For these animals the effluent provides an ideal medium for feeding and burrowing, while the absence of predators allows populations to increase largely unchecked.

The success of tubificids in these environments is due to their ability to respire at very low oxygen tensions. Figure 3.18 shows that the respiration rate of *Tubifex tubifex* and *Branchiura sowerbyi* is almost unaffected by dissolved oxygen concentrations down to 20 per cent of air saturation. Tubificids contain the pigment haemoglobin, which has a high affinity for oxygen. The haemoglobin in *T. tubifex* exhibits a negative Bohr effect, which means that oxygen can be taken up at low pH, when the carbon dioxide content of the water is high, a frequent occurrence in organically polluted waters. The pigment functions in the transport of oxygen, but it appears not to store oxygen for use during prolonged periods of anoxia. Tubificids are known to survive in anaerobic conditions for up to four weeks and they may metabolize glycogen anaerobically at this time. Tubificids can also feed, defaecate and lay eggs at low oxygen tensions. Naidid worms may also respond to organic pollution by large increases in numbers, especially on the stony substrata which are favoured by the group. *Nais elinguis* appears particularly tolerant of pollution and may increase in numbers twenty-fold below a sewage outfall.

As the water below a discharge of organic material becomes more oxygenated, tubificids decrease in abundance and are replaced by the midge larva

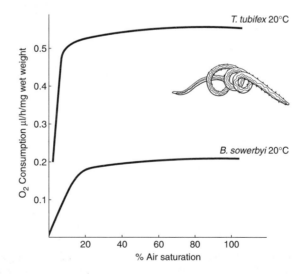

Fig. 3.18. Rate of oxygen consumption in relation to dissolved oxygen concentration, as percentage of air saturation, in two species of tubificid worm, *Tubifex tubifex* and *Branchiura sowerbyi* (adapted from Aston, 1973).

(blood worm) *Chironomus riparius*. *C. riparius* cannot withstand oxygen conditions as low as those tolerated by tubificids. The *Chironomus* zone contains species other than *C. riparius*, including the carnivorous Tanypodinae, which feed on tubificids and small chironomids. Pupal exuviae (the 'skins' which remain floating on the surface of the water after the adult midges have hatched) have been used to characterize the distribution of chironomids in streams. Figure 3.19 shows how *C. riparius* dominates the chironomid community below a sewage outfall, being absent in the clean water immediately above. As the water self-purifies downstream the proportion of *C. riparius* decreases until it is absent 1 km below the discharge. Chironomid pupal exuviae have been used widely in the United Kingdom to detect point sources of organic pollution (Wilson, 1994).

The life cycle of *C. riparius*, with many broods each year, together with the selection of sites by ovipositing females, gives the species numerous opportunities during a year to invade a site which has become suitable as a result,

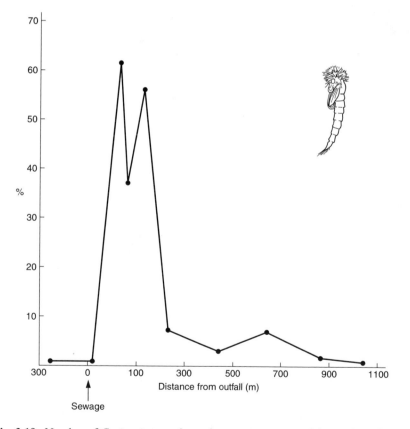

Fig. 3.19. Number of *C. riparius* pupal exuviae as a percentage of the total number of chironomid exuviae at stations above and below a sewage outfall on the River Chew in southwest England (adapted from Wilson and McGill, 1977).

for example, of temporary pollution caused by organic run-off from a farm-yard. Chironomid densities may exceed $30\,000\,\mathrm{m}^{-2}$ downstream of an organic discharge (Davies and Hawkes, 1981).

Chironomus, like tubificids, has haemoglobin in its blood and the content may reach up to 25 per cent of the value for human blood. The haemoglobin has a molecular weight of 34 400, half that in mammalian blood, and the pigment content of the blood increases when the water is poorly aerated. It acts as a carrier mainly when the oxygen tension of the water is low, at a time when the amount of oxygen required by the animal cannot be supplied by physical solution. *Chironomus riparius* lives in a tube, which is kept oxygenated by the undulatory movements of the animal's body. This extends the layer of oxygenated mud into the sediments and increases their rate of oxidation.

Below the chironomid zone, the isopod *Asellus aquaticus* becomes numer-ous (Fig. 3.12D, p. 75), especially where beds of *Cladophora* occur. The body length and weight of *Asellus* was found to be significantly lower in polluted sites, while females produced fewer eggs (Tolba and Holdich, 1981). Leeches, molluscs and the alder fly *Sialis lutaria* are also often abundant in this zone. The amphipod *Gammarus pulex* is very much more sensitive to organic pol-lution. *Gammarus* is killed within five hours at oxygen concentrations in the water of $1\,\mathrm{mg\,l}^{-1}\,\mathrm{O_2}$. Its distribution in rivers is largely determined by the number of hours during the night that the oxygen concentration falls below this critical concentration (Hawkes and Davies, 1971).

As the self-purification process continues, other species of invertebrates ap-pear in the community. The species most sensitive to organic pollution are the stoneflies (Plecoptera) and to a lesser extent the mayflies (Ephemeroptera), which are often absent even at mild levels of pollution. *Amphinemura sulcicollis* is more tolerant than other stoneflies of organic pollution and *Baetis rhodani* and *Caenis horaria* are more tolerant than other mayflies.

The distribution of selected invertebrate species in the chronically polluted River Mersey catchment (p. 81) is shown in Fig. 3.20. The upper stretches of the Goyt support a rich invertebrate fauna, including stoneflies and mayflies, and in the absence of pollution, physical conditions would be suitable for most of the length of the river. Below the discharge of effluent from a textile factory, however, only oligochaete worms and chironomids occur. Aeration provided by the steep gradient of the river allows species such as *Asellus* and the mayfly *Baetis* to recolonize, and lower down small numbers of other mayflies and stoneflies survive by inoculation from a clean tributary.

The invertebrate fauna of the River Tame steadily deteriorates downstream as effluents from various sewage works are added to the river. The Etherow has a generally richer fauna and that of the Mersey has been showing some im-provement with upgrading of sewage works, but the last stretch of the river, just above the tidal limit, supports only tubificids and chironomids.

Field experiments have been conducted to investigate the effect of episodes of deoxygenation on stream invertebrates, events associated, for example,

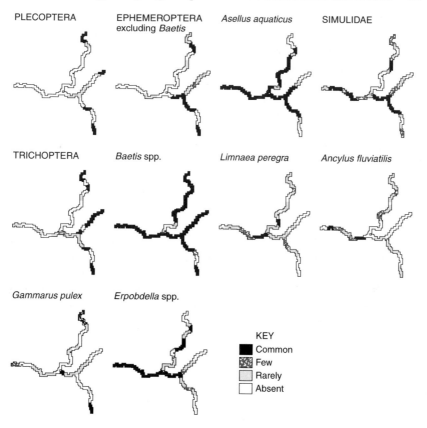

Fig. 3.20. Distribution of selected invertebrate taxa in the Mersey catchment, 1978 1986 (from Holland and Harding, 1984).

with accidents at farms (p. 50). Sodium sulphite was added to a stream to reduce oxygen to around $1 \, \text{mg} \, l^{-1}$ for a period of 6 hours (Fig. 3.21). Drift densities of invertebrates, normally less than 0.5 individuals per m^3 during daylight, increased within one hour to 43 individuals per m^3, mainly of mayflies

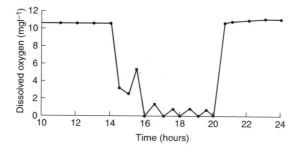

Fig. 3.21. Dissolved oxygen in stream water following treatment with sodium sulphite to mimic an organic pollution episode (from Edwards *et al.*, 1991).

and stoneflies. A later peak included *Gammarus*, simulids and chironomids. Invertebrate taxa in the stream decreased from 30 to 14 but recovery was relatively rapid, with initial richness and numbers being achieved within 6 to 12 weeks (Fig. 3.22).

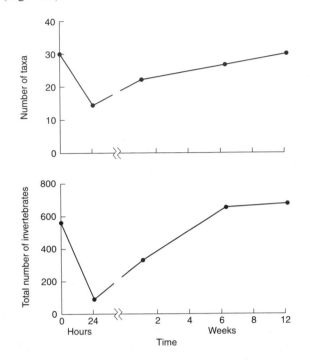

Fig. 3.22. Species richness and abundance of invertebrates in a small stream following a pulse of deoxygenation (from Edwards *et al.*, 1991).

Fish Fishes are considerably more mobile than any of the organisms discussed so far and they can sometimes avoid pollution incidents. The generalized relationship between the oxygen consumption of a fish and the oxygen tension of the water is shown in Fig. 3.23. The standard metabolic rate (equivalent to the basal metabolic rate) is independent of the oxygen concentration of water down to the incipient lethal level, below which survival becomes difficult. The rise in the standard rate that sometimes occurs before the incipient lethal level is reached is due to the increased respiratory activity necessary at low oxygen tensions. The active metabolic rate depends on oxygen tension to some fairly high critical level (the incipient limiting level) after which it is independent of oxygen tension.

When the oxygen supply becomes deficient, a fish breathes more rapidly and the amplitude of the respiratory movements increases. Oxygen lack and carbon dioxide excess, both of which occur with organic pollution, increase the ventilation volume of fishes and, at lower levels of oxygen, the cardiac output is reduced. This lowers the rate of passage of blood through the gills, so

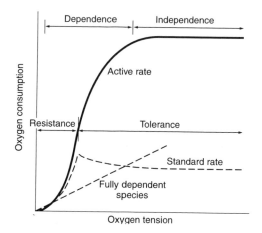

Fig. 3.23. Relationships between the active and standard rates of oxygen consumption of fishes and the oxygen tension of the water (from Warren, 1971).

allowing a longer period of time for uptake of oxygen, and also conserves oxygen by reducing muscular work. The percentage volume of red corpuscles in the blood may also be increased as the fish increases its urine flow to reduce the volume of the blood. The zone of resistance is reached when the oxygen tension in the water is so low that the homeostatic mechanisms of the fish are no longer able to maintain the oxygen tension in the afferent blood and the standard metabolism begins to fall. The effects of low oxygen on the biology of fish have been reviewed by Alabaster and Lloyd (1982).

Severe organic pollution renders rivers fishless for considerable distances downstream of the discharge. Fish begin to reappear in the *Cladophora/ Asellus* zone, the stickleback *Gasterosteus aculeatus* often being the first species to occur. Many pollutional and environmental factors affect the abilities of fishes to take up oxygen, so that their reactions to organic pollution in field situations are complex. Ammonia, often present in organic effluents and highly toxic to fish, is a frequent problem.

Organic pollution tends to be most severe in the lower reaches of rivers and in estuaries, and this can cause particular problems for migratory fish such as salmon (*Salmo salar*) and sea trout (*Salmo trutta*), which have high oxygen requirements. These fishes may be prevented by severe pollution downstream from reaching their breeding grounds in the headwaters of rivers, even though conditions are perfectly satisfactory there.

The improvement in water quality of the River Thames in London following upgrading of sewage treatment works has already been described (p. 57). The Thames used to support an abundant community of fish, including salmon, and even as late as 1828, some 400 fishermen earned their living from the tidal reaches of the river. Fish disappeared in the middle years of that century but, following the first attempts at sewage treatment, they began to re-establish from 1890. A further deterioration took place from about 1915, and for around 45 years the tidal Thames was devoid of fish over a stretch of at least 60 km. Following further improvements in sewage treatment facilities in the 1960s, regular monitoring of the species and numbers of fish in the river began. Figure 3.24 shows the number of species recorded. By 1978 almost one hundred species had been noted and attempts are being made to re-establish migratory stocks of salmon. More than two hundred salmon returned to the river to spawn in 1992. One problem has been the occasional pollution of the river during storms in summer, when the old sewerage system, taking run-off as well as sewage, is unable to cope. A barge producing 30 t of oxygen per day has been sited where the oxygen sag is greatest and this is sufficient to prevent fish kills at critical periods.

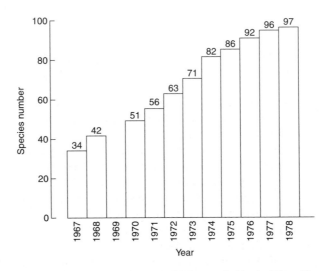

Fig. 3.24. Cumulative number of species of fish recorded in the River Thames between Fulham and Gravesend (from Wood, 1982).

Conclusion

Organic contamination of freshwaters was the first type of pollution to be recognized and studied and the first for which treatment facilities were developed, dating back to the last century. Where effort is put into providing or upgrading sewage treatment facilities the improvements in river water quality can be dramatic. An example is provided by the River Clyde, flowing through the industrialized central lowlands of Scotland (Fig. 3.25). In 1968 substantial stretches of the river were grossly polluted with organic and industrial effluents but, following investment in pollution abatement measures, no stretches could be so described by 1988. Nevertheless there is still heavy pollution in the lower river, while amounts of PCB in the food chain (not considered in this biological classification) remain unacceptably high (Mason *et al.*, 1992).

Many rivers remain grossly polluted, even in the developed world. In

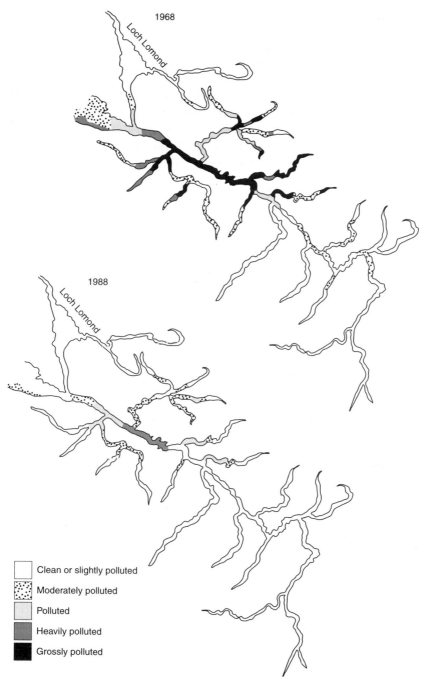

Fig. 3.25. Biological classification of water quality in the River Clyde, Scotland, 1968 and 1988 (adapted from Hammerton, 1994).

Belgium, for example, less than 10 per cent of the sewage receives treatment and only now is the capital, Brussels, building its first full sewage treatment works (Kinnersley, 1994). With the more open political system evolving in Eastern Europe, it is becoming clear that immense pollution problems are present there. Organic pollution becomes most acute in the underdeveloped world, where the treatment both of sewage effluents and drinking-water supplies is rudimentary or non-existent, faecal contamination of the latter leading to chronic problems of disease and periodic epidemics. But organic pollution and the diseases associated with it are the consequence of a lack of financial resources or a lack of political will, or both, rather than of a lack of scientific and technological know-how.

Chapter 4

EUTROPHICATION

The terms *oligotrophic* and *eutrophic* were originally used in the early years of this century to describe nutrient conditions in the development of peat bogs. When they were introduced to the limnological world, lakes were classified as having oligotrophic water if they were clear in summer and eutrophic water if they were turbid owing to the presence of algae. Considerable confusion over these terms has developed as other definitions have arisen. In particular, oligotrophic lakes have been defined as having a low productivity and eutrophic lakes as having a high productivity. Although in general terms high nutrient levels and high productivity go together this may not always be the case. Marl lakes, for instance, have highly calcareous, nutrient-rich water (eutrophic), with a heavy precipitation of carbonates. Under these conditions phosphates and many micronutrients form insoluble compounds which precipitate to the bottom of the lake. Thus algal productivity is low and marl lakes may be considered as oligotrophic.

Eutrophication will be defined as the *enrichment of waters by inorganic plant nutrients*. Nutrients are usually nitrogen and phosphorus and result in an increase in primary production. The income of nutrients may be considered as *artificial* or *cultural* eutrophication if the increase is due to human activities, or *natural* eutrophication if the rate of increase is caused by a non-human process, such as a forest fire. From our definition of pollution in Chapter 1 (p. 1) only artificial eutrophication will concern us here.

The general characteristics of oligotrophic and eutrophic waters are given in Table 4.1. Lakes with intermediate characteristics are described as *mesotrophic*. A fourth general category of lake includes those which receive large amounts of organic matter of terrestrial plant origin – *dystrophic* or brown-water lakes, so called because of the heavily stained water. They usually have a low productivity of plankton.

Henderson-Sellers and Markland (1987) and Harper (1992) provide detailed reviews of eutrophication processes.

Sources of nutrients

A number of compounds and elements (e.g. silicon, manganese, vitamins) may at times be limiting to algal growth, but it is only excesses of nitrogen and

Table 4.1   The general characteristics of oligotrophic and eutrophic lakes

	Oligotrophic	*Eutrophic*
Depth	Deeper	Shallower
Summer oxygen in hypolimnion	Present	Low, sometimes absent
Algae	High species diversity, with low density and productivity, often dominated by Chlorophyceae	Low species diversity with high density and productivity, often dominated by cyanobacteria
Blooms	Rare	Frequent
Plant nutrient flux	Low	High
Animal production	Low	High
Fish	Salmonids (e.g. trout, char) and coregonids (whitefish) often dominant	Coarse fish (e.g. perch, roach, carp) often dominant

phosphorus that give rise to algal nuisances. In the majority of lakes phosphorus is normally the limiting element because the relative proportions of biologically available phosphorus in the water compared to other elements are often less than the relative proportions of these elements in algae. An increase in phosphorus will therefore result in an increase in productivity. In many oligotrophic waters, however, it appears that nitrogen is limiting. If nitrogen becomes limiting in eutrophic conditions some cyanobacteria (also known as blue-green algae or cyanophytes) are able to fix atmospheric nitrogen and will grow provided phosphorus is not limiting. Cyanobacteria can also store large amounts of phosphorus within their cells, which allows them to survive when phosphorus may become limiting in late summer (Thompson and Rhee, 1994).

The majority of polluting nutrients enter watercourses and lakes in effluents from sewage treatment works, in untreated sewage, from farming activities and from precipitation. Local eutrophication may also occur where livestock are given access to the water for drinking or where large congregations of waterfowl gather. Sources might be *discrete*, such as a specific sewage outfall, or *diffuse*, such as from farmland within the catchment.

Table 4.2 gives a general summary of the quantities of nitrogen and phosphorus that are derived from different types of land use in the United States. Note the considerable increase in the export of phosphate as forest is converted to agricultural use and agricultural land is subject to urban development. The relative importance of sources of nutrients will vary, of course, with the type of land use in the catchment.

Urban sources

Nutrients from urban sources may be derived from domestic sewage, industrial wastes and storm drainage. The contribution of nitrogen and phosphorus

Table 4.2 The quantities of nitrogen and phosphorus derived from various types of land use in the United States (from Lee *et al.*, 1978)

Land use	Total phosphorus (g m^{-2} yr^{-1})	Total nitrogen (g m^{-2} yr^{-1})
Urban	0.1	0.5
Rural agriculture	0.05	0.5
Forest	0.01	0.3
Other sources		
Rainfall	0.02	0.8
Dry deposition	0.08	1.6

per person averages 10.8 g N per capita per day and 2.18 g P per capita per day, though there is a considerable range. In the 1940s detergents were developed containing sodium tripolyphosphate, which softens water by neutralizing calcium and keeps dirt in suspension once it has washed off the clothes. Between 1950 and 1970 detergent consumption increased more than five times in the United States and more than seven times in Great Britain, and detergent is now an important source of phosphorus in domestic sewage, often making up more than half the total phosphorus in effluents.

Industrial sources of nutrients may be locally important, depending on the type of industry, the volume of effluent and the amount of treatment it receives. For instance, the brewing industry, which releases about 11 000 m^3 effluent each day into rivers in England and Wales, produces an effluent containing some 156 mg l^{-1} N and 20 mg l^{-1} P. Food processing generally and those concerns such as the wool industry that require substantial washing procedures, are likely to produce effluents containing high concentrations of nitrogen and phosphorus.

Rural sources

Rural sources of nutrients include those from agriculture, from forest management and from rural dwellings, of which the first is the most universally important. Rural dwellings generally dispose of their sewage into septic tanks which might cause local pollution. Summer houses, which frequently have primitive sewage disposal facilities, are often built on lakesides.

Nutrients are lost from farmland in three ways:

1. by drainage water percolating through the soil leaching soluble plant nutrients;
2. by inefficient return to the land of the excreta of stock;
3. by the erosion of surface soils or by the movement of fine soil particles into subsoil drainage systems.

The use of fertilizers has vastly increased during this century and in the United States the total inorganic fertilizer used is equivalent to 40 kg/person/ year. Nitrogen and phosphorus behave differently in soils. The nitrate anion is fairly mobile because of the predominantly negative charge on soil particles, so that it is readily leached if it is not taken up by plants. By contrast, phosphate is precipitated as insoluble ion (calcium or aluminium phosphate) and then released only slowly.

The solubility of nitrate means that agriculture is a major contributor to nitrate loadings in freshwaters and often up to half the nitrogen applied to crops is lost to groundwater. Agriculture accounts for 71 per cent of the mass flow of nitrogen in the River Great Ouse in the English Midlands, compared with only 6 per cent for phosphorus, the balance being from sewage effluents.

The concentration of nitrate in rivers follows closely the volume of water flowing. Levels are low during summer, even when fertilizer is being added, because the growing plants utilize nitrogen as soon as it becomes available. There is also little net downward movement of water in the soil during the summer because of high rates of evaporation and transpiration. With a decrease in transpiration and evaporation in autumn and winter, nitrate is leached from the soil and levels in rivers rise. The rate of loss declines again in late winter because soluble nitrate reserves are depleted and low temperatures reduce the rate of nitrification. The annual nitrate level in rivers closely follows the annual levels of fertilizer application within the catchment and this has been steadily increasing over the past three decades (Fig. 4.1). For instance in eight stretches of river in the English Midlands annual increases ranged from $0.07-0.22\,\mathrm{mg\,l^{-1}\,yr^{-1}}$, greatest rates of increase occurring in the most urbanized stretches of river (José, 1989). However in the rural and intensively arable eastern region of England, rates of increase are higher still, of the order of $0.25-0.28\,\mathrm{mg\,l^{-1}\,yr^{-1}}$ (Warn and Page, 1984).

The loss of phosphate by leaching from agricultural land is negligible, so that the input to freshwaters is largely by erosion. Arable farming increases the

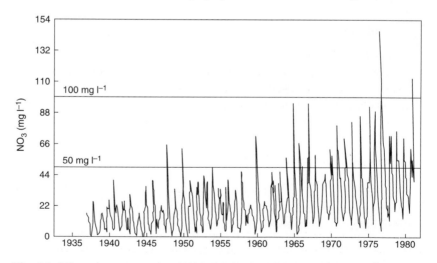

Fig. 4.1. Nitrogen concentration (1937–82) in the water of the River Stour, eastern England, close to the abstraction point for a water supply reservoir (from Gray, 1994).

natural rate of erosion of land because the soil is often bare for several months of the year. Losses of phosphorus to water equivalent to 60 per cent of the fertilizer applied to the land have been reported. Much of this phosphorus is tightly bound to the soil particles and is not immediately biologically available when it reaches freshwaters, but the solubility of phosphate is enhanced when the soil particles become incorporated into anaerobic muds.

The other chief source of nutrients from the agricultural industry is animal farming, especially where this involves intensive rearing. The increase in size of livestock units and the problem of waste disposal has already been described (p. 50). Labour costs for handling farmyard manure are high and areas of intensive livestock rearing are often distant from areas of arable farming, where the wastes could be used on the land. The amount of phosphorus excreted by British livestock each year is four times that excreted by the human population.

The management of forests may have local effects on the nutrient loading of rivers. The experiments in the Hubbard Brook Watershed in the United States, where a forest was cut and left on site, with regrowth prevented by the application of herbicides, showed that nitrate increased some fifty times compared with uncut controls (Likens *et al.*, 1970). Most forest practices, of course, are nowhere near as extreme as this. In some countries of the world, forests are regularly fertilized and this may result in local eutrophication. Phosphate is added to newly established conifer plantations in Britain and with the increase in afforestation this could result in an overall increase in productivity in upland catchments. This can be observed in northwest Scotland where naturally oligotrophic streams, if flowing from plantations, frequently have dense growths of filamentous algae in spring, smothering the stoney bottoms which may be important spawning grounds for salmonids.

Two case histories

Before dealing with the general effects of eutrophication I want to describe conditions in two lakes which have been studied in detail: Lake Washington in North America and Lough Neagh in the British Isles.

Lake Washington

Lake Washington (Fig. 4.2), in the northwestern United States, has a surface area of 87.6 km^2 and a maximum depth of 76.5 m. Water, which is low in nutrients, enters the lake via the Cedar River from the south and the outflow is the Sammamish River, which flows into Lake Sammamish. Lake Washington is connected to the Pacific Ocean at Puget Sound, but a series of locks holds the lake level higher. The city of Seattle (population 489 000) lies between Puget Sound and the western shore of the lake. The events in Lake

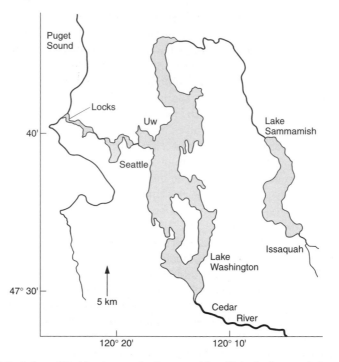

Fig. 4.2. Lakes Washington and Sammamish, United States (adapted from Edmondson, 1969).

Washington have been detailed in a series of papers by Edmondson (e.g. 1969, 1970, 1971, 1972a, b, 1979).

Early in this century raw sewage was disposed of directly into Lake Washington and, by 1926, a population of 50 000 was adding raw sewage to the lake through 30 outfalls. By 1936 a series of interceptors and tunnels had been constructed to divert Seattle's sewage to the sea at Puget Sound, and this reduced pollution within the lake. However, Seattle began to expand along the edges of the lake and development took place on the east side. A series of works was built to cope with the expansion and discharged secondarily treated sewage into the lake, there being ten sewage works by 1954. In addition, feeder streams into the lake were being contaminated with drainage from septic tanks. Sewage was responsible for 56 per cent of the total input of phosphorus to the lake.

Work during 1950 showed that conditions in Lake Washington were very different from those revealed in an earlier survey in 1933, and during the next few years conditions deteriorated sufficiently to attract considerable public attention to the problems. There was a greater summer oxygen deficit in the hypolimnion in 1950 compared with 1933 while the winter phosphate concentration, which gives a good index of supply to the plankton, had shown a marked increase. The summer densities of phytoplankton increased severalfold during the 1950s and in 1955 a dense bloom of the cyanobacterium *Oscillatoria rubescens* developed. A *bloom* can be defined as *an aggregation of plankton sufficiently dense to be readily visible*. Lakeside bathing beaches were periodically closed due to pollution, but the situation was far worse in Puget Sound, which was receiving Seattle's untreated sewage.

The amenity value of Lake Washington, close to a large urban centre, had seriously declined and this gave rise to a vociferous movement to prevent further deterioration. It was decided to divert the majority of sewage from the lake to Puget Sound and at the same time to improve the quality of the effluent entering the Sound to reduce pollution there. Diversion began in March 1963, with one third of the sewage being transferred to Puget Sound. Ninety-nine per cent of the sewage had been diverted by March 1967 and the project was completed a year later at a total cost of $125 million.

Lake Washington responded quickly to the reduction in nutrients. Winter levels of phosphate began to decline rapidly after 1965. Levels of nitrogen declined more slowly because agriculture is a major contributor of nitrogen to the lake. The level of chlorophyll *a* had fallen by 1970 to one fifth the level in 1963 and there were corresponding increases in the transparency of the water (Fig. 4.3). Edmondson (1991) described the lake as being in excellent condition in 1990 for a waterbody in an urban area. The transparency of the water was rarely less than 5 m and usually much greater, while, although small numbers of cyanobacteria occur, algae are no longer a nuisance.

The observations on Lake Washington illustrate clearly the relationship

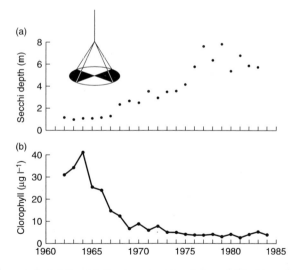

Fig. 4.3. Changes in (a) Secchi disc transparency (m), and (b) chlorophyll in surface water (µg l^{-1}) in Lake Washington 1962–84 (adapted from Gulati, 1989). Note: a Secchi disc is 20 cm in diameter and divided into four segments, two blank and two white. It is lowered into the water on a line until it just ceases to be visible – the Secchi depth.

between an increase in nutrients and an increase in undesirable biological productivity, but they also show that eutrophication can be reversed if a major source of nutrients is removed. Edmondson (1991) provides an overview of Lake Washington, including the social and political problems encountered during its restoration.

Lough Neagh

Lough Neagh (Fig. 4.4), in Northern Ireland, has a surface area of 383 km^2 and a mean depth of 8.6 m, so it is much larger, but much shallower, than Lake Washington. It is fed by six major rivers and the Lower River Bann drains the lough to the sea. Lough Neagh is an important source of water for Ulster; there is also a commercial fishery and extensive recreational use.

Fig. 4.4. Lough Neagh and its catchment area in Northern Ireland.

In 1967 a bloom of the cyanobacterium *Anabaena flos-aquae* disrupted the treatment process in the waterworks, killed large numbers of eels (*Anguilla anguilla*) held in cages and interfered with recreation (Smith, 1993). This encouraged a detailed study of the lake, which was shown to be highly eutrophic

(Wood and Smith, 1993). The nitrate and phosphate levels in Kinnego Bay in Lough Neagh were some three times higher than the maximum recorded from Lake Washington, while the level of chlorophyll *a* was some eight times higher. In the main body of the lough the values were somewhat lower than this, though it is still one of the most productive lakes in the world.

A study of the diatoms in sediment cores demonstrated a progressive enrichment of the lake from the end of the nineteenth century (Battarbee and Carter, 1993). Species changed, *Cyclotella*, for example, replacing *Stephanodiscus*, while diatom abundance increased, especially since about 1960. There have been three main effects on the phytoplankton of the lake due to eutrophication. The majority of desmid species have disappeared, while there is a large diatom peak which utilizes all of the available silica. Two cyanobacteria, *Oscillatoria redekei* and *O. agardhii* are almost perpetually present. Sometimes the *Oscillatoria* crop fails in the summer, to be replaced, after a period of clear water, by the nitrogen-fixing *Anabaena* or *Aphanizomenon*. However, *Anabaena* has never again come to dominate the lake as it did in 1967 (Gibson, 1993).

Soluble reactive phosphorus (SRP) was found to be the critical nutrient controlling algal growth. The sources of SRP are shown in Fig. 4.5. Some 40 per cent is from background sources, mainly land drainage, and no correlation was found between the use of phosphorus fertilizer in the catchment and the annual concentration of phosphorus in inflowing rivers, though such a relationship does exist for nitrogen. Sewage treatment works (STWs) accounted for 54 per cent of the inflow of SRP, the balance being met by effluent from creameries

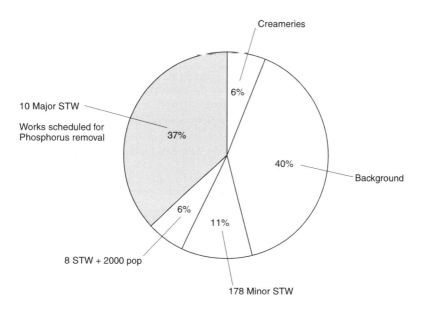

Fig. 4.5. Predicted budget of soluble reactive phosphorus (SRP) to Lough Neagh (from Smith, 1993).

in the catchment. Ten major STWs accounted for 37 per cent of the SRP, but all the treatment works combined were responsible for only 10 per cent of the nitrate loading to the rivers, the balance deriving from land drainage. Phosphorus then is the only critical nutrient which could be reduced at the treatment works.

At the ten major treatment works, the majority of SRP in the incoming sewage is to be precipitated out using an alumino–ferric liquor, such that final effluents should contain no more than $1\,mg\,l^{-1}$ total P. The first works to be treated was that at Ballymena, which resulted in a 50 per cent decrease in SRP in the River Main. There was no such decrease in the adjacent River Moyola, whose treatment works had not been dosed. Significant reductions in nutrient loadings to Lough Neagh are anticipated as the programme proceeds (Smith, 1993).

Wood and Smith (1993) should be consulted for a detailed consideration of the water quality and ecology of Lough Neagh.

The general effects of eutrophication

We have examined two enriched lakes in some detail and can now generalize about the effects of an increase in the rate of nutrient income on waterbodies. The effects of eutrophication on the receiving ecosystem and the particular problems these cause to man are summarized in Table 4.3.

Eutrophication causes marked changes in the biota. The Norfolk Broads area of eastern England can serve as an initial example. The Norfolk Broads are a group of relatively small, shallow lakes formed during medieval times by the flooding of peat diggings. They are calcareous and naturally eutrophic, but the water used to be generally clear and dominated by a rich flora of charo-

Table 4.3   The effects of eutrophication on the receiving ecosystem and the problems to man associated with these effects

Effects
1. Species diversity often decreases and the dominant biota change
2. Plant and animal biomass increases
3. Turbidity increases
4. Rate of sedimentation increases, shortening the life-span of the lake
5. Anoxic conditions may develop

Problems
1. Treatment of potable water may be difficult and the supply may have an unacceptable taste or odour
2. The water may be injurious to health
3. The amenity value of the water may decrease
4. Increased vegetation may impede water flow and navigation
5. Commercially important species (such as salmonids and coregonids (whitefish)) may disappear

phytes (large multicellular algae also called stoneworts) and aquatic flowering plants (macrophytes). The plants support a diverse fauna of invertebrates, some of them rare in Britain. The area as a whole is outstanding for its wildlife interest, but at the same time maintains a large tourist industry, based mainly on boating holidays and angling. A detailed description of the region is provided by George (1992).

Conditions began to deteriorate in the Broads during the late 1950s but the area was not subject to any scientific study. A survey in 1972 and 1973 of 28 broads showed that 11 were completely devoid of aquatic macrophytes or had only a poor macrophyte growth, consisting mainly of floating-leaved water lilies (Mason and Bryant, 1975). A later study of 42 broads found only four with clear water and a flora of submerged charophytes (phase 1). A further six had a flora dominated by submerged species which can grow rapidly to the surface and form mats (phase 2a). Such species include *Ceratophyllum demersum* and *Potamogeton pectinatus*, often associated with floating-leaved water lilies. At a later stage only water lilies remain (phase 2b). Thirty-two broads were devoid of aquatic plants (phase 3).

These phases represent stages in the eutrophication process, phases 2b and 3 falling into a group of broads with high phosphorus, and group 2a being intermediate (Moss, 1983). These changes are clearly illustrated in Fig. 4.6. With the loss of macrophytes permanent algal blooms developed and the transparency of the water in summer at one site, Barton Broad, measured with a Secchi disc, was only 11 cm. Broads with no submerged macrophytes had a poorly developed benthic fauna dominated by tubificid worms and chironomids. A detailed study of the benthos of an unpolluted broad and a culturally enriched broad (Mason, 1977a) showed that the unpolluted site (Upton Broad) had 40 taxa in the benthos, 17 of them occurring commonly. The enriched site (Alderfen Broad) had 22 taxa, only 7 of them occurring commonly. In the Norfolk Broads there has been a change from a community dominated by macrophytes to one dominated by phytoplankton.

We can now consider in more detail the effects of eutrophication on the biota of freshwaters.

Plankton

The types of phytoplankton associated with lakes of different trophic status are listed in Table 4.4. Picoplankton and desmids are particularly important in lakes low in nutrients, while cyanobacteria often dominate lakes with high nutrient concentrations. The diatom flora also changes. Oligotrophic lakes have a diatom flora that is often dominated by species of *Cyclotella* and *Tabellaria*, while eutrophic lakes have *Asterionella*, *Fragilaria crotonensis*, *Stephanodiscus astraea* and *Melosira granulata* as dominants. It should be emphasized, however, that we are looking at a continuum, not a series of sharp divisions of trophic state, and the phytoplankton communities in any one lake must be viewed accordingly.

Fig. 4.6. (a) A turf pond dug some 15 years ago and isolated from river water (phase 2a) and (b) a broad from which macrophytes have disappeared (phase 3) (photographs by the author).

As an example, the seasonal changes in phytoplankton biomass of the hypertrophic Ardleigh Reservoir, in eastern England, are shown in Fig. 4.7. The reservoir is shallow and is characterized by high concentrations of phosphorus, up to $0.78\,\text{mg}\,\text{l}^{-1}$. Diatoms (Bacillariophyta) dominated in winter and early spring, to be succeeded by a mixture of cryptomonads (Cryptophyta) and green algae (Chlorophyta), the latter group becoming dominant in late spring.

Table 4.4 Characteristic algal associations of oligotrophic and eutrophic lakes

	Algal group	*Examples*
Oligotrophic lakes	Picoplankton (often small cyanobacteria)	*Synechococcus*
	Desmid plankton	*Staurdesmus, Staurastrum*
	Chrysophycean plankton	*Dinobryon*
Mesotrophic lakes	Diatom plankton	*Cyclotella, Tabellaria*
	Dinoflagellate plankton	*Peridinium, Ceratium*
	Chlorococcal plankton	*Oocystis, Eudorina*
Eutrophic lakes	Diatom plankton	*Asterionella, Fragilaria, Stephanodiscus, Melosira*
	Dinoflagellate plankton	*Peridinium, Glenodinium*
	Chlorococcal plankton	*Scenedesmus*
	Cyanobacterial plankton	*Aphanizomenon, Anabaena, Microcystis*

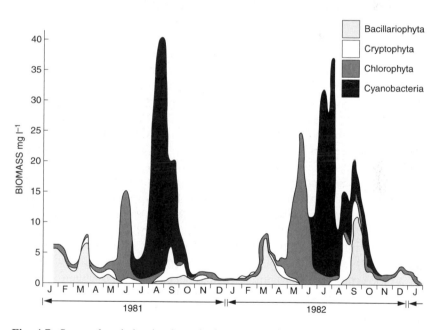

Fig. 4.7. Seasonal variation in phytoplankton composition and total biomass (mg l⁻¹ wet weight) in Ardleigh Reservoir (adapted from Abdul-Hussein and Mason, 1988).

In early summer the cyanobacteria *Anabaena* and *Aphanizomenon* joined the chlorophytes, while during high summer the cyanobacterium *Microcystis aeruginosa* dominated, the intense bloom of this species subsiding in early autumn to be replaced by cryptophytes and later by diatoms. *Microcystis* is unable to fix atmospheric nitrogen and is dependent on nitrate and ammonia released by the processes of decomposition in the waterbody. There was a significant relationship between ammonia and *Microcystis* in Ardleigh Reservoir. This seasonal change in the phytoplankton appears fairly typical of eutrophic waters. Turbulence is important in determining the dominant phytoplankton group, for cyanobacteria are outcompeted in turbulent waterbodies even though conditions would otherwise be suitable for them (Steinberg and Hartmann, 1988).

Once established, bloom-forming cyanobacteria influence their surrounding environment, physically, chemically and biologically, in ways which favour their continued persistence. Vincent (1989), for example, reported how a large bloom of *Anabaena* in Lake Rotongaio, New Zealand, restricted the euphotic zone to less than 1.5 m. *Anabaena*, with low maintenance energy requirements and an efficient light-harvesting mechanism, could cope with this poor light regime. *Anabaena* also strongly absorbed solar radiation, which raised surface water temperatures and further promoted cyanobacteria, which have a high temperature optimum for growth.

The thermal gradient created apparently restricted the propagation of turbulence down through the water column, favouring buoyant species of flagellates at the expense of diatoms, which sank in the still water. It also reduced the upward transfer of ammonium ions from the hypolimnion, putting all but nitrogen-fixing species at a disadvantage. *Anabaena*, by contrast, was able to take up low concentrations of ammonium ions and ammonia over a broad range of pH, while the increased pH in the water induced by the *Anabaena* bloom may have additionally favoured cyanobacteria over other phytoplankton groups through decreased CO_2 availability at high pH. Cyanobacteria are largely resistant to grazing by zooplankton. They also produce specialized cells which can survive prolonged periods of unfavourable conditions. Therefore once nutrient conditions have developed which favour the establishment of cyanobacteria (high dissolved inorganic phosphorus and low dissolved inorganic nitrogen concentrations in Lake Rotangaio) positive feedback mechanisms enable populations to grow rapidly and persist at high densities for long periods.

Diatoms are particularly interesting because they have siliceous frustules (outer walls), which fall to the lake floor on the death of the cell. By examining a time series of sections through a sediment core it is possible to describe the changes that have occurred in the waters of a lake. Moss (1972) has examined a sediment core from Gull Lake, Michigan. When the earlier sediments were laid down, the diatoms *Cyclotella michiganiana* and *Stephanodiscus niagarae* were dominant. From 20 cm in the sediment upwards they were joined by *Cyclotella comta* and *Melosira italica*. At 15 cm in the sediment a marked change occurred, with those species already present becoming much more abundant and *Asterionella formosa* appearing. At 10 cm *Fragilaria crotonensis* appeared, while the abundance of the other species continued to increase. The total diatom density near the surface of the sediment was fifty times greater than in the lower layers. The sediment record thus reflected the increase in nutrient supply to the lake both in terms of a change in diatom species and in an increase in diatom abundance (see also p. 101).

Cores through the sediment of several of the Norfolk Broads have shown similar changes as eutrophication progressed. There is historical evidence for eutrophication from agricultural fertilizers and from domestic buildings, with their requirement for effluent disposal (Moss, 1983). Increased algal growth has led to an increased rate of sedimentation. In Barton Broad the rate of sedimentation doubled in the 1950s (from $1.2–3.1 \, mm \, yr^{-1}$ to $5 \, mm \, yr^{-1}$), doubled again in the 1960s, and increased to $12 \, mm \, yr^{-1}$ in the 1970s (Moss, 1980).

The concentration of chlorophyll a in the water is often taken as an index of the biomass of algae present, and together with total phosphorus and water transparency (Secchi depth) has been used to classify the trophic status of lakes (Table 4.5). Enrichment also affects the rate of primary production. The mean daily rates of primary production are $30–100 \, mg \, C \, m^{-2} \, d^{-1}$ for oligotrophic lakes and $300–3000 \, mg \, C \, m^{-2} \, d^{-1}$ for eutrophic lakes (Rohde, 1969), being equivalent to annual rates of $7–25 \, g \, C \, m^{-2}$ and $75–700 \, g \, C \, m^{-2}$ respectively.

A typical pattern of seasonal change in the zooplankton is shown in Fig. 4.8. A spring increase in algae is followed by an increase in herbivores. They overeat their food supply and phytoplankton decline, followed by the zooplankton. There is usually a clear-water phase before algae again increase in numbers, followed by a further expansion in zooplankton numbers. With any rise in primary productivity we could reasonably expect an increase in the zooplankton. This does not necessarily occur however, for different types of algae are utilized to different extents by zooplankton. Changes in the availability of suitable algal food are responsible for much of the summer fluctuation in zooplankton numbers apparent in Fig. 4.8. Both the quantity of food and its quality affect the ingestion rate of zooplankton (Benndorf and Horn, 1985). Cyanobacteria are assimilated very inefficiently compared with green algae

Table 4.5 OECD boundary values for fixed trophic classification system (from Ryding and Rast, 1989, after OECD 1982)

Trophic category	TP	mean Chl	maximum Chl	mean Secchi	minimum Secchi
Ultra-oligotrophic	<4.0	<1.0	<2.5	>12.0	>6.0
Oligotrophic	<10.0	<2.5	<8.0	>6.0	>3.0
Mesotrophic	10–35	2.5–8	8–25	6–3	3–1.5
Eutrophic	35–100	8–25	25–75	3–1.5	1.5–0.7
Hypertrophic	>100	>25	>75	<1.5	<0.7

Explanation of terms:
TP = mean annual in-lake total phosphorus concentration ($\mu g \, l^{-1}$);
mean Chl = mean annual chlorophyll a concentration in surface waters ($\mu g \, l^{-1}$);
maximum Chl = peak annual chlorophyll a concentration in surface waters ($\mu g \, l^{-1}$);
mean Secchi = mean annual Secchi depth transparency (m);
minimum Secchi = minimum annual Secchi depth transparency (m).

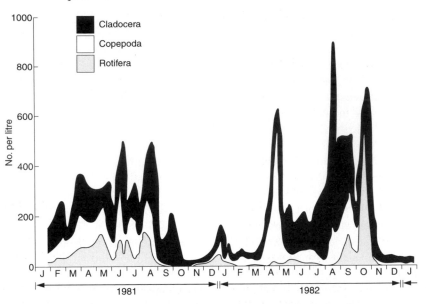

Fig. 4.8. Numbers of Cladocera, Copepoda and Rotifera in the zooplankton of Ardleigh Reservoir (from Abdul-Hussein, 1985).

and diatoms. The bottom-dwelling *Chydorus sphaericus* can eat cyanobacteria and its population often increases markedly during phytoplankton blooms. Larger cladocerans, such as *Daphnia*, filter much more efficiently than smaller plankton, such as *Bosmina*, and this greater feeding efficiency may suppress smaller zooplankton when larger animals are abundant. However, predation by fish is usually on larger forms, so that smaller zooplankton may dominate (see below).

Macrophytes

The loss of macrophytes from the Norfolk Broads was described earlier in this chapter, and similar changes have occurred in other lake systems, such as in the Netherlands (Van Liere *et al.*, 1984). The loss of aquatic plants is often abrupt, luxuriant beds disappearing in only one or two seasons. The decline in charaphytes and rooted macrophytes such as *Najas marina* is easiest to explain. There is a direct relationship between nutrient loading and increased growth of both phytoplankton and periphyton (the microflora and fauna attached to parts of rooted aquatic plants) and also of non-rooted macrophytes, with a concomitant decline in rooted, submerged macrophytes (Hough *et al.*, 1989). The low-growing rooted macrophytes are effectively shaded out, and in the Norfolk Broads this group was the first to disappear. But why have species such as *Ceratophyllum* and *Myriophyllum*, at best only partially rooted and forming dense mats at the surface of lakes, also disappeared? Similarly those species, such as water lilies, which are rooted but have floating leaves are eventually also lost. Macrophyte-dominated or phytoplankton-dominated states of waterbodies may exist over a wide range of high nutrient concentrations (Irvine *et al.*, 1989) and indeed, species such as *Ceratophyllum* and *Myriophyllum* have been proposed as suitable species for use in waste-water treatment lagoons, where nutrient levels will be extremely high (Polprasert, 1989). In experimental ponds there was no significant relationship between phosphorus loading or concentrations and the biomass of aquatic plants (Balls *et al.*, 1989).

It appears that, over a range of nutrient concentrations, shallow lakes may have two alternative stable states, either clear water and dominated by aquatic vegetation or turbid water characterized by a high algal biomass (Irvine *et al.*, 1989; Scheffer *et al.*, 1993). The macrophyte community may be stabilized by luxury uptake of nutrients (thus making nutrients unavailable to plankton), allelopathy (the secretion of chemicals which prevent plankton growth), shedding of leaves with heavy epiphyte burdens, sheltering of large populations of grazers which eat phytoplankton, and the provision of refuges for these grazers from predation by fish. The phytoplankton community may be stabilized by an early growing season (shading the later growth of macrophytes); easier acquisition of carbon dioxide, especially when pH is high late in the season; increasing the vulnerability of herbivores to predation in the unstructured environment; and the production of large, inedible algae. In many lakes the stability of the macrophyte-dominated community has, however, been broken, with a switch to communities dominated by algae. Possible reasons for this will be discussed below.

Nutrient-rich lakes are surrounded by emergent macrophytes, such as reed-mace or cat-tails (*Typha*) and reed (*Phragmites australis*). Over the same timescale as the loss of aquatic macrophytes, reedbeds have also regressed. In the Norfolk Broads, Boorman and Fuller (1981) calculated that there were 216 ha of reedswamp in 1880, 121 ha in 1946 and 49 ha in 1977. Declines in reed area have been described from more than 35 lakes in Europe (Ostendorp, 1989). For the Norfolk Broads, Boar *et al.* (1989) have suggested that at nitrate concentrations above $6 \, \text{mg} \, l^{-1}$, which occur commonly in Broads' water, there is a decrease in the rhizome to shoot ratio of reed. This may make floating reedbeds unable to withstand mechanical disturbances caused by wave action. Ostendorp (1989) has pointed out, however, that reed decline is not restricted to polluted (eutrophic) lakes but also occurs in mesotrophic lakes, while reeds can thrive in hypertrophic lakes and waste water treatment plants (p. 68) with nitrate concentrations as high as $10 \, \text{mg} \, l^{-1}$. It may be significant that, over the same timescale, large areas of intertidal saltmarsh, a habitat broadly similar to that of reedswamp, have also regressed (Long and Mason, 1983). The discussion is continued below (p. 110).

In flowing waters it is not macrophyte loss but excessive macrophyte growth that occurs. This may include the large, filamentous alga *Cladophora*, especially in waters rich in sewage effluents (see p. 79) and angiosperms such as species of *Ranunculus*, or in the tropics, water hyacinth *Eichhornia crassipes*.

Benthos

The reduction in diversity that occurs in the benthic fauna with enrichment has already been referred to but there may also be changes in the seasonal pattern of occurrence, with consequences for organisms higher in the food chain. The eutrophic but unpolluted Upton Broad, mentioned earlier, was dominated by larvae of the midge *Tanytarsus holochlorus*, with lesser numbers of caddis larvae, mayfly nymphs and snails, whereas the culturally eutrophic Alderfen Broad was dominated by the tubificid worm *Potamothrix hammoniensis* and the midge larva *Chironomus plumosus*. The benthic populations in Upton Broad were highly seasonal and there was little biomass present in the autumn and winter. Alderfen Broad was markedly less seasonal, with a substantial biomass of benthic animals in autumn and winter. This is illustrated in terms of community respiration in Fig. 4.9. The fish in Upton Broad would probably have very little food from late summer to mid-spring and indeed the populations of fish were small and recruitment poor compared with high populations and good recruitment in Alderfen Broad.

Fish

The general changes in the fish fauna with eutrophication are summarized in Fig. 4.10, using information from 51 European lakes. Four stages in eutrophication are recognized from changes in fish yield. Coregonid fish dominate in oligotrophic lakes and their yield increases in the early stages of enrichment and then declines. There is some increase in the percids in the intermediate stages of eutrophication but they then decline. The yield of cyprinid fish increases sharply at intermediate stages of eutrophication and falls off sharply in highly eutrophic waters.

In Dutch lakes, eutrophication has led to the dominance of bream (*Abramis brama*) in the fish community (Lammens, 1989). Bream grow rapidly compared to other fish species and quickly become too large to be eaten by the chief predatory fish, zander (*Stizostedion lucioperca*). They exert considerable grazing pressure on zooplankton and accelerate the eutrophication process (see next section). Large bream also stir and mix the sediments, which may provide further difficulties for the re-establishment of macrophytes.

Community interactions and eutrophication

A naturally eutrophic lake has an abundant and diverse community of aquatic macrophytes in the littoral zone, while in shallow lakes the plants will occupy

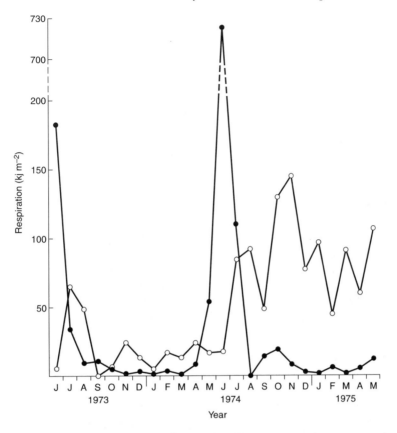

Fig. 4.9. Monthly respiration (kJ m^{-2}) of the benthic macroinvertebrate community at Alderfen Broad (○) and Upton Broad (●), eastern England, June 1973–May 1975 (from Mason, 1977a).

much of the waterbody. The canopy provides a refuge for zooplankton, especially those larger species which would otherwise be vulnerable to predation by fish. Large zooplankton, such as *Daphnia*, are efficient grazers of phytoplankton so that water remains clear, providing ideal conditions for macrophyte growth. Fertilization experiments have generally failed to reduce the preponderance of macrophytes (other than low-growing, rooted species) in these diverse systems (Moss and Leah, 1982; Irvine *et al.*, 1989) showing that the ecosystem is very robust. Why then have such ecosystems collapsed in the Norfolk Broads and elsewhere?

Stansfield *et al.* (1989) have suggested that the switch was operated by a reduction in large zooplankton caused by the widespread introduction of organochlorine insecticides into agriculture in the late 1950s and 1960s, when chemicals were used with little control. Cladocera are rather susceptible to pesticides. Sediment core analysis indicated that *Daphnia* became scarce at this time. With the absence of *Daphnia*, large populations of phytoplankton

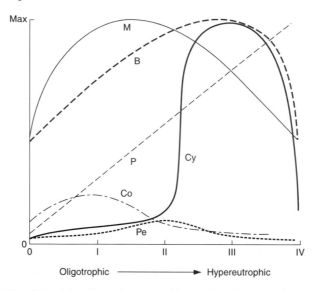

Fig. 4.10. Trends in lakes becoming eutrophic. B, density of benthos; Co, yield of whitefish (coregonids); Cy, yield of cyprinid fish; M, submerged macrophytes; P, phosphorus content of water, density of plankton and turbidity; Pe, yield of percid fish; 0–IV, stages of eutrophication (from Hartmann, 1977).

and epiphytes developed to shade out macrophytes. Any later recovery in the zooplankton population would be prevented because, no longer having a macrophyte refuge, they were intensively preyed upon by fish. The community would remain dominated by small forms, such as *Bosmina*, which are less efficient grazers, and feedback mechanisms would maintain the system in the phytoplankton stage. Daytime refuges, in the form of water lilies, have been shown to be essential for the survival of large-bodied Cladocera in the Broads (Timms and Moss, 1984).

Would the sudden elimination of zooplankton and the consequent algal blooms have been sufficient to stifle completely those vigorous macrophytes, such as *Ceratophyllum*, which grow rapidly through the waterbody to form surface mats of vegetation, and those like water-lilies which grow from a large rhizome to put out floating leaves, immune to the shading of phytoplankton? Those waters that carried increased concentrations of nutrients and raised levels of novel insecticides into the Broads in the 1950s and throughout the 1960s would have also carried various new herbicide formulations, applied to control weeds on agricultural land and indeed to control water plants in drainage ditches within the region. It is tempting to speculate that a herbicide, or more likely a cocktail of herbicides, reduced the vigour and hence the competitive ability of macrophytes at a time when conditions were optimal for algal blooms to develop. These herbicides may also have had a role in the decline of reedbeds and indeed of coastal saltmarshes. Herbicides, however, unlike organochlorines, break down quickly in lake sediments, so any part that they may have played in macrophyte decline must remain conjectural.

Problems for man

The changes in nutrient levels and biology outlined above can create a number of problems directly affecting human activities (Table 4.3) and these can be summarized into three main areas: problems associated with water purification, supply and consumption; problems associated with aesthetic and recreational activities; and problems associated with the management of watercourses and lakes.

Water purification

It is essential that a water treatment plant can produce a finished product of water of a consistently high quality, irrespective of the size of demand (Gray, 1994). The increase in phytoplankton with eutrophication frequently causes severe problems in water purification. Water is treated either by double slow sand filtration preceded by passage through rapid sand filters or microstrainers, or by coagulation and sedimentation followed by rapid sand filtration. In the first method large algae are removed at the primary filter, which may become rapidly blocked at high algal densities or high densities of zooplankton, whereas large quantities of small algae can overload the slow sand filters. The coagulation and sedimentation in the second method of treatment are more efficient at removing large numbers of small algae, so that the residual large algae may block the rapid sand filters. Blocked filters seriously reduce the throughput of water at the treatment works and occasionally a reservoir has to be taken out of service temporarily to clean the filters. Where the water supply for potable and industrial use in an area comes mainly from one large source the blockage of filters in the treatment works can be potentially very serious.

The smallest algal cells often pass through the filters, producing a turbid final water. The cells may decompose in the distribution pipes and some breakdown products, notably mucopolysaccharides, chelate with iron and aluminium that is added to the treatment, leading to increased metal levels passing to the supply. The decomposing algae also stimulate the growth of bacteria. Fungi and invertebrates can feed on these biofilms growing within the distribution network. The resulting water coming from the tap can have an unpleasant appearance, taste and odour and may regularly contain chironomids and *Asellus*, leading to many complaints from consumers! Most reservoirs in eastern England have suffered periodically from an inability to treat water adequately owing to high concentrations of plankton. This has led to the closure of various treatment works for periods of up to two months, six months and over a year (Hayes and Greene, 1984).

Nitrates

A water supply with a high nitrate level presents a potential health risk. In particular infants under six months of age may develop *methaemoglobinaemia* by drinking bottle-fed milk that is high in nitrates. Babies have gastric juices with a very low pH, which favours the reduction of nitrate ions to nitrite. Nitrite ions readily pass into the bloodstream, where they oxidize the ferrous ions in the haemoglobin molecules, reducing the oxygen carrying capacity of the blood. Above 25 per cent methaemoglobin there is a blueing of the skin (cyanosis) and associated symptoms. Death occurs at levels of between 60 per cent and 85 per cent methaemoglobin. There have been several thousand cases notified worldwide, many of them fatal. It appears that methaemoglobinaemia only occurs when bacteriologically impure water with nitrate levels approaching $100\,mg\,l^{-1}$ is supplied, and the problem is non-existent with treated, piped water. The disease is largely associated with bottle-fed infants using impure well-water with high levels of nitrates (Gray, 1994). In the UK the last reported case was in 1972.

The European Health Standards for drinking water recommend that nitrate concentrations should not exceed $50\,mg\,NO_3\,l^{-1}$. Water remains acceptable at nitrate levels of $50-100\,mg\,NO_3\,l^{-1}$ but is unacceptable at levels above $100\,mg\,NO_3\,l^{-1}$. Standards in the United States are more stringent, with nitrate levels higher than $45\,mg\,NO_3\,l^{-1}$ unacceptable. Water derived from river intakes in lowland England frequently has nitrate levels exceeding the maximum recommended level ($100\,mg\,NO_3\,l^{-1}$) and has to be mixed with water low in nitrates before it enters the public supply. Alternatively, bottled water, low in nitrates, is supplied to mothers who are bottle-feeding young infants.

High concentrations of nitrates (greater than $100\,mg\,NO_3\,l^{-1}$) in water may result in the formation of nitrosamines in the stomach. Nitrosamines have caused stomach cancer in animal experiments, but epidemiological studies have, as yet, failed to find a link between exposure to nitrates and cancer.

Amenity value

The aesthetic and recreational value of eutrophic waters is often reduced (Fig. 4.11). There may be interference with fishing, sailing and swimming as a result of the production of surface scums during algal blooms and of the growth of algae and macrophytes on shores. The smell resulting from decaying algae washed ashore is often highly offensive. The dense swarms of midges which emerge from eutrophic lakes can be a considerable nuisance to lakeside visitors and residents, while the insecticides used to control midges cause damage to other organisms, including fishes, in the lake.

Management problems, other than water treatment, revolve around excessive weed growth and fisheries. The biomass of aquatic macrophytes increases with nutrient input and plants spread over lakes, rivers and canals making

Fig. 4.11. A bloom of *Aphanizomenon* (photograph by the author).

navigation difficult and hindering recreation. Large growths of macrophytes impede the flow of water and increase the risk of flooding during storms. Many waterways have to be manually cleared of weeds, often twice a year at great expense.

It has already been recorded (p. 110) that increases in enrichment cause changes in the fish community. Salmonids and coregonids, which are high-quality food fish, are replaced by cyprinids, which are of low quality, though their biomass is usually higher. This change in dominant species is due to de-oxygenation of the hypolimnetic water. Low oxygen conditions may develop in eutrophic waters when algae and macrophytes die and decompose, and this will also affect the numbers and species of fishes present. An increase in pH, associated with increased growth of plants, may also kill fish.

Algal and cyanobacterial toxins

High densities of algae may produce toxins that are lethal to animals. The alga *Prymnesium parvum*, for example, grows well in nutrient-rich, slightly brackish waters and produces a toxin that is very potent to fish. It has caused problems in commercial fish-ponds in Israel and has severely damaged a first-class recreational fishery at Hickling Broad, eastern England (Wortley and Phillips, 1987). Fish kills have occurred in Hickling Broad since 1969 and the large populations of *Prymnesium* appear to be related to the increase in total phosphorus in the water. Large populations of cells are necessary to produce sufficient toxins to kill fishes, and high toxin activity occurs when phosphorus becomes limiting, because the formation of phospholipids in cell membranes is disrupted, allowing leakage of the toxin.

Toxins produced by cyanobacteria may also pose risks to human health, livestock and wildlife. Species such as *Microcystis*, *Aphanizomenon* and *Anabaena* can produce potent poisons, as measured in mouse bioassays. They can induce rapid and fatal liver damage at low concentrations and may also be neurotoxic, causing paralysis and death in as little as five minutes. No human deaths have been recorded, but there are a number of cases of algal toxins being implicated in illness. Worldwide there are many cases of mortalities of livestock and wildlife. In September 1989, at Rutland Water in eastern England, eight dogs and a number of sheep died rapidly after drinking reservoir water, which was experiencing a bloom of *Microcystis*. Filtrates from algal scums contained the toxin microcystin at concentrations of 0.1– 1.7 $\mu g\,l^{-1}$ (Environmental Data Services, 1989) and public access, including angling and watersports, to all reservoirs in the region was banned for several weeks. Toxins do not always occur in blooms of cyanobacteria and concentrations when they do occur can be highly variable with time. Sampling indicates that there is a 45–75 per cent chance of an individual bloom being toxic (Lawton and Codd, 1991). This makes them very difficult to predict, detect and monitor.

Problems also arise with the bacterium *Clostridium botulinum*, which grows in the sediment of shallow, eutrophic lakes and releases a toxin during periods of hot weather. Birds are especially susceptible to botulism in shallow waters and serious losses of commercially valuable wildfowl periodically occur, especially in the United States (Eklund and Dowell, 1987).

These wide ranging effects of increases in nutrient loadings on receiving waterbodies show why eutrophication is such an important problem today.

Experiments within lakes

Much of the information concerning eutrophication has been gained by comparing observations made on polluted and unpolluted waters or by comparing data collected from one site at successive intervals. It is often difficult, however, to determine the cause–effect relationships in such a complex situation where many factors may vary simultaneously. The opposite approach is to conduct laboratory experiments, where cause–effect relationships can be verified by direct test, but it is often dangerous to apply results obtained under precise laboratory conditions to field situations. The manipulation of waters in the field lies between these two approaches and has proved highly illuminating in elucidating aspects of the eutrophication process.

Part of a lake may be isolated and nutrients added, using conditions in the water outside the enclosure as a control, or there may be a series of experimental and control enclosures (Ravera, 1989). Some enclosures may be very large. For example Lund tubes (Lund, 1978), used in the Lake District of

England, were made of butyl rubber, with a diameter of 45.5 m and a depth of 15 m. Such enclosures are too large and expensive to replicate. Smaller enclosures are generally made of polythene. One problem with enclosures is that the walls develop a microflora and fauna which can markedly alter the nutrient balance within, such problems increasing in smaller enclosures with large surface to volume ratios.

Scientists at the Freshwater Institute of the Fisheries Research Board of Canada have observed the effects of nutrient additions to whole lakes. Some 46 small lakes, within an area containing several hundred in Ontario, have been set aside for experimental research on eutrophication (Experimental Lakes Area). Lake 227 has a surface area of 5 ha and a maximum depth of 10 m. Phosphorus and nitrogen were added weekly from 1969 to 1972, the annual addition amounting to $0.48 \, g \, P \, m^{-2} \, yr^{-1}$ and $6.29 \, g \, N \, m^{-2} \, yr^{-1}$, these additions increasing the natural inputs by about five times.

The transparency of water, measured with a Secchi disc, became on average less in each year (Fig. 4.12). In 1969 the maximum biomass of phytoplankton was $16\,160 \, mg \, m^{-3}$, compared with $5000 \, mg \, m^{-3}$ in 1968, before fertilization. In 1970 the peak standing crop had more than doubled to $35\,000 \, mg \, m^{-3}$. Poor weather conditions in 1971 prevented further increase in the maximum biomass, but the peak biomass in 1972 increased to $63\,000 \, mg \, m^{-3}$.

Before fertilization the phytoplankton of Lake 227 was dominated by Chrysophyceae (golden-brown algae). In 1969, Chlorophyta (green algae) dominated in the summer. In 1970, cyanobacteria, which had never occurred before fertilization, dominated in the late summer. In 1971, cyanobacteria (especially *Oscillatoria*) again showed periods of dominance, although the

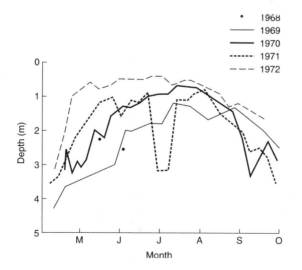

Fig. 4.12. Seasonal changes in the Secchi disk visibility in Lake 227, Ontario, Canada, 1968–72 (from Schindler *et al.*, 1973).

situation was complicated by unusually cold weather in July and early August. In 1972, cyanobacteria and Chlorophyta were again abundant, with the latter remaining dominant. The phytoplankton production in Lake 227 was some ten times higher than that occurring in natural lakes in the area (Schindler *et al.*, 1973; Schindler and Fee, 1973).

In Lake 227, algal blooms consisted primarily of chlorophytes, with cyanobacteria only occasionally dominant. By contrast a lower rate of fertiliz-ation to another lake, 226, consistently resulted in blooms of nitrogen-fixing cyanobacteria (Fig. 4.13). As well as the lower level of nutrient addition, Lake 226 was receiving a different ratio of N:P (5:1 by weight) compared with Lake 227 (14:1). It therefore appeared that the onset of a cyanobacterial bloom was triggered by a scarcity of nitrogen relative to phosphorus, rather than merely a

Fig. 4.13. Experimental Lake 226, Ontario. The far lake has been fertilized with phos-phorus, nitrogen and carbon, and supports an intense bloom of the nitrogen-fixing cyanobacterium *Anabaena spiroides*. The near lake received nitrogen and carbon only and has exhibited no algal bloom (photograph by Dr K. Mills).

phosphorus input. This was tested in Lake 227 in 1975 when nitrogen additions were reduced while phosphorus loading was kept constant (Schindler, 1988a). Cyanobacteria dominated the lake in summer, fixing atmospheric nitrogen to compensate for deficiencies in nitrogen supply.

These observations on whole-lake fertilization were supplemented by experiments in replicated enclosures (large polythene tubes) in which nutrients were added singly or in various combinations (Schindler *et al.*, 1971). The addition of either organic or inorganic carbon resulted in no increase in algal standing crop. The addition of nitrogen and phosphorus together always produced large standing crops of algae. When added singly, phosphorus always caused some increase in standing crop, whereas the addition of nitrogen alone never elicited a response. So, both whole-lake fertilization and manipulations in enclosures demonstrate that it is the concentration of phosphorus which is controlling the standing crop of phytoplankton. Limiting the supply of phosphorus to lakes would therefore appear to be the key to overcoming problems caused by eutrophication.

The problems of replication in whole-lake experiments have been discussed by Carpenter (1989). It is recommended that, because of limitations of experimental systems and high costs, experiments should consist of a minimum of one experimental and one paired reference system. A series of such experiments, performed at different times and in many locations, will provide more ecological insight than a replicated experiment in a single region.

Modelling eutrophication

Many lakes are exhibiting eutrophication as a result of human activities, and it is clearly impossible to study all of these in detail. Relationships which would allow the prediction of changes caused by rates of enrichment and which require the measurement of as few parameters as possible would obviously be extremely valuable to the managers of freshwater ecosystems.

Models of eutrophication vary widely in their complexity (e.g. Golterman, 1991). Modelling approaches to eutrophication are described in Jørgensen (1980) and Henderson-Sellers and Markland (1987). Vollenweider (1975) has emphasized that satisfactory models should meet three essential criteria. They should be general, realistic and precise. In practice one of these criteria usually has to be sacrificed in order to maximize the others.

Models of eutrophication have concentrated especially on the relationships of nitrogen and phosphorus to algal production, with particular emphasis on phosphorus because this nutrient can explain much of the variance in chlorophyll levels in lakes (Heyman and Lundgren, 1988).

The best known conceptual model is that of Vollenweider (1969, 1975):

$$\overline{P} = \frac{L}{\overline{z}\,(r_s + r_f)}$$

where $\bar{P}$ is the concentration of total phosphorus (g m^{-3}), often measured at the time of lake overturn in spring; L is the phosphorus loading (g m^{-2} lake area per year); $\bar{z}$ is the mean depth of the lake (m); r_s is the sedimentation rate coefficient (the fraction of P lost per year to the sediments); and r_f is the hydraulic flushing rate (the number of times the water in the lake is replaced each year).

This input–output model assumes that the lake is in a steady state, with loadings and flushing rates not changing with time, and is uniformly mixed. Most of the terms in the model can be measured using straightforward techniques, only r_s presenting special difficulties. The model has been used to produce loading values of nitrogen and phosphorus which would be permissible for lakes of different depths, and loading values which would cause deterioration of the lake (Table 4.6). The ratio of nitrogen to phosphorus in the water can also provide useful information. If the N:P ratio exceeds 16:1, phosphorus is likely to be limiting to algal growth. If the ratio is less than 16:1, nitrogen may be limiting.

Table 4.6 Vollenweider's permissible loading levels for total nitrogen and total phosphorus (biochemically active) (g m^{-2} yr^{-1})

Mean depth (m)	Permissible loading		Dangerous loading	
	N	P	N	P
5	1.0	0.07	2.0	0.13
10	1.5	0.10	3.0	0.20
50	4.0	0.25	8.0	0.50
100	6.0	0.40	12.0	0.80
150	7.5	0.50	15.0	1.00
200	9.0	0.60	18.0	1.20

The simple relationship between loading and depth has been further developed by including such parameters as the internal loading (such as from the sediments) and the length of the shoreline. From a study of many lakes worldwide empirical relationships between phosphorus loading, chlorophyll *a* concentration and transparency (measured simply using a Secchi disc) have been calculated (OECD, 1982), as shown in Fig. 4.14.

In Ontario, Canada, there is a great demand for summer cottages and other developments on lakesides, and their often rudimentary sanitation is a potential source of nutrients to sensitive lakes. A model was developed by Dillon and Rigler (1975) to predict the effect of development on lakes. The model consists of a sequence of equations relating nutrient loadings to watershed characteristics, lake nutrient concentrations to loading and biological responses to nutrient concentrations. The approach is illustrated in Fig. 4.15. The *total phosphorus* supplied to a lake from natural sources (J_N) derives from the catchment (J_E) and direct precipitation (J_{PR}). An estimate for the export of

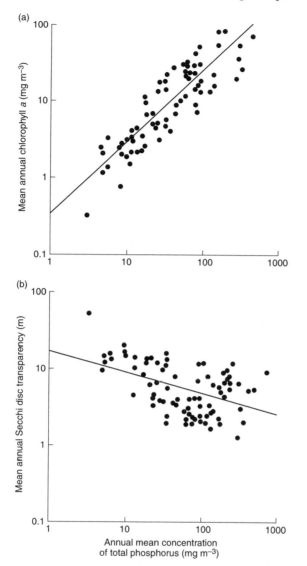

Fig. 4.14. Relationships between total phosphorus and a) chlorophyll *a*; b) Secchi depth from an international study of lake eutrophication (adapted from OECD, 1982).

phosphorus from a catchment can be made by examining the land use (e.g. the proportions of forestry, arable and pasture) and the geology (e.g. whether igneous or sedimentary). The export coefficient is multiplied by the area of the catchment and these are summed for all tributaries to the lake to obtain J_N. J_{PR} can be determined from long-term average rainfall values and the average concentration of phosphorus in rainwater.

The total natural phosphorus supplied to a lake in a year is:

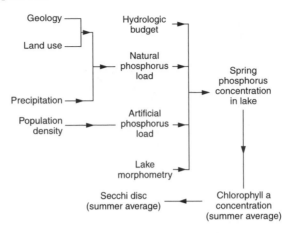

Fig. 4.15. Scheme of empirical models used to assess the effects of development on the trophic status of lakes (from Dillon and Rigler, 1975).

$$J_N = J_E + J_{PR} \, \text{mg yr}^{-1}$$

and the total natural loading (L_N) is J_N divided by the surface area of the lake (A_O) in $\text{mg m}^{-2} \text{yr}^{-1}$.

The artificial loading (J_A) to the lake will depend on a number of factors, such as the number of cottages, caravans etc., their methods of sewage disposal, the number of inhabitants and the time they spend in the cottage each year. Dillon and Rigler assumed that the average North American supplies 0.8 kg P to the environment each year. An estimate of the number and usage of cottages requires a survey.

The total supply of phosphorus to the lake (J_T) is:

$$J_T = J_N + J_A \, \text{mg yr}^{-1}$$

and the total loading (L_T) is:

$$L_T = J_T/A_O \, \text{mg m}^{-2} \text{yr}^{-1}$$

To predict the spring phosphorus concentration ($\bar{P}$) in the lake, Dillon and Rigler used the Vollenweider model described earlier. The sedimentation rate (r_s) was obtained indirectly through the retention coefficient, which is highly correlated with the loading per unit area and hence easily predicted.

Knowing the spring phosphorus concentration ($\bar{P}$), the summer average chlorophyll *a* concentration can be predicted:

$$\log_{10} (\text{chl}.a) = 1.45 \log_{10} (\bar{P}) - 1.14 \, \text{mg m}^{-3}$$

Chlorophyll *a* can then be related to transparency (the Secchi disc reading, Fig. 4.16).

Changes in the loading of phosphorus to a lake result in changes in the phosphorus concentration and hence water quality, but the response is gradual rather than immediate, and follows an exponential relationship. The response

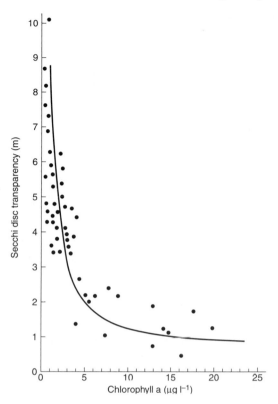

Fig. 4.16. Relationship between Secchi disc depth and chlorophyll *a* concentration for a number of lakes in southern Ontario. The values for each lake are based on means collected over the period of stratification (June–September) (from Dillon and Rigler, 1975).

time can be described by the half-life of the change in concentration ($t_{1/2}$) which depends only on the rate coefficients:

$$t_{1/2} = \ln 2(r_s + r_f)$$

Lakes with rapid flushing times will have short half-lives and hence response times, while lakes with slow flushing rates will respond only slowly to changes in loading.

The Dillon–Rigler model is mostly used by managers in the reverse direction. A minimum summer lake transparency or maximum chlorophyll *a* concentration which is considered acceptable would be decided. The maximum permissible loading for phosphorus from artificial sources (L_A) would be determined using the model and this would be translated into the maximum allowable development in terms of cottages, etc.

It would obviously not be feasible in terms of resources or time to scientifically evaluate every lake in Ontario where development was anticipated, so the model is valuable in that reasonable predictions of effects of development can be made with little or no field work. While some of the parameters used in the model apply only to Ontario, similar models have wide application in the management of water resources. Refinements of the Dillon–Rigler model allow lake managers to predict the effect of lakeside development throughout an entire watershed, including lakes situated downstream (Hutchinson *et al.*, 1991). Schindler (1987a) considers that the development of loading models to control phosphorus inputs is the greatest single success of ecological knowledge over environmental problems.

Controlling eutrophication

Table 4.5 lists the guidelines for classifying the trophic status of a waterbody following a survey. In Ardleigh Reservoir, whose plankton was described on p. 104, peak phosphorus concentrations were some 250 times the minimum concentration for assigning a waterbody to the eutrophic category, while peak concentrations of nitrogen were 10 times the minimum concentration. We have seen that the productivity of freshwaters is most often limited by phosphorus. Eutrophication may therefore be most effectively controlled by reducing the load of phosphorus, and there are a number of reasons why phosphorus control is likely to be more effective than nitrogen control. Phosphate is present in only trace amounts in oligotrophic lakes and the inflow streams are low in phosphate where they are not influenced by human activities. In contrast, the inflow streams may contain large quantities of nitrates. Nitrates are leached readily from agricultural lands whereas phosphates tend to be tightly bound (p. 97). Some cyanobacteria and bacteria can fix gaseous nitrogen so will not be limited by nitrogen provided other essential nutrients are available. It is also relatively easier and cheaper to remove phosphorus than nitrogen during the sewage treatment process. Nevertheless, where water is reused for potable supply it may be necessary to reduce the level of nitrogen for public health or legal reasons (p. 114).

Rast and Holland (1988) have provided a basic, practical framework for the development of a eutrophication management strategy. It consists of the following steps:

1. Identify eutrophication problem and establish management goals.
2. Assess the extent of available information about the lake or reservoir.
3. Identify available, feasible eutrophication control methods.
4. Analyse all costs and expected benefits of alternative management strategies.
5. Analyse adequacy of existing institutional and regulatory framework for implementing alternative management strategies.
6. Select desired control strategy and disseminate plan to interested parties.
7. Use institutional mechanisms to minimize future eutrophication problems.

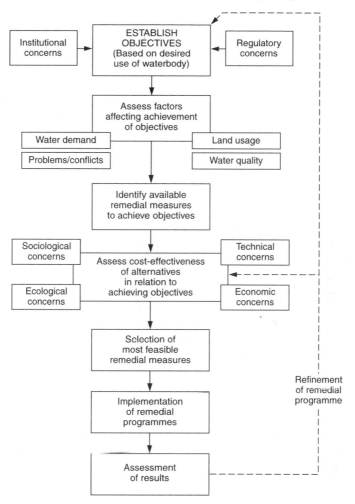

Fig. 4.17. Sequence of decisions to be made in the development and implementation of eutrophication control programmes (from Rast and Holland, 1988).

A schematic representation of these steps is illustrated in Fig. 4.17.
Restoration methods may be divided into two classes:

1. Reduction of external loading of nutrients.
2. Manipulations within the lake ecosystem.

In practice it is often necessary to apply both techniques simultaneously to ensure some measure of success. Ryding and Rast (1989) and Cooke (1993) describe in detail the steps needed for eutrophication control.

Reduction in external loading

The discharge of nutrient-rich sewage and animal wastes into watercourses and lakes is often the major source of enrichment problems, while detergents are often the most important source of phosphorus in waste waters. In the United States 40–70 per cent of phosphorus within sewage came from phosphate detergents and the output from municipal sewage plants may contain up to 70 per cent of the total input of phosphorus to a lake. The elimination of phosphorus as an essential component of detergents could therefore remove about 50 per cent of the total phosphorus entering some lakes. Effective phosphate-free detergents have been developed and some states and cities in the United States have banned or restricted the phosphorus content of detergents.

Primary sewage treatment removes only about 5–15 per cent of nutrients and secondary treatment only 30–50 per cent, so that a tertiary treatment process is necessary to remove the majority of phosphorus contained within sewage. Treatment may involve chemical removal, physical removal or biological removal.

Phosphates are precipitated, a process known as phosphate stripping, using coagulants of lime or compounds of aluminium or iron. The precipitate is then separated in a sedimentation unit. The process is 90–95 per cent efficient at removing phosphorus. Biological removal uses the ability of some micro-organisms to take up phosphorus in excess of their immediate nutritional requirements and store it within the cells in the form of polyphosphates. These can then be removed from the system with the sludge. Biological removal of phosphorus can be achieved by modifying existing processes for nitrogen removal.

Phosphorus may also be removed by passing water into treatment ponds. Much of the element is absorbed on to sediments which will settle out in the pond. If the water is retained for three or more days nutrients will be taken up by algal blooms, which will die and transfer phosphorus to the sediment. It may also be possible to use simple floating plants, such as duckweeds (*Lemna*) or water fern (*Azolla*) which can be regularly harvested. Details are provided by Viessman and Hammer (1993).

The operating costs of tertiary treatment may be equivalent to those of secondary treatment, thus doubling the overall costs of plants removing phosphate. It is therefore essential to know accurately the proportion of the total phosphorus loading that is due to sewage effluents and the efficiency of the proposed tertiary treatment plant before it is installed. A nutrient budget for Barton Broad, eastern England, showed that 80 per cent of a total phosphorus loading of $18–22 \, \text{mg} \, P \, m^{-2} \, yr^{-1}$ was derived from sewage sources, and it was considered that an 80 per cent reduction in the phosphorus content of sewage effluents was required to permit the re-establishment of aquatic macrophytes (Osborne, 1981). Subsequently phosphorus removal was initiated by adding ferric sulphate to final effluents from sewage treatment works upstream of Barton Broad, and this eventually reduced the annual phosphorus load to the lake by 90 per cent. Total phosphorus concentrations within the lake, however, remain high, summer values exceeding $100\,\mu g\,P\,l^{-1}$, while the annual mean chlorophyll concentration exceeds $100\,\mu g\,l^{-1}$ and the Secchi depth transparency is only 30 cm in summer (Phillips and Kerrison, 1991). Internal loading can be as high as $130\,mg\,P\,m^{-2}\,d^{-1}$, compared with an external load of only $12\,mg\,P\,m^{-2}\,d^{-1}$ (Phillips *et al.*, 1994). Macrophytes have not returned. An alternative to tertiary sewage treatment is to divert nutrient-rich wastes away from vulnerable lakes, which can only be done if there is somewhere, usually the sea, to put the diverted water. Lake Washington (p. 97) was one example where this procedure was applied successfully. It may also be possible, where the majority of external loading is derived from one source as it usually is in a pumped-storage reservoir scheme, to treat the inflow water in a holding lagoon before it enters the main waterbody. In Ardleigh Reservoir, 90 per cent of the load of soluble reactive phosphorus was derived from water pumped from the River Colne, which has a number of sewage treatment works upstream (Redshaw *et al.*, 1988), and this water is now being treated with ferric sulphate, to precipitate out phosphorus, before it enters the reservoir.

It is also possible to reduce nutrient loadings to lakes by altering land-use practices in the catchment. The prevention of erosion, changes in the timing and reduction in the frequency of fertilizer applications and the development of methods for dealing with animal waste are some of the measures that can be taken. Modifying cropping systems, especially ensuring that land is not bare in winter, or putting land down to permanent pasture can also be undertaken, and this latter is being encouraged in so-called nitrate sensitive areas in the United Kingdom, where groundwater is heavily contaminated with nitrate. The planting of buffer strips between farmland and watercourse may also be effective and such a scheme has been recommended to protect the deteriorating Slapton Ley nature reserve in southwest England (Wilson *et al.*, 1993). Ryding and Rast (1989) provide a detailed assessment of methods to control external sources of nutrients.

Cullen and Forsberg (1988) recognized three main types of response of lakes to a reduction in the external load of phosphorus:

1. Reduction in lake phosphorus and chlorophyll concentration sufficient to change the trophic category of the water, e.g. from hypertrophic to eutrophic or eutrophic to mesotrophic.
2. Reduction in lake phosphorus and chlorophyll concentration which is insufficient to change the trophic status, though lakes become less eutrophic.
3. Reduction in lake phosphorus that results in little or no reduction in chlorophyll or algal biomass, though nuisance species may be reduced.

In a survey of 43 lakes where point sources of phosphorus had been reduced, Cullen and Forsberg consider that 15 fell into type 1, 9 into type 2 and 19 into type 3, that is, in 44 per cent of cases there had been no significant improvement.

In the Netherlands, for example, the external phosphorus load to the

Loosdrecht lakes ecosystem was substantially reduced in 1984 but there was no improvement in water quality (Liere *et al.*, 1992). In Esthwaite Water in the English Lake District, the removal, from 1986, of the phosphorus from the sewage effluent contributing 47–67 per cent of the total P loading, resulted in no marked changes in water quality and no reduction in the cyanobacterial dominance of the phytoplankton (Heaney *et al.*, 1992). Reduction of phosphorus loadings, usually in excess of 70 per cent, to 27 Danish lakes resulted in very little improvement in conditions (Jeppeson *et al.*, 1991). Barton Broad was mentioned above. Therefore, within-lake measures *in addition* to external phosphorus control are often essential to achieve the desired objectives of lake restoration.

Manipulations within the lake ecosystem

In stratified eutrophic lakes oxygen depletion occurs in the hypolimnion, which may result in the death of fish. Anaerobic conditions at the sediment surface result in the release of phosphate which then becomes available to algae at overturn, resulting in more photosynthesis and more organic matter to deplete oxygen further in the hypolimnion. This cycle can be broken by aeration. Destratification also increases the turbulence in the water, which may affect the ability of the phytoplankton to grow, especially the nuisance cyanobacteria. The techniques of destratification and re-aeration are described in detail by Henderson-Sellers and Markland (1987) and Harper (1992).

In the Biesbosch Reservoirs in the Netherlands, fed by polluted and highly eutrophic water from the River Meuse, the injection of air at the bottom of the reservoirs prevents thermal stratification and hence a serious deterioration in water quality is avoided. It also mixes algae over a sufficient depth so that light, rather than nutrients, limits growth. This, combined with grazing by zooplankton, maintains the biomass at acceptable levels (Oskam and Breeman, 1992).

One of the major problems following a reduction in external phosphorus is an internal loading of phosphorus from the sediments. There is normally a net flow of phosphorus to the sediments but release occurs under conditions of low oxygen. Aerobic release may also occur, probably dependent on desorption of phosphorus from inorganic complexes and the mineralization of phosphorus associated with organic material, especially where sediments are in contact with overlying water low in orthophosphate. In Ardleigh Reservoir, the net sediment release of soluble reactive phosphorus between July and September was equal to $23\,\mathrm{mg\,P\,m^{-2}\,d^{-1}}$, equivalent to 33 per cent of the mean annual loading of the nutrient (Redshaw *et al.*, 1988). At other times of the year the sediments acted as a sink, but release coincided with the development of large algal blooms. In the Norfolk Broads internal loading has been recorded as high as $278\,\mathrm{mg\,P\,m^{-2}\,d^{-1}}$ (Phillips *et al.*, 1994). Reviews of internal loading from sediments are provided by Boström (1984), Forsberg

(1989) and Marsden (1989). Phosphorus cycling between sediments and water is complex and is still poorly understood. It is clear, however, that it is not only a chemical process dependent on redox and pH, but microbial processes are also involved. Microorganisms may release or bind phosphorus through metabolic reactions, extracellular release and cell lysis. Brunberg and Boström (1992) have shown that colonies of *Microcystis* in lake sediments stimulate bacterial activity, eventually resulting in a release of phosphorus to the water column. As *Microcystis* colonies may last for several years in the sediments, they may be important for delaying the recovery of lakes.

One approach to preventing the release of phosphorus from the sediment is chemically to 'seal' the nutrient in. Foxcote Reservoir, in midland England, is highly eutrophic and periodically the water has been untreatable for up to six months of the year. Following laboratory tests, pumping of river water to the reservoir was stopped in early April and the waterbody was dosed, from a boat, with 9000 l of ferric sulphate liquor (3.5 mg Fe l^{-1}). Within 5 minutes of dosing the soluble reactive phosphorus in the water dropped from 0.025 mg l^{-1} to 0.001 mg l^{-1}. A layer of ferric hydroxide floc, about 1 cm thick, formed over the reservoir sediment, acting as a barrier for phosphorus release (Redshaw, 1983). In June pumping of river water recommenced but it was treated with ferric sulphate before it was released into the reservoir. By the summer following the treatment the algal populations had declined sharply (Fig. 4.18) and in the next year a dense growth of macrophytes had developed in the reservoir. The sudden reduction in the phosphorus loading switched the waterbody from phytoplankton dominated to macrophyte dominated.

Jaeger (1994) reported a 90 per cent decline in maximum chlorophyll levels in a small glacial lake following treatment with ferric chloride. Boers *et al.* (1994) observed a similar rapid decline in phosphorus and chlorophyll following the treatment of a shallow Dutch lake with ferric chloride. Their

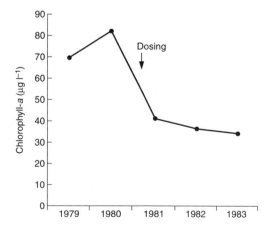

Fig. 4.18. Peak algal standing crops (μg chlorophyll *a* l^{-1}) in Foxcote Reservoir before and after treatment with ferric sulphate (from data in Hayes *et al.*, 1984).

success was only transitory, however, due mainly, it was thought, to the short residency time of the water in the lake (35 days) and the high external phosphorus loading. This emphasizes the need to control external P sources before undertaking within-lake treatment. Aluminium salts have also been used with some success to coagulate and seal phosphorus into sediments (Cooke, 1993); the treatment effectiveness in six Washington State (USA) lakes ranged from 50–80 per cent and lasted for at least five years (Welch and Shrieve, 1994). The metal is potentially toxic, however, and is best avoided in any water likely to be used for public supply.

The alternative, generally more expensive, technique is to remove the sediment entirely, thus both taking away the internal source of phosphorus and deepening the lake. Moss *et al.* (1986) described the restoration of two small lakes in the Norfolk Broads system. In Alderfen Broad, diversion of an inflow stream in 1979 led to effective isolation from the main source of phosphorus. At Cockshoot Broad a dam was constructed to isolate the lake from the nutrient-rich river (Fig. 4.19) and the sediment was pumped out.

In the first two years after isolation the phytoplankton in Alderfen Broad was greatly reduced, the water became clear and a dense growth of *Ceratophyllum demersum* developed. Over the following three years, however, the plants gradually disappeared to be replaced by a large phytoplankton bloom, including cyanobacteria, as phosphorus was released from the sedi-

Fig. 4.19. The dam across the channel linking Cockshoot Broad to the River Bure. Nutrient-rich river water is excluded and the phosphorus-rich sediment was pumped out of the broad (photograph by the author).

ments. It is not known why phosphorus release did not occur in the early years after diversion but it may have been the result of decreased input of organic matter from reduced spring plankton populations. Once plants re-established, the organic matter produced by them may have led to the increase in phosphorus release from the sediments (Phillips, 1992), the plants being at least partly responsible for their own demise. Poor fish recruitment and adult fish mortality in the late 1980s led to an increase in large-bodied cladocerans and a return of *Ceratophyllum*, which produced a substantial biomass over the period 1988–91, despite high phosphorus concentrations (Perrow *et al.*, 1994). The community in Alderfen Broad remains unstable and the sediment has been removed to reduce internal loading and deepen the lake, though a successful outcome is not yet guaranteed.

At Cockshoot, after isolation and sediment removal, phytoplankton growth declined, the water became clear and a diverse assemblage of aquatic plants developed in part of the broad, possibly helped by a scarcity of planktivorous fish (many of which had been removed by netting). Most of the open water, however, has not been colonized by plants despite relatively low phosphorus levels and the absence of release from the sediments. The reasons are not known but it may be due to a combination of factors – the lack of sufficient source material of macrophytes, excessive grazing by a water-bird, the coot (*Fulica atra*), or phytoplankton still at sufficiently high density to prevent adequate light transmission to the bottom of the lake (Phillips, 1992). This restoration has been only partially successful.

Other techniques to prevent recycling of nutrients or to accelerate the outflow of nutrients have included sealing lake bottoms (with polythene sheeting), selectively discharging hypolimnetic water in water supply reservoirs, or diluting and/or flushing with water from an oligotrophic source. The removal of nutrient-rich macrophytes, algae or fish (biotic harvesting) has also been attempted.

Recently barley straw has been used successfully to reduce the growth of filamentous and planktonic algae in canals and ponds. Reductions of 90 per cent in algae growing on microscope slides downstream of the straw were achieved, with no change in nutrients (Welch *et al.*, 1990; Ridge and Barrett, 1992). The mechanism of action remains unknown (Pillinger *et al.*, 1992) but the method has considerable potential and barley straw is in great surplus in many cereal-growing areas.

Biomanipulation

The restoration techniques so far discussed have been largely concerned with controlling the sources of phosphorus which fuel algal growth, so-called *bottom-up* techniques. On p. 109 I described the complex interactions occurring between the members of the aquatic community, and how a breakdown in one of these interactions could lead to the rapid collapse of the system, terminating in a stable, phytoplankton-dominated community. Once phosphorus reduction has been achieved it may be possible to recreate the complex community to control excessive algal growth by *top-down* methods or *biomanipulation*. Biomanipulation can be defined as *the manipulation of the food web of aquatic ecosystems to increase the numbers of grazers on algae* (Moss, 1992).

Fish may be significant predators of larger zooplankton, especially in the absence of cover (p. 112) and such predation can result in intense algal blooms. The addition of increasing numbers of blue-gill sunfish (*Lepomis macrochirus*) to enclosures in a lake led to an increase in chlorophyll *a* concentration and a decrease in transparency in proportion to fish numbers (Lynch and Shapiro, 1980). In another experiment, Round Lake, Minnesota, was treated with rotenone, a fish poison, in autumn 1980. The majority of the fish, mainly planktivores, were killed and the lake was then stocked with piscivorous species. Over the following two years there were decreases in chlorophyll *a* and total phosphorus in the water and increases in transparency, the body size of zooplankton, and of *Daphnia* (Shapiro and Wright, 1984). Figure 4.20 illustrates the reduction in algal primary production in a Swedish lake, Lilla Stockelidsvatten, following treatment with rotenone in November 1973. The improvement in conditions was quite dramatic and Secchi disc transparency more than doubled. Successful results in reducing algal growth following fish removal have also been achieved in a water supply reservoir (Benndorf, 1987). Fig. 4.21 dramatically illustrates the effects of fish removal on water quality in the University of Essex lakes.

In the Norfolk Broads experiments with enclosures have been undertaken (Irvine *et al.*, 1991; Phillips and Kerrison, 1991). Hoveton Great Broad (Fig. 4.22) is fairly large (31 ha), shallow (1–1.5 m) and largely devoid of macrophytes. Pens of 4 m² were constructed which allowed for free movement of water and phytoplankton, but prevented movement of zooplankton and excluded fish. The phytoplankton community in the pens and open water were

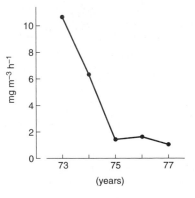

Fig. 4.20. Algal primary production (mean midday values) in Lake Lilla Stockelidsvatten, following removal of fish by rotenone in November 1973 (from De Bernardi, 1989, after Henrikson *et al.*, 1980).

Fig. 4.21. Two small, connected lakes on the campus of the University of Essex. The upper lake (a) developed a dense algal bloom in August 1994 as it did in previous years. The lower lake (b) had a large population (30 000) of rudd and bream removed in autumn 1993 after several years of study; in contrast to previous years no algal bloom developed in 1994, the water remained clear and macrophytes began to establish. Swards of macrophytes, mainly *Potamogton crispus*, dominated the lake in summer 1995 (photographs by the author).

Fig. 4.22. Aerial view of the experimental enclosures in Hoveton Great Broad, eastern England. A barrier which excludes fish but allows water movement can be seen, together with a series of circular experimental enclosures (photograph courtesy of Dr G. L. Phillips, National Rivers Authority).

similar throughout the summer but densities of *Daphnia* were much higher in the enclosures. They also had higher egg ratios and mean brood sizes because of the greater proportion of larger animals in the pens. This suggests that zooplankton populations in the lake are controlled by size-selective predation by fish and not by algal production. It is suggested that the removal of fish would extend a brief clear water phase, caused by zooplankton grazing, throughout the summer, allowing macrophytes to establish.

Dutch lakes are shallow and very similar to the Norfolk Broads. They have suffered a similar loss of macrophytic vegetation with eutrophication. Once macrophytes have gone, populations of the predatory pike (*Esox lucius*) decline. Pike need refuges to reduce the risk of intraspecific predation and also cover for hunting. In waters with dense vegetation populations of young pike are high and they effectively control the abundance of young bream. In the absence of pike, bream come to dominate the fish community, eating zooplankton, stirring up the bottom sediments and accelerating eutrophication (Hosper, 1989). Lake Bleiswijkse Zoom was divided into two compartments. In April 1987 the majority of planktivorous and bottom-feeding fish (at a density of $650\,\mathrm{kg\,ha^{-1}}$) were removed from one compartment and fry of the predator zander were introduced. The other compartment was used as a reference. Removal of fish resulted in low concentrations of chlorophyll *a*, total phosphorus, total nitrogen and suspended solids, while Secchi disc transparency increased from 20 cm to 110 cm (Meijer *et al.*, 1989). By June charaphytes (the

Phase 1 community of the Broads, p. 103) had become abundant, creating a habitat for the re-establishment of the major predator, the pike. The success in biomanipulation is due not only to increased grazing pressures on algae but also to a decreased availability of phosphorus. This results from the sedimentation of detritus as the increased growth of macrophytes reduces turbulence in the water (Boers *et al.*, 1991).

As well as food web effects, fish may also have direct effects on macrophytes. The removal of fish, predominantly bream, from a gravel pit resulted in a ninety-fold increase in plant cover and an increase in plant biomass from $1 \mathrm{g\,m^{-2}}$ to $47 \mathrm{g\,m^{-2}}$. Zooplankton numbers remained high and unchanged. The re-introduction of fish into an enclosure resulted in a decline of plant biomass from $47 \mathrm{g\,m^{-2}}$ to $5 \mathrm{g\,m^{-2}}$ (Wright and Phillips, 1992). Fish both inhibited the initial establishment of seedlings and reduced the growth of existing stands. Bottom-feeding fish like bream also resuspend the sediments, reducing the transparency of shallow lakes, while the excretory products of large populations of fish add significant amounts of nutrients to the water (Breukelaar *et al.*, 1994).

Fish population manipulation as a technique to control eutrophication is a recent approach and its long-term success is by no means guaranteed. Lakes are often dominated by single, unstable populations of macrophytes. For example, following manipulation in Lake Vaeng, Denmark, in 1986–88, *Potamogeton crispus* became dominant as Secchi depth doubled. By 1990, however, *Elodea canadensis*, which is not native to Europe, became exclusively dominant in the lake (Lauridsen *et al.*, 1994). There is also evidence that the clear-water phase resulting from fish removal may only be temporary. For example Lake Bleiswijkse Zoom (see above) gradually increased in turbidity from the second year of biomanipulation, while the condition in other lakes deteriorated after five years (Meijer *et al.*, 1994). Planktivorous fish populations build up despite the presence of predators. Biomanipulation has much promise but is not yet a proven technique.

Dutch workers have suggested that the zebra mussel (*Dreissena polymorpha*) could be used along with fish manipulations to control algal growth (Reeders *et al.*, 1989). Populations of mussels in two lakes were sufficient to filter the entire water mass once or twice per month. The effect of zebra mussels is shown dramatically in Lake Erie, North America, which was invaded by this non-native in 1988. The number of algal cells in the lake in the mid-1980s was only 20 per cent of the population in the mid-1960s, due to changes in phosphorus loading. With the invasion of the mussels, however, there was a decline of more than 90 per cent in the algal cells in 1988–90, and by 1991 the chlorophyll levels were more typical of oligotrophic waters. Densities of the diatom *Fragilaria* fell from 360 cells $\mathrm{ml^{-1}}$ in 1987 to 14 cells $\mathrm{ml^{-1}}$ in 1989 and less than 1 cell $\mathrm{ml^{-1}}$ in 1990 (Nicholls and Hopkins, 1993). Zebra mussels are unselective feeders, any unwanted material being deposited in a mucus stream as pseudofaeces, so that all algal species are reduced in numbers. This contrasts with the size-selective feeding of zooplankton.

Economically it is the eutrophication of water supply reservoirs which is of most concern. Very often these reservoirs also support recreational fisheries and are intensively stocked. In Ardleigh Reservoir, referred to several times previously, the zooplankton community is dominated by small-bodied forms, while *Daphnia hyalina* has a small adult size relative to populations in similar waters (Mason and Abdul-Hussein, 1991). The reservoir is managed as a put-and-take fishery for rainbow trout, which are unable to breed in such eutrophic waters. Trout stomachs contain many zooplankton. From the discussion above a reduction in fish populations would seem to be a sensible measure as part of the restoration process, which at present consists of treating the inflow water with ferric salts, with only limited success. However the costs of water treatment problems have never been weighed against the benefits of the revenue received from angling, and even the countenancing of such an analysis is seen as politically unacceptable. In many such reservoir fisheries pike are destroyed whenever possible and even fish-eating birds, especially the cormorant (*Phalacrocorax carbo*), are tolerated only because they are legally protected.

Lakes can be restored successfully, especially when a reduction in phosphorus loading is followed by biomanipulation. Biomanipulation is likely to be most effective in shallow waters where macrophytes are a major component of the ecosystem, or in artificial waterbodies where fish removal is a feasible and acceptable option. Each lake must be considered as an individual, however, and will require its own prescription, so that a detailed limnological study is always necessary. Adequate finance and political will are also essential. That so many attempts at restoration have, at least, been only marginally successful is due in part to a lack of imagination by lake managers, and to the often grudging way such schemes are sanctioned by the authorities following prolonged public campaigns.

Chapter 5

ACIDIFICATION

Acid rain and the acidification of freshwaters has received considerable attention over the past 20 years, but the problem is not new. There were observations of lakes in Scandinavia losing their fish populations as early as the 1920s, while studies of diatom remains in sediment cores from lakes in southwest Scotland have indicated that acidification began around the middle of the last century. There has certainly been an acceleration of the process in the last three decades. For example, of 87 lakes in southern Norway surveyed in both the periods 1923–49 and 1970–80, 24 per cent had a pH below 5.5 in the earlier period, compared to 47 per cent in the later.

Schindler (1988b), Wellburn (1994) and Howells (1990) provide syntheses of the causes and consequences of acid rain, while Morris *et al.* (1989) and Steinberg and Wright (1994) contain review papers dealing specifically with freshwater ecosystems.

Sources of acidity

The main pollutants responsible for acid rain are sulphur dioxide (SO_2) and the oxides of nitrogen (NOx). Figure 5.1 illustrates the rate of emission of these gases in the United Kingdom during the present century. Currently some 1.8 million t of sulphur and 0.6 million t of nitrogen are emitted annually. Around 60 per cent of the SO_2 is derived from power stations and 30 per cent from industrial plants, the amount of emission having been reduced since 1970 with improved air pollution control legislation. By contrast the emission of nitrogen has continued to rise, with 45 per cent derived from power stations and 30 per cent from the exhausts of vehicles (Mason, J., 1989). There is also evidence that ammonium compounds, mainly from the excreta of intensively farmed livestock, are increasingly involved in acidification.

When SO_2 and NOx reach the atmosphere they react with moisture and undergo oxidation, resulting in the formation of sulphuric and nitric acids, which exist mainly in the clouds and fall to earth in rain or snow (*wet deposition*). Conversion rates are very rapid, being approximately 100 per cent per hour in summer and 20 per cent per hour in winter. Alternatively, in a dry atmosphere complex photochemical reactions, involving highly reactive oxidiz-

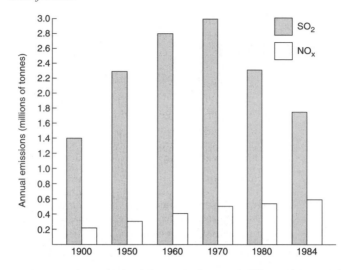

Fig. 5.1. Annual emissions of sulphur and nitrogen (millions of tonnes) in the UK (from data in Mason, J., 1989).

ing agents such as ozone, result in the production of sulphuric and nitric acids, conversion rates being approximately 16 per cent per day in summer and 3 per cent per day in winter, much slower than the reactions in a moist atmosphere. The acids from these reactions reach the surface of the earth in gaseous or particulate form (*dry deposition*). The production of acids from precursor gases seems to be limited by the availability of oxidizing agents, hydrocarbons and ultraviolet radiation from the sun. There is evidence that the whole atmosphere in the northern hemisphere is more reactive than it was a few decades ago, and this accelerates the process of acidification. A review of chemical reactions in the atmosphere is provided by Clarke (1992).

Uncontaminated rainwater, in equilibrium with atmospheric carbon dioxide, has a pH of 5.6. Almost everywhere in the world the pH of rain and snow is lower than this. Fig. 5.2a illustrates the annual fall-out of sulphur in Europe, the highest concentrations being in central England, central Europe and the Alps. The fall-out of nitrogen follows a similar pattern. This pattern is reflected in the average pH of rainfall (Fig. 5.2b), with eastern Britain, parts of Scandinavia and central Europe averaging less than pH 4.3, more than ten times more acid than 'natural' rainfall (pH being measured on a log scale). The average annual rainfall of the United Kingdom during 1978/80 was between pH 4.5 and 4.2 almost everywhere. North America is similarly affected, with large areas of the northeast suffering acidification, as well as the southeast, midwest and far west. Some 2500 lakes and 36 000 km of streams in the United States have been identified as either acidic or sensitive to acidification from acid rain (Downey *et al.*, 1994). The lowest pH recorded in a rainfall sample in Pennsylvania in 1978 was 2.32! (Lynch and Corbett, 1980).

Sulphuric acid contributes roughly 70 per cent to the mean total annual

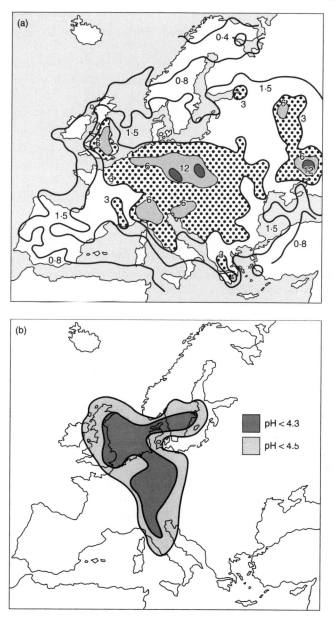

Fig. 5.2. Total fallout of (a) sulphur ($g\,m^{-2}\,yr^{-1}$) over Europe and (b) the pH of rainfall (adapted from Pearce, 1987, and OECD, 1977).

acidity of precipitation in northwest Europe and about 60 per cent in eastern North America, most of the remainder being due to nitric acid. Although much of the pollution falls fairly locally in the area of production (Fig. 5.2a), some of it will be transported by winds for several thousand kilometres before it eventually falls in rain. It is calculated, for example, that about 17 per cent of the acid falling on Norway derives from Britain and about 20 per cent of that falling on Sweden is from Eastern Europe.

The input of acidic materials to freshwaters may be from three sources: directly from the atmosphere, indirectly from the atmosphere via run-off in the catchment, and from the generation of acidity within the catchment (e.g. by soil acidification). The impact of acid precipitation on freshwaters is dependent on the surrounding geology and soils, which determine the capacity of the water to neutralize acids. Regions with a calcareous geology are not sensitive to acidification and even small amounts of limestone in a drainage basin exert considerable influence in areas which would otherwise be very vulnerable (Henriksen, 1989). Acidification is most likely to occur where the bedrock is granite or gneiss, with thin soils which have insufficient base cations freely available to neutralize the acidity which is deposited. Areas in Europe and North America with sensitive geology and soils are shown in Fig. 5.3.

The acids naturally present in soil, organic and carbonic acids, are so-called weak acids, and although they contribute significantly to the overall acidity of the soil, they do not dissociate into their respective anions and cations to the same extent as the strong acids (H^+, SO_4^{2-} and NO_3^-), which are derived from acid precipitation. The transfer of acidity from soil to surface waters requires mobile negative ions to bind to the acid hydrogen ions. As the rate of supply of dissociated hydrogen ions increases in soil water from acid precipitation, the rate of cation exchange is increased: that is, cations such as Na^+, K^+ and Mg^{2+} are displaced from exchange sites on soil particles. These can then move through the soil, provided that mobile ions are present to transport them, and acid precipitation provides these in the form of SO_4^{2-}. The nitrate ions could serve the same purpose, but most are taken up by vegetation as soils are generally nitrate deficient. Strong acids also mobilize the aluminium ion (Al^{3+}), which is of particular significance because of its toxicity. Heavy metals may also be mobilized (Veselý, 1994). The sulphate ion thus efficiently transfers acidity from soils to surface waters.

Land use influences the rate of acidification of freshwaters, forestry being especially important. Fig. 5.4 illustrates the headwaters of the River Severn in central Wales, which winds through the plantation forest of Hafren. The pH in the river falls to a minimum of 4.4, and minimum pH values below 5.5 are recorded for several kilometres below the forest. By contrast in a tributary of the Severn, the River Dulas, which drains open moorland, a minimum pH of 5.0 was recorded near the source, but most of the river had a minimum pH above 5.5 (Mason and Macdonald, 1987). Vegetation scavenges both dry deposition and pollutants held in mists and fogs (*occult deposition*) very efficiently and the rainfall reaching the soil beneath vegetation is more acid than incoming rain, because it contains both these scavenged pollutants and materials leached out of the plants. Conifers appear to be especially efficient at scavenging pollutants. Data from the Llyn Brianne catchment in mid-Wales is shown in Table 5.1. Both acidity and sulphate deposition increase beneath

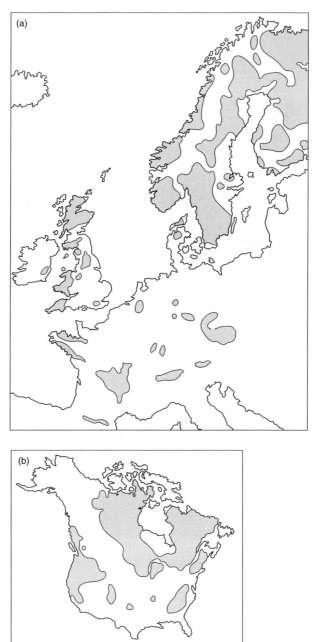

Fig. 5.3. Areas of (a) Europe and (b) North America with geology and soils rendering freshwaters liable to acidification.

Fig. 5.4. Upper reaches of the River Severn in mid-Wales showing extensive afforestation with conifers in the catchment, resulting in marked acidification of the water (photograph by the author).

spruce. The higher rate of evapotranspiration from trees compared with grassland results in a reduced amount of run-off of water through soils, with a greater concentration of pollutants. Trees also take up nutrients for growth, including ions which could potentially neutralize acids. The accumulation of litter as forests develop also leads to the natural acidification of soils. For example, significant acidification of soils over only 15 years took place in broad-leaved plantations on former arable soil, despite their buffering capacity (Hughes-Clarke and Mason, 1992).

The drainage channels cut through plantation forests expose sulphur and nitrogen to the atmosphere, acids being produced by oxidation. The relationship between forest cover and stream chemistry is illustrated in Fig. 5.5. Comparing an unafforested catchment and one with 30 per cent forest cover, mainly of Sitka spruce, a sharp decline in pH is apparent while both mean sul-

Table 5.1   Acidity and sulphate reaching soils beneath different types of vegetation at Llyn Brianne, Wales (adapted from Gee and Stoner, 1988)

	Mean pH	*Mean SO$_4$ ($\mu equiv\ l^{-1}$)*
Ambient precipitation	4.60	54
Moor grass	4.70	124
Oak	4.70	115
12-year-old spruce	4.27	144
25-year-old spruce	4.32	296

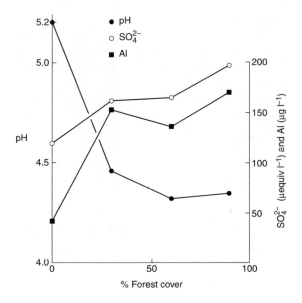

Fig. 5.5. Effect of afforestation on mean pH, sulphate ($\mu\,eq\,l^{-1}$) and aluminium ($\mu g\,l^{-1}$) in streams in adjacent catchments in Scotland (adapted from data in Harriman and Wells, 1985).

phate and mean aluminium increase markedly. In an extensive survey in Wales, pH was found to decrease and aluminium concentration of stream water increased with increasing percentage of plantation forest cover within catchments (Ormerod *et al.*, 1989).

The practice of liming upland grasslands to raise productivity increased in the United Kingdom from a total usage of less than 20 000 t per annum in the 1940s to more than 60 000 t in the late 1950s. It has since gradually declined, with a change in the liming subsidy, to around 30 000 t per annum. Ormerod and Edwards (1985) have suggested that liming could have had a substantial effect on the mineral concentrations of soft water streams, for increased acidification in some Welsh streams since the mid-1960s parallels a decline in the application of lime to agricultural land.

Long- and short-term changes in acidity

It is extremely difficult to determine long-term trends in acidification by examining series of water quality records. Historically these have been taken for other purposes and usually in lowland, buffered reaches of rivers. Furthermore, analytical techniques have been refined in recent years. In particular, accurate pH measurement in waters of low conductivity is difficult and earlier analysts would have been unaware of this. In an analysis of 75 data sets from the United Kingdom, Ellis and Hunt (1986) could find evidence of a downward trend in only six. The most convincing was the River Ystwyth in west Wales, where mean pH fell from 6.5 to 5.5 over the period 1970–84. Figure 5.6 shows a comparison of historical (1930–34) and recent (1979) pH values, measured colorimetrically for a series of lakes in the Adirondack Mountains in the United States. There were fewer lakes with pH above 7.0 and more with pH below 6.0 in the later survey, acidification being greater in those lakes which had a pH originally above 6.5. In southern Norway, of 87 lakes surveyed in both the periods 1923–49 and 1970–80, 24 per cent had a pH below 5.5 in the earlier period compared to 47 per cent in the second period.

In Chapter 4 the analysis of diatom remains in sediment cores in order to track progressive eutrophication was described (p. 106). The same technique has been used to follow the course of acidification in lakes. In lakes in Galloway progressive acidification began around 1850. Species of diatom typical of acid waters, such as *Tabellaria binalis* and *T. quadriseptata*, began to increase, while species indicating less acid conditions, such as *Anomoeoneis vitrea* and *Fragilaria virescens*, gradually declined in abundance. Species indicative of strongly acid sites (pH < 5) dominated in various sites from be-

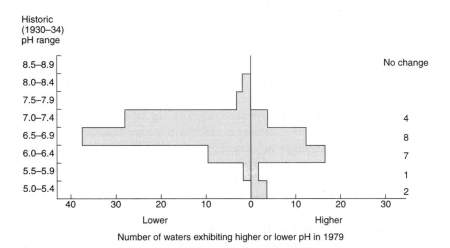

Fig. 5.6. Comparison of historical and recent pH values for 138 lakes in the Adirondack Mountains, United States (adapted from Havas *et al.*, 1984, after Pfeiffer and Festa, 1980).

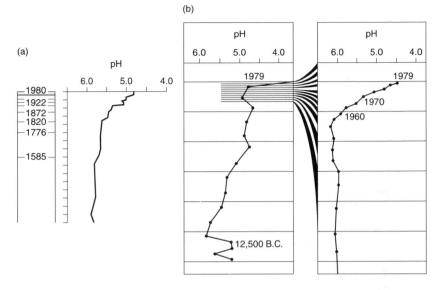

Fig. 5.7. Historical reconstruction of pH using sediment diatom assemblages of (a) Round Loch of Glenhead, southwest Scotland, and (b) Lake Gårdsön, Sweden. In the latter, the species composition dates back to the last Ice Age, with the changes since the beginning of this century shown in more detail (adapted from Henriksen, 1989, after Flower and Battarbee, 1983, and Renberg and Hedberg, 1982).

tween 1930 and 1950 (Flower *et al.*, 1987). Fig. 5.7 illustrates the inferred pH for the Round Loch of Glenhead, southwest Scotland, and Lake Gårdsjön, Sweden, reconstructed from diatom communities in sediment cores. The pH in the Scottish loch had dropped from pH 5.5 to 4.5 during the past 130 years, with the most dramatic change in recent decades. In Lake Gårdsjön changes became detectable in the early 1950s, but accelerated during the 1960s and 1970s. From a review of a number of reconstructions using diatoms it has been concluded that acidification will occur only when the ratio of calcium in water ($\mu\,eq\,l^{-1}$) to sulphur deposition ($g\,m^{-2}\,y^{-1}$) is less than 60:1 (Battarbee, 1994; Battarbee and Charles, 1994).

There are also marked short-term changes in the acidity of freshwaters. Seasonal changes occur: the summer composition of streamwater is closer to that of groundwater, having a higher pH. Lower pH values will be commoner in winter, as the composition of surface run-off will more closely reflect that of precipitation. In addition, brief pulses, or *episodes*, of high acidity may occur during heavy rainfall or snow-melt. During dry periods, acids will be deposited and accumulate on vegetation, to be washed off during heavy rain, increasing the input of acid to surface waters. Similarly acids accumulate in lying snow, to be rapidly mobilized during a thaw. An example of these acid episodes is illustrated for the Vikedal River, southwestern Norway, in Fig. 5.8. A continuous monitor was installed in the river and large fluctuations in pH were related to changes in water flow. During heavy rainfalls, pH fell by more

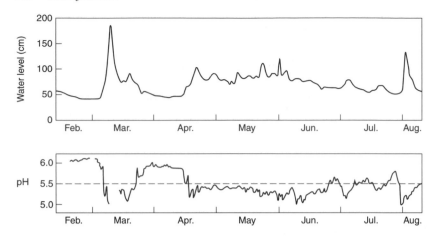

Fig. 5.8. Variations in pH and water level in the Vikedal River, Norway, February–July, 1983 (after Henriksen *et al.*, 1984).

than half a unit within three hours. A survey of episodic acid events in Canada caused by snow-melt produced maximum pH depressions of between 0.4 and 2.6 units, the minimum pH value recorded being 3.2 in one lake. Alkalinity was depressed by 100–200 μ eq l^{-1} (Tranter *et al.*, 1994). The major cause of these episodic events was sulphate ions, deriving ultimately from anthropogenic sources. These episodes can be highly damaging to the river ecosystem, fish kills being one regular consequence (Muniz, 1991), but routine monitoring may be insufficiently frequent to detect them.

Effects of acidification on the aquatic environment

Water chemistry

The generalized changes occurring during the acidification of lakes are shown in Fig. 5.9. The buffering capacity of the water will be determined by the concentration of bicarbonate ions and the process of acidification can be seen as taking place in three stages (Henriksen, 1989). In the first stage, the dissolved bicarbonate buffers the inputs of strong acids:

$$H^+ + HCO_3^- \rightarrow H_2O + CO_2$$

The pH generally remains above 6 and the plant and animal communities remain stable as the lake loses alkalinity. In the second stage, transition lakes, the bicarbonate buffer may be lost during long periods of acid inputs, resulting in large fluctuations in pH and periodic fish kills. In the final stage the loss of alkalinity is complete and the lake retains a low but stable pH, usually below 5, while metal levels, especially aluminium, may be elevated, resulting

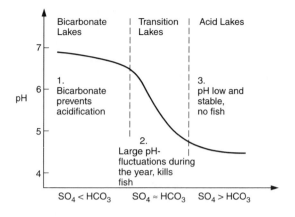

Fig. 5.9. The acidification process in lakes.

in the extermination of fish populations. Those lakes which have a high buffering capacity will never reach the permanently acid state.

Primary producers

Acidification results in a decrease in diversity of the phytoplankton. Acidic lakes have 10–20 species, compared to 30–80 species in circumneutral oligotrophic lakes (Muniz, 1991). Dinoflagellates tend to dominate many acid lakes, but despite the decrease in species there is little evidence of a decrease in biomass or photosynthesis if the supply of phosphorus is maintained (Olaveson and Nalewajko, 1994). Some species survive acidic conditions because they have evolved acid and aluminium tolerance in phosphate metabolism (Smith, 1990).

The attached algae (*periphyton*) show a somewhat different response. Although diversity generally declines there is a proliferation of growth, due to a reduction in grazing and an increase in the clarity of acid waters. Periphytic growth begins to increase as pH drops to below 6 and is extensive below pH 5.5 (France and Wellbourn, 1992). Filamentous algae such as *Mougeotia* and *Zygnema* often dominate, while mats of cyanobacteria may form in conditions too acid to support phytoplanktonic cyanobacteria. In Welsh streams, the green algae *Mougeotia*, *Ulothrix* and *Stigeoclonium* were most frequent at low pH (Ormerod and Wade, 1990).

Species richness of macrophytes is less in lakes of lower pH (Jackson and Charles, 1988). Changes in the macrophytes of lakes have included a reduction in the dominant *Lobelia*, especially at pH of 4 or less, and an increase in the dominance of *Sphagnum* (Grahn, 1986) and it has been suggested that increases in *Sphagnum* may accelerate the acidification process because the moss has a high ion-exchange capacity in its cell walls (Hendrey and Vertucci, 1980). Ormerod *et al.* (1987) have studied the macroflora of acid streams in

Wales and believe that the presence or absence of indicator species can be used in the bankside assessment of stream acidity. The presence of the liverworts *Scapania undulata* and/or *Nardia compressa*, and the absence of the macroscopic alga *Lemanea*, indicate waters in the pH range 4.9–5.2. With the moss *Fontinalis squammosa* present, but *Lemanea* absent, the pH range is 5.6–5.8, and with both these two present the pH is likely to be in the range 5.8–6.2. Stream water is likely to be above pH 6.2 where *Lemanea* is present, but *F. squammosa* is absent.

Decomposers

Some studies have indicated that in acid waters there is a reduced rate of decomposition and, in laboratory experiments, the rate of oxygen consumption by microorganisms using birch leaves as a substrate was halved when the pH dropped from 7.0 to 5.2. There is a shift in dominance from bacteria to fungi (Haines, 1981; Perry *et al.*, 1987). It has been suggested that, with increasing acidification, there will tend to be a build-up of organic matter, such as leaves and twigs, which could lead to a reduction in mineralization. The consequent decrease in limiting nutrients such as phosphorus could then lead to a reduction in primary production. This scenario is not universally supported, however, and several studies have failed to show a decrease in decomposition with acidification.

Invertebrates

Not only is there a change in the phytoplankton of lakes, but there are also changes in the zooplankton, a topic examined in detail by Brett (1989). Figure 5.10 illustrates the relationship between species richness of the crustacean fauna and pH in over 70 acidic waterbodies in moorland areas of Great Britain. There is a marked decrease in the number of species with increased acidity, but no such relationship occurred with calcium. The greatly reduced scatter of points below pH 5 indicates that pH is the controlling factor on the community in acidic waters, whereas above pH 5 other factors are involved. Common dominant species in acidic lakes in North America include *Diaptomus minutus*, *Bosmina longirostris* and the rotifer *Keratella taurocephala*, these species often representing the entire zooplankton community (Mierle *et al.*, 1986; Brett, 1989). *Eudiaptomus gracilis*, *Eubosmina longispina* and *Keratella serrulata* are typical species of acidic lakes in Scandinavia. Species of *Daphnia* and *Cyclops* are mostly scarce or absent in highly acid lakes, which tend to have an increased representation of generalist species.

A number of factors may explain the changes in zooplankton community structure in lakes after acidification. High concentrations of H^+ and metals impose physiological stresses. For example, Havas and Likens (1985) studied the combined effects of acidity and aluminium on mortality and sodium balance

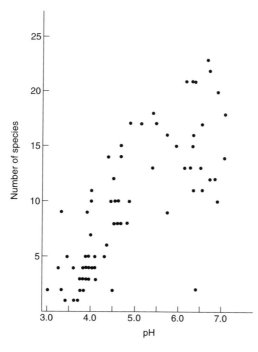

Fig. 5.10. Relationship between the number of crustacean species in acidic water-bodies and pH (adapted from Fryer, 1980).

in *Daphnia magna*. Both H^+ and Al interfered with sodium balance. At pH 6.5, raised Al (1.02 mg l^{-1}) led to a decreased rate of Na^+ influx of 46 per cent and an increased rate of outflux of 25 per cent, there being a net loss of sodium. At pH 4.5, Na^+ influx was inhibited by 73 per cent, compared with treatments at pH 6.5, whether or not Al was present. However Al decreased Na^+ outflux by 31 per cent at pH 4.5, so reducing the net loss of sodium and temporarily prolonging the survival of *Daphnia*. Physiological stress may be most important in highly acid environments where the community has been reduced to just a few species, a shift in competitive fitness favouring those species that are most tolerant of acidity, frequently generalists.

Other important factors are shifts in the structure of phytoplankton populations, for example towards inedible species, changes in the availability of bacterioplankton and detritus, and changes in the predator community, including the elimination of fish (Brett, 1989). Havens and DeCosta (1985) describe how, in acidified enclosures, both the abundance and mean body size of *Bosmina longirostris* increased relative to the controls, which they attributed to a decline in its predator, *Mesocyclops edax*.

Marked changes also occur in the benthos of lakes during acidification. Figure 5.11 illustrates the tolerance limits of 17 common species of crustaceans, bivalves and snails to pH, based on a survey of 1500 Norwegian lakes. Snails and bivalves, with calcareous shells, largely disappeared below pH 6, though

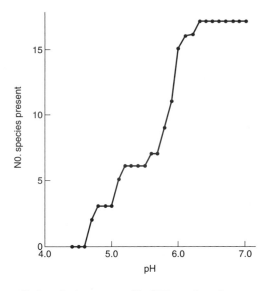

Fig. 5.11. Lower limits of tolerance to pH of 17 species of crustaceans, bivalves and snails, based on a survey of 1500 Norwegian lakes. All species were present in waters greater than pH 6.3 (adapted from data in Økland and Økland, 1980).

some small mussels survived down to pH 4.7 and some snails to pH 5.2, albeit at low densities. The crustaceans *Lepidurus arcticus* and *Gammarus lacustris*, important food items for fish, were sensitive to acidity, whereas *Asellus aquaticus* was found down to pH 5.2 and less frequently down to pH 4.8.

In North America the most abundant amphipod of lakes, *Hyalella azteca*, is acid sensitive, being found, for example, in 69 of 70 lakes in Ontario with positive alkalinity, but in none of 9 lakes with negative alkalinity (that is, acidified) (Stephenson and Mackie, 1986). Three of five species of crayfish of soft waters in North America are sensitive to acidification, with delayed hardening of the exoskeleton, parasitism, egg mortality and recruitment failures being reported (Mierle *et al.*, 1986). The physiology of crayfish in acid waters is reviewed by McMahon and Stuart (1989).

Studying headwater streams of the River Tywi in west Wales, Stoner *et al.* (1984) found that, where pH was greater than 5.5 and hardness greater than 8 mg l^{-1}, the invertebrate community consisted of 60–78 taxa, whereas in streams with a mean pH less than 5.5 and hardness less than 10 mg l^{-1}, only 23–37 invertebrate taxa were present. In a study of 104 upland streams in Wales, pH and aluminium concentration strongly influenced species richness and the structure of the macroinvertebrate community (Wade *et al.*, 1989). Species such as *Gammarus pulex*, *Ancylus fluviatilis* and *Hydropsyche* spp., as well as several herbivorous mayflies, were absent from acidic streams, even though potential food sources were present (Ormerod *et al.*, 1987). In the Adirondack Mountains of New York State streams with pH in the range 5.8–7.2 had relatively diverse invertebrate assemblages, the mayfly *Ephemerella funeralis* and the beetle *Oulimnius latiusculus* dominating. In the more acidic streams (range 4.4–5.0) these two species were absent and the community had less than half as many taxa. Stoneflies dominated (Simpson *et al.*, 1985). This impoverishment of the invertebrate community with acidification of streams is a general phenomenon (Sutcliffe and Hildrew, 1989). In contrast, the relationships between acidification and population density and biomass of macro-invertebrates are weak and inconsistent.

As with the plankton, the decline in richness of the stream fauna could be a combination of physiological stress, a change in food supply and a reduction in predators. Aquatic invertebrates need to actively take up sodium, chloride, potassium and calcium ions for survival and uptake is dependent on external concentrations. In acid waters ion concentrations may be too low, while hydrogen and aluminium ions may become dominant in the water. Being small and mobile, these may be transported inwards instead of essential ions, disturbing the normal internal equilibrium and possibly leading to a fatal loss of vital ions from blood and tissue (Sutcliffe and Hildrew, 1989). A net loss of sodium ions at low pH, upsetting osmoregulation, has been reported for stoneflies and mayflies (Twitchen, 1990; Frick and Herrmann, 1990), while increases in aluminium have been shown to decrease sodium ions in body fluids. In artificial streams, the mortality of a number of species was enhanced at pH 4 in the presence of elevated aluminium. The toxic effect of aluminium was less in the presence of organic matter (Burton and Allan, 1986). Dissolved organic carbon probably complexes with aluminium, reducing its toxicity, and its presence is extremely important in influencing the survival of stream biota under acidic conditions. Crustaceans and gastropods appear especially susceptible to low ion concentration in the presence of acidity, probably because they are more permeable, while all arthropods are vulnerable when they moult, a time when permeability is greatly increased.

There is currently little evidence to support the view that food availability and a reduction in predation pressure may influence the fauna of acid streams. A decrease in the decomposition of leaf packs and a substantial decline in the invertebrate populations within them was reported at pH 4 (Burton *et al.*, 1985), but the decline appeared to be due to invertebrate mortality rather than a change in food quality. Sutcliffe and Hildrew (1989) review the limited evidence suggesting that food quality for grazing invertebrates may be reduced in acid streams.

Metals are generally more soluble at low pH and laboratory experiments have often demonstrated increased metal uptake by invertebrates under acid conditions. However, to date, field observations lend little support to laboratory findings (Wren and Stephenson, 1991; Gerhardt, 1993).

Fish

The first reports of acidification came from fisheries inspectors in Norway in the early 1900s, who reported kills of Atlantic salmon (*Salmo salar*). Salmon showed a sharp decline in numbers, beginning in the 1970s, in seven southern rivers receiving acid precipitation, whereas there has been no overall change in rivers not receiving acid precipitation (Fig. 5.12). Brown trout (*Salmo trutta*) began disappearing from mountain lakes in Norway in the 1920s and 1930s and by the 1970s fishless lakes were reported from many regions in southern Scandinavia (Fig. 5.13). A survey in 1986 found that the number of fishless lakes in southern and southwestern Norway had doubled since a previous survey in 1971–75 (Henriksen *et al.*, 1989).

The evidence that fish populations were being reduced by acidification

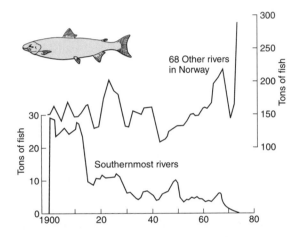

Fig. 5.12. Average catch of salmon in seven rivers in southern Norway receiving acid precipitation and 68 other rivers that do not receive acid precipitation (from Henriksen, 1989).

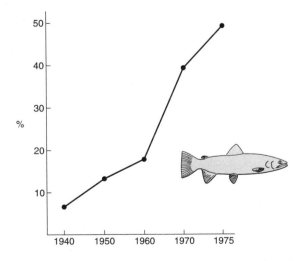

Fig. 5.13. Percentage of lakes (*n* = 2850) in Scandinavia which have lost their population of brown trout.

came much later in other countries. Haines and Baker (1986) estimated that 200–400 lakes in the Adirondack Mountains may have lost their fish populations due to acidification. Acidification has affected fish populations in several east coast states and eastern provinces of Canada.

Historical data on changes in fish populations are largely lacking in Great Britain. In Loch Fleet, southwest Scotland, catches of trout by rod and line have been recorded since the 1920s. A sharp decrease in catches occurred in the early 1950s (Fig. 5.14) and no fish were caught after 1975 (Turnpenny *et*

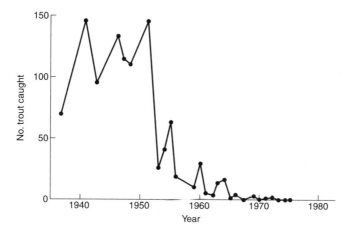

Fig. 5.14. Number of trout caught by rod and line in Loch Fleet, southwest Scotland (from Turnpenny, 1989).

al., 1988). Declines in fisheries due to acidification are widespread in this region of Scotland (Harriman *et al.*, 1987), though this appears to be the only area of the country seriously affected, despite the susceptibility of Scottish waters owing to geology and soil type. There is, however, a widespread decline in stream fisheries in Scotland associated with extensive conifer plantations (Egglishaw *et al.*, 1986), which increase the acidity of the drainage basin. Fisheries in upland areas of central and west Wales have also declined, especially in afforested catchments (Stoner and Gee, 1985).

Turnpenny (1989) has examined the occurrence of fish species in streams of various pH ranges in Great Britain. Brown trout were found in only 28 per cent of streams with pH less than 5, but in 95 per cent of streams with pH greater than 6.5. No streams with pH less than 5.5 contained salmon. Population density of salmonids in streams with pH less than 5 was 16.1 per cent that of streams with pH greater than 6.5, while biomass was greatly reduced. Eels (*Anguilla anguilla*) were widely distributed with respect to acidity, but biomass was three to four times lower in streams with pH less than 5.5 than those with pH greater than 6. Minnows (*Phoxinus phoxinus*) and bullheads (*Cottus gobio*) were scarce in streams with pH less than 6.5. In lakes in Finland, roach (*Rutilus rutilus*) were found to be the most sensitive species, disappearing from a number of study sites in the 1980s. Even the least sensitive species, whitefish (*Coregonus peled*) and perch (*Perca fluviatilis*) showed signs of stress and reproductive impairment in both male and female. Laboratory experiments with newly hatched fry showed sensitivity to pH and aluminium in the order roach > zander (*Stizostedion lucioperca*) > whitefish, perch > pike (*Esox lucius*), the last species tolerating one pH unit lower than roach (Rask, 1992; Vuorinen and Vuorinen, 1992).

Acid waters are associated with two types of fishery problem, fish kills during episodes of acidity and the gradual decline of fish populations in waters known to be acid. The *generalized* short-term effects of acidity on fish are given in Table 5.2, though this does not take into account the influence of associated events, such as increases in aluminium.

Table 5.2   Generalized short-term effects of acidity upon fish (from Wellburn, 1988)

pH range	Effect
6.5–9	No effect
6.0–6.4	Unlikely to be harmful except when carbon dioxide levels are very high ($>1000\,mg\,l^{-1}$)
5.0–5.9	Not especially harmful except when carbon dioxide levels are high ($>20\,mg\,l^{-1}$) or ferric ions are present
4.5–4.9	Harmful to the eggs of salmon and trout species (salmonids) and to adult fish when levels of Ca^{2+}, Na^+ and Cl^- are low
4.0–4.4	Harmful to adult fish of many types which have not been progressively acclimated to low pH
3.5–3.9	Lethal to salmonids, although acclimated roach can survive for longer
3.0–3.4	Most fish are killed within hours at these levels

The effect of acidity on fish is mediated by way of the gills, five major functions being affected: ion regulation, osmoregulation, acid–base balance, nitrogen excretion and respiration (Brakke *et al.*, 1994). High levels of sodium and chloride ions are found in the blood plasma of fish, and ions which are lost from the urine or from the gills must be replaced by active transport across the gills against a large concentration gradient. When calcium is present in the water it reduces the egress of sodium and chloride ions and the ingress of hydrogen ions. The excessive loss of ions such as sodium, which cannot be replaced quickly enough by active transport, is the main cause of mortality in acid waters. When the concentrations of sodium and chloride ions in the blood plasma fall by about a third, the body cells swell and extracellular fluids become more concentrated. Potassium may be lost from the cells to compensate for this, but if it is not eliminated from the body quickly, depolarization of nerve and muscle cells occurs, resulting in uncontrolled twitching of the fish before it dies.

Aluminium is toxic to fish in the pH range 5.0–5.5 . Aluminium ions interfere with the regulation by calcium of gill permeability, therefore enhancing the loss of sodium in the critical pH range. Aluminium ions also cause clogging of the gills with mucus and interfere with respiration. Electrolyte loss and gill dysfunction affect especially the younger life stages of the fish (Appelberg *et al.*, 1992). High concentrations of other metals, such as copper and zinc, may also sometimes prove toxic in acid waters. In one study a small lake was divided into two, one half being acidified. Yellow perch (*Perca flavescens*) accumulated 16 per cent more mercury over two years in the acidified section. Processes within the acidified half seemed to result in greater methylation of mercury (Wiener *et al.*, 1990).

Eggs may be especially sensitive to acidity. The relationship between pH levels and calcium levels and the survival of freshly fertilized trout eggs is shown in Fig. 5.15. Survival after eight days was 100 per cent at pH 5.1,

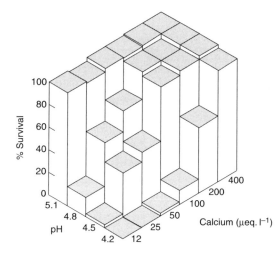

Fig. 5.15. Percentage survival of freshly fertilized brown trout eggs after 8 days in a range of pH and calcium concentrations (from Brown and Sadler, 1989).

irrespective of calcium concentration, and at the highest concentration of calcium (400 μ eq l^{-1}) irrespective of pH. At low pH (4.2) and low calcium (less than 50 μ eq l^{-1}) all eggs died. In experiments with brook trout (*Salvelinus fontinalis*) eggs, Hunn *et al.* (1987) found that mortality exceeded 80 per cent at pH 4.5, was 15–18 per cent at pH 5.5 and was less than 2 per cent at pH 7.5. Embryo mortality was reduced at the lowest pH in the presence of aluminium. All larvae died within 30 days at pH 4.5, none surviving for more than ten days. After fertilization, which itself may be influenced by acidity, fish embryos appear to be susceptible through their entire developmental period, though the times immediately following fertilization and shortly before and during hatching appear to be especially critical (Rosseland, 1986). Low pH reduces the activity of the hatching enzyme chorionase, which dissolves the inner wall of the egg. Hatching success is therefore reduced.

The effect of acidification on fish growth is somewhat confused, some workers reporting reduced growth rates and others finding no reduction. In

Atlantic salmon, the period of smoltification, when the fish change physiologically and behaviourally for their downstream migration, appears to be a supersensitive time to acidification (Rosseland, 1986). Skeletal deformities have also been found in fish from acid waters (e.g. Campbell *et al.*, 1986), due probably to a malfunction in calcium metabolism.

In conclusion, fish populations may disappear from acidified waters because of periodic mortality caused by acid episodes, or because mortality in the susceptible early stages of development and growth may mean that the population fails to maintain itself or grow. This will result in a gradual population decline and extinction even in waters where conditions are not overtly toxic to adult fish (Rosseland, 1986).

Higher vertebrates

As with fish, reproductive failure is the major effect of acid water on amphibians. In general they appear more tolerant than fish. Aston *et al.* (1987) reported high densities of spawn of the common frog (*Rana temporaria*) in waters of pH down to 4.2. Experiments showed that the survival of embryos to normal free-swimming larvae was not affected by pH in the absence of aluminium. The survival of embryos decreased, however, with increasing aluminium concentrations (Tyler-Jones *et al.*, 1989). For 26 species of amphibian the lethal pH, causing 100 per cent mortality of embryos, ranged from 3.4–4.5 (Freda, 1986). Andrén *et al.* (1988) found that egg mortality and the embryonic development time of three species of frog (*Rana arvalis, R. temporaria* and *R. dalmatina*) increased when pH declined. Raised levels of aluminium had no effect on the eggs, but pH and aluminium both influenced larval mortality, larval deformities being recorded. *R. arvalis* was the most tolerant. Salamanders, which are much more aquatic than frogs and toads, are more sensitive to reduced pH (Mierle *et al.*, 1986). Changes in the food supply are also likely to be significant and Freda (1986) concludes that pond acidity is having a major effect on the local distribution and abundance of amphibians in North America.

Dippers (*Cinclus cinclus*) are song birds that live along swiftly flowing streams where they feed mainly on aquatic invertebrates, such as mayfly and caddis larvae, which they collect by 'flying' underwater or foraging on the stream bottom. There is a direct relationship between overall dipper abundance and stream pH in Wales (Fig. 5.16) and their breeding density (number of pairs per 10 km) was also highly correlated with the abundance of mayfly and caddis. Dippers were scarcest along those streams with low pH (less than 5.7–6.0) and raised concentrations of aluminium (greater than 0.08–0.10 g m^{-3}). On the River Irfon, historical data indicates a decline of 1.7 pH units between the 1960s and 1984, while the dipper population has fallen by 70–80 per cent. Dippers breeding along acidic streams showed a delayed start to the breeding season, smaller clutch size, reduced egg mass, thinner egg

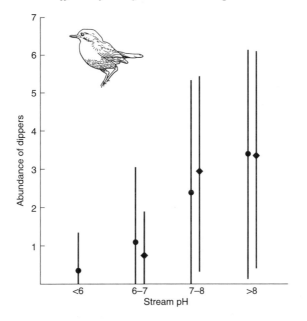

Fig. 5.16. Abundance of dippers (mean with standard deviation) along Welsh streams of different pH during autumn 1986 (dots) and 1987 (diamonds). Abundance is measured as catching per unit effort of ringing activity (adapted from Ormerod and Tyler, 1993).

shells, reduced brood size, slower nestling growth and poor adult condition. Mayfly and caddisfly larvae are scarce in these acid streams and dippers spend longer foraging and deliver food at a slower rate to nestlings. Fledgling survival is also reduced (Tyler and Ormerod, 1992; Ormerod and Tyler, 1993).

Fish-eating great northern divers or common loons (*Gavia immer*) in Canada suffer a high level of brood mortalities on acidic lakes, which is thought to be the result of a shortage of suitable food (Alvo *et al.*, 1988). In contrast lake birds that feed on benthic insects may benefit from a reduction in competition with fish, an example being the goldeneye duck (*Bucephala clangula*) (Eriksson, 1987).

Those areas of Great Britain liable to suffer acidification of freshwaters (Fig. 5.3) are also the regions which currently hold thriving populations of otters (*Lutra lutra*) (p. 11), so they are of especial importance for the conservation of this threatened species. The River Severn, in Wales, holds a good population, but otters do not much use the top 15 km of the river, which flows through forestry plantations (Fig. 5.5). They frequent a tributary, the Dulas, which rises in open moorland, almost to the source. The minimum pH of the Severn was less than 5.5 for the top 15 km or so of its length, whereas only the top 2 km of the Dulas had a minimum pH as low as this (p. 140). Fish were almost entirely absent from the top of the Severn, whereas good populations existed in the Dulas. This lack of fish, due to acidification, in the headwaters of the Severn prevented otters living there (Mason and Macdonald, 1987).

In a broader study of three regions of Scotland, 72 one-kilometre stretches of river bank were surveyed and signs of otters were found at all of them. There were, however, highly significant correlations between the use made by otters of stretches and both pH and conductivity of the water. The effect was greatest in southwest Scotland, the region with the most acidified waters. Fewer signs of the presence of otters were found at those sites where minimum pH was low enough to adversely affect fish populations (Mason and Macdonald, 1989).

Humans

As mentioned previously, acid waters bring toxic metals into solution. These may then accumulate in fish to be eaten by humans, mercury being of special concern (p. 167). Alternatively heavy metals may enter the drinking water supply, while acid water also dissolves metals from the pipes of the water distribution system (WHO Working Group, 1986). Particular concern has been expressed over lead and aluminium. The drinking water of a number of cities is distributed through lead pipes and increased lead concentrations may result in blood and nervous disorders, a reduced intellectual development in children and behavioural abnormalities (p. 176). High concentrations of aluminium have been linked to osteomalacia, a bone-wasting disease, and to Alzheimer's disease, premature senile dementia.

Demonstrating the effects of acidification by ecosystem manipulation

In the previous chapter (p. 117) I described how an entire lake was fertilized to chart the progress of eutrophication. In the same Experimental Lakes Area of Canada, Lake 223 was acidified over an eight-year period, causing dramatic changes in ecosystem function (Schindler *et al.*, 1985; Schindler, 1988b). After a two-year preliminary study, sulphuric acid was added from 1976 to gradually decrease the pH from an initial 6.8 to an eventual 5.0 in 1981. The pH was then maintained at 5.0-5.1 from 1981–1983. Over the period there was no overall change in either production or decomposition processes in the lake, but marked changes took place in the species composition. In the phytoplankton, *Dinobryon* decreased and *Gymnodinium* and *Merismopedia* increased below pH 5.5, whereas at pH 5.6 a filamentous alga, *Mougeotia*, not previously recorded in the lake, formed thick mats in the littoral (shallow-water) areas. The zooplankton community shifted from one dominated by copepods to one dominated by cladocerans. The cladoceran *Daphnia galeata mendotae* was largely replaced by *Daphnia catawba* and *Holopedium gibberum*, most probably owing to a reduction in fish predation. The overall biomass of the zooplankton, however, remained relatively constant.

Among the first animals to disappear from Lake 223 were opossum shrimps (*Mysis relicta*) and fathead minnows (*Pimephales promelas*). The estimated population of shrimps before the experiment was almost seven million, but at pH 5.9 they had all but disappeared. Fathead minnows failed to reproduce and died out a year after the shrimps. Both white sucker (*Catostomus commersoni*) and lake trout (*Salvelinus namaycush*) produced young in abundance at pH 6.4, though recruitment in both species failed at pH 5 and no 0-group (first year) fish of any species were observed in the following year. The condition of adult fish was poor. Cannibalism was noted in trout as many of their food items were killed. At pH 5.6 the exoskeleton of crayfish (*Oronectes virilis*) began to lose its calcium and animals became infected with a microsporozoan parasite. They were absent by the end of the experiment, as were leeches and the mayfly *Hexagenia*. Some taxa, such as chironomid midges, did well as they were released from the pressure of predation in the simplified ecosystem. As no species of fish reproduced at values of pH below 5.4, it was predicted that the lake would become fishless within ten years, based on knowledge of the natural mortalities of long-lived species. In addition to species replacement and an overall loss of diversity, acidification resulted in modifications in animal behaviour, alterations in predator–prey and parasite–host relationships, and changes in physiology and reproduction. Ecosystem stress resulted in the decline of many species before they were affected by direct toxicity of hydrogen ions (Schindler, 1988b).

The changes in Lake 223 were caused solely by increases in hydrogen ion concentration and not by any secondary effects such as aluminium toxicity. This might explain some unexpected results, such as the lack of decline in rates of primary production, decomposition or nutrient cycling. Furthermore, the minimum pH was 5.0 in Lake 223, whereas many acidified lakes have mean pH much lower than this.

The effects of adding acid and aluminium to a small headwater stream in central Wales were studied by Ormerod *et al.* (1987). Sulphuric acid was added over a 24-h period to reduce pH from 7.0 to 4.2–4.5, the upstream stretch acting as a reference reach. Some 200 m downstream aluminium sulphate was added continuously to increase the concentration from background ($0.05\,\mathrm{g\,m^{-3}}$) to 0.3–$0.4\,\mathrm{g\,m^{-3}}$. Toxicity tests were carried out *in situ* using six species of macroinvertebrates, brown trout and salmon. Three species of invertebrates showed no increase in mortality, while three others showed a 25 per cent increase in mortality in both treatment zones. Brown trout and salmon showed a 7–10 per cent mortality in the acid zone and a 50–87 per cent mortality in the aluminium zone, salmon succumbing more rapidly than trout. Downstream drift of blackfly larvae (Simulidae) increased in both acid and aluminium zones, while several species increased in drift in aluminium zones, the most pronounced response being that of the mayfly (*Baetis rhodani*), which was the only species showing a significant decline in benthic density at the end of the experiment. In an experimental study in California, where *Baetis* was also susceptible to acid, a high proportion of drifting invertebrates (45–100 per cent, depending on species) were found to be dead (Kratz *et al.*, 1994). In a further experiment in Wales citrate was added at a downstream site to bind aluminium. The drift density of *Baetis rhodani* increased six-fold above the control stretch in acid and labile aluminium treatments, but was not affected in the zone of organically-bound aluminium (Weatherley *et al.*, 1988). Greater rates of invertebrate drift during aluminium additions than during acid additions have also been reported from an experimentally dosed stream in New Hampshire, in the United States, by Hall *et al.* (1987), who consider that fluctuations in aluminium concentration at pHs not toxic to invertebrates may be an important factor regulating their distribution and abundance in poorly buffered streams. The Welsh experiments are reviewed by Weatherley *et al.* (1990).

Long-term experiments in acidifying whole catchments have been conducted in Norway (Wright *et al.*, 1994). Eight years of acid additions resulted in marked changes in the chemistry of run-off. Sulphate and base cations (sodium, potassium, calcium and magnesium) increased while the acid neutralizing capacity decreased. The run-off was acidic, rich in aluminium ions and toxic to fish.

Reversing acidification

In Lake 223, described above, sulphuric acid additions were decreased in 1984, allowing the pH to recover from 5.0 to 5.4, while in 1985 it increased further to 5.5–5.6 owing to unusually wet weather (Schindler, 1987b). White suckers and pearl dace (*Semotilus margarita*), which had survived without breeding, spawned when the pH reached 5.4, and the zooplankton species *Holopedium gibberum* and *Daphnia catawba* declined, presumably owing to heavy predation by larval fish. Lake trout recovered their overall condition, but evidence of breeding was not obtained in the year of recovery. The crayfish, opossum shrimp and fathead minnow had not returned by 1985 and the pH was not considered sufficiently high to support them. Nevertheless, signs of recovery in the lake fauna were seen rapidly after the rise in pH.

There are two possible ways of reversing the acidification of freshwaters – reducing emissions and adding lime. In Norway, a small catchment was covered with a roof to exclude acid rain, clean water being applied beneath the roof. Nitrate and sulphate run-off quickly declined relative to controls (Wright, 1987).

Almost all of the countries in Europe, together with Canada and the United States, committed themselves in 1983 to reduce sulphur emissions by 30 per cent within a decade, though this has not been achieved. Flue gas desulphurization is the favoured method of cutting emissions at power stations. Limestone slurry and air are injected to convert sulphur dioxide to gypsum

($CaSo_4$). A major problem is that sources of limestone are often situated in areas of outstanding landscape and the movement of limestone and gypsum on the scale required involves many thousands of lorry journeys each year. In southern Norway it has been predicted that a 30 per cent reduction in sulphur emission could reduce the number of acid lakes by 20 per cent, while a 50 per cent reduction may reduce the number by 35 per cent (Henriksen *et al.*, 1989). For the River Towy catchment in Wales, modelling has predicted that a 50 per cent reduction in atmospheric deposition would prevent further deterioration in soft-water streams draining moorland, but in streams draining conifer plantations the rate of deterioration would only be slowed down, mainly because of the continued leaching of toxic aluminium from forest soils (Edwards, 1989; Ormerod *et al.*, 1990). Similarly for southwest Scotland predictions from a mathematical model have suggested that a 50 per cent reduction in acid emissions is necessary to prevent further increases in stream acidity on moorlands, while afforested catchments will require greater reductions (Whitehead *et al.*, 1987).

This blanket approach to pollutant reduction is unsatisfactory because its effectiveness cannot be measured. A different strategy is to determine the *critical load* or concentration, which is exceeded when the rate of deposition of pollutant causes harmful effects on a receptor (Bull, 1991; Lükewille, 1994). This requires the mapping of sources of pollutants (such as power stations), mapping of load deposition, of sensitive receptors and of areas where critical loads are exceeded. By modelling, it is possible to calculate the reduction in emissions required to reduce deposition to acceptable levels in particular sensitive areas, while the costs relative to the benefits can also be considered.

Improvements in acidic waters resulting from a reduction in emissions currently remain somewhat academic, but the liming of waters has been practised for some years. As early as 1926, in Norway, intake water to fish hatcheries has been successfully limed to prevent mortality. In Sweden about 4000 lakes have been limed since 1976, at an annual expenditure of US$17 million, similar activities on a smaller scale in Norway costing $1.5 million each year (Henriksen, 1989). Pulverized limestone ($CaCO_3$), hydrated lime ($Ca(OH)_2$) and quicklime (CaO) may be used, but pulverized limestone is preferred because, though less reactive than the other compounds, it is less caustic and cheaper. The effectiveness of liming will depend on the retention time of water in a lake. In lakes with short retention times liming must be carried out annually or biannually, whereas with retention times of 2–5 years, reacidification following liming may take 5–10 years (Wright, 1985). Lake Hovvatn in southern Norway was limed in 1981 and over 11 000 brown trout were stocked over the next four years. Over this period the lake re-acidified to levels considered critical to fish and severe mortality occurred a year later. Growth rates reduced as acidification progressed (Barlaup *et al.*, 1994). The addition of lime to rivers and streams reduces the toxic effect of acidification to salmonids (Rosseland *et al.*, 1986; Weatherley *et al.*, 1989b), though there is the possibility of precipitating out aluminium salts which may then prove toxic at higher pH levels. Fish kills have occurred in lakes following liming when aluminium levels remained high (Leivestad *et al.*, 1987) and, as reported above, aluminium is especially toxic to fish in the pH range 5.0–5.5, so that fish mortality could be increased during the liming process. In the Appalachian Mountains of Virginia a low-cost direct liming has been practised in trout streams to offset trends in acidification (Downey *et al.*, 1994). Single treatments with limestone (cost up to $1500 per treatment) were sufficient to maintain native trout populations.

For long-term results the liming of catchments rather than water might prove more effective. Subcatchments of Loch Fleet, in southwest Scotland, were subjected to experimental liming and, from data collected over 3–5 years, it was predicted that limestone, applied as a slurry or as dry powder at rates of $20\,t\,ha^{-1}$ could maintain acceptable water quality for 15 years (Howells *et al.*, 1992). Fish were re-stocked, bred and maintained good condition (Turnpenny, 1992). However concern has been expressed over damage to terrestrial vegetation, especially *Sphagnum*, while the efficacy elsewhere has been questioned (Weatherley and Ormerod, 1992), particularly in view of the expense.

Because of the widespread nature of the acidification problem, and the expense of treating acidified waters (many operations for example use helicopters to apply doses), liming is likely to offer only a local solution to protect fisheries resources of particular commercial or recreational value. As stressed by Henriksen (1989) it is the *causes*, not the *symptoms* which need to be attacked, and this requires an international committment to reducing sulphur emissions.

Acid mine drainage

Drainage water from mines causes very considerable pollution problems over wide areas. In the United States, some 19 300 km of watercourses and 73 000 ha of lakes are adversely affected by mining pollution (Kleinmann and Hedin, 1990). In the Clark Fork River complex in Montana, waste material (tailings) covers $35\,km^2$ and is estimated to contain 9000 t arsenic, 200 t cadmium, 90 000 t copper, 20 000 t lead, 200 t silver and 50 000 t zinc (Sell, 1992). As well as severe adverse environmental effects, still detectable in rivers 200 km downstream, the surrounding area has one of the highest mortality rates in people aged between 35 and 74 in the United States.

Substantial lengths of river in many industrial countries are affected with acid water from both coal and metal mines. There are also severe problems in developing countries. In the Amazon, for example, prospecting for gold has resulted in rivers and fish being severely contaminated with mercury used in the refining process.

Abandoned mines are more of a problem than working mines and this is causing great concern in the UK with the closure of many coal mines during the early 1990s. When coal is mined, substantial quantities of the mineral pyrite, a crystal composed of reduced iron and sulphur (FeS_2), are often exposed to the oxidizing action of air, water and chemosynthetic bacteria, and these utilize the energy obtained from the conversion of the inorganic sulphur to sulphate and sulphuric acid. The reactions are as follows:

$$2FeS_2 + 2H_2O + 7O_2 \rightarrow 2FeSO_4 + 2H_2SO_4$$
$$2FeSO_4 + O_2 + 2H_2SO_4 \rightarrow 2Fe(SO_4)_3 + 2H_2O$$
$$Fe(SO_4) + 6H_2O \rightarrow 2Fe(OH)_3 + 3H_2SO_4$$

At higher pH pyrite can be oxidized in the presence of ferric iron:

$$FeS_2 + 14Fe^{3+} + 8H_2O \rightarrow 15Fe^{2+} + 2SO_4^{2-} + 16H^+$$

The conversion of ferrous sulphate to ferric sulphate occurs very slowly below pH 4, but it is rapid in the presence of microbial catalysts such as iron-oxidizing bacteria, which can increase the rate of oxidation a million-fold.

The most numerous bacteria in acid mine waters are those which oxidize sulphur (*Thiobacillus thiooxidans*) and iron (*Thiobacillus ferrooxidans*), and their densities may be as high as $10^9 \, ml^{-1}$. Walsh (1978) reported that iron-oxidizing *Metallogenium* are active in mine waters of between pH 3.5 and 4.5 and create the lower acidic conditions (below pH 3.5) suitable for the functioning of *Thiobacillus ferrooxidans*. At pH greater than 4.5, iron oxidation proceeds rapidly without the need for biological catalysis. When living in streams, *Thiobacillus* are frequently embedded in filamentous streamers, white or cream in colour, which are composed of a network of polymer fibres secreted by the bacteria (Wakao *et al.*, 1985). Harrison (1984) provides a review of the biology of *Thiobacillus* and its associates.

The type of water issuing from mines depends on the kind of mine and the surrounding geology. Where drainage is in the pH range 2–4.5, aluminium may be as high as $2000 \, mg \, l^{-1}$, though at higher pH it is usually never more than $20 \, mg \, l^{-1}$. At this low pH range, ferrous iron may be as high as $10\,000 \, mg \, l^{-1}$ and there will be no ferric iron. The high acidity of mine water also brings heavy metals into solution, adding to the pollution problems (Kelly, 1988). Those organisms in the receiving stream responsible for breaking down organic matter may be destroyed by the high acidity so that the self-purification process is inhibited and oxygen depleted water, often with suspended solids, extends much further downstream than would otherwise be the case. Washings from active mines also result in a heavy load of suspended matter. As the pH increases downstream, due to the dilution of mine water with stream water and run-off, the iron is oxidized by the dissolved oxygen in the stream, which turns a bright orange colour. Ferric hydroxide precipitates out on to the stream bottom, covering the substratum with a brown slime and smothering benthic algae and macrophytes.

Hargreaves *et al.* (1975) have described the photosynthetic flora from 15 sites in England, 13 associated with coal mines, having a pH of less than 3.0. All of the sites also had high levels of one or more heavy metals and of silica, and most had high concentrations of phosphate, ammonia and nitrate. Twenty-four species of photosynthetic plants were found in flowing waters, with an additional four species restricted to pool sites. The alga *Euglena mutabilis* proved to be the most widespread species and was also the most abundant, sometimes forming 80 per cent cover. Some species, although widespread, could not survive the lowest pH and species richness was greatest in the range pH 2.5–3.0.

Four species were found to have been recorded both in the English survey and in studies of acid streams in the United States, namely *Euglena mutabilis*, *Lepocinclis ovum*, *Eunotia exigua* and *Ulothrix zonata*. In acid-polluted tundra pools in northern Canada, *E. mutabilis* has been recorded at pH 1.8 (Sheath *et al.*, 1982). The presence of heavy metals may influence the composition of the species. For example *Eunotia exigua* is more likely to be found where copper and zinc concentrations are increased (Whitton and Diaz, 1981).

Several studies have been made of the effects of acid mine drainage on invertebrate communities. In a pool with pH 3.2 and contaminated with metals in a mining complex in Canada, the fauna was made up of 99.5 per cent Chironomidae (Wickham *et al.*, 1987) and this domination of acid mine waters by chironomids appears characteristic. Cranefly larvae (Tipulidae) and alderfly larvae (Megaloptera) are also frequently common. Crustaceans are usually absent.

In a study of 46 metal-contaminated sites in 12 streams in Cornwall, south-west England, copper was found to have most influence on the invertebrate community, followed by aluminium. Interactions between toxic metals and pH, alkalinity, hardness and dissolved organic matter added to the complexity of the response of the invertebrate community. At the most severely contaminated sites, with copper exceeding $500\,\mu\mathrm{g}\,l^{-1}$, the community was reduced to just four species – a flatworm (*Phagocata vitta*), two chironomids (*Chaetocladius melaleucus* and *Eukiefferiella claripennis*) and a net-spinning caddis (*Plectrocnemia conspersa*) (Gower *et al.*, 1994).

Typically, acid waters below a discharge result in a low diversity of species, but with large populations, released from competition and from predation by fish. Further downstream, in the zone of neutralization, the precipitation of iron, which alters the characteristics of the substratum and clogs the gills of invertebrates, results in both low diversity and low population density.

Fish are very severely affected by acid mine drainage and low pH. Letterman and Mitsch (1978) recorded ten species of fish, with a biomass of $22\,812\,\mathrm{kg}\,\mathrm{ha}^{-1}$, upstream of a mine discharge in Ben's Creek, Pennsylvania. Downstream of the discharge only six species occurred, with a biomass of $11\,\mathrm{kg}\,\mathrm{ha}^{-1}$. The overall populations above and below the discharge were $31\,056\,\mathrm{ha}^{-1}$ and $963\,\mathrm{ha}^{-1}$ respectively.

If the source of pollution is water percolating through mine waste heaps, it can often be treated successfully by covering the tip with clay and top soil. Water coming from springs or underground water sources tends to be more acidic, with higher loads of metals, and is much more difficult to control. It can be neutralized with lime and the metals flocculated out but this is very expensive and frequently ineffective. Artificial wetlands can be constructed to purify mine waste water, generally by bacterial action, and there are over 400 of these in the United States (Fennessy and Mitsch, 1989; Kleinmann and Hedin, 1990).

The succession of *Metallogenium* to *Thiobacillus ferrooxidans* which occurs as the pH drops may provide a biological approach to preventing pollution. *T. ferrooxidans* activity results in an exponential increase over a limited time period in the production of ferric iron. If the onset of *T. ferrooxidans* activity can be delayed, even by a small percentage increase in the mine water residence time, the majority of the highly polluting later cycles of the pyrite degradation cycle could be avoided. In the laboratory acidity and iron release can be controlled by perfusing mine water with artificial mine water enriched with ferrous iron, which inhibits *Metallogenium* growth. The inhibition of *Metallogenium* causes a reduction in the activity of later members in the succession, such as *T. ferrooxidans*.

For a detailed survey of the impact of mining on freshwaters Kelly (1988) should be consulted.

Chapter 6

HEAVY METALS AND ORGANOCHLORINES

The majority of pollutants discussed so far are damaging to natural ecosystems and man's resources, but only those diseases transmitted by faecal contamination of water have caused widespread human mortality. Although many compounds have been shown to exert toxic effects in animal experiments, it has proved extraordinarily difficult to show that concentrations present in water can be directly harmful to humans. That does not mean that they have no effect, but that it is extremely difficult to conclusively demonstrate that particular pollutants can be damaging to human health. When chemicals are known to have toxic effects in laboratory animals, stringent standards are recommended for permitted concentrations in water (p. 275), although their scientific basis is often suspect.

There is, however, considerable evidence that some heavy metals and organochlorines may be harmful to health at levels recorded in the environment. They are conservative pollutants in that they are not broken down, or are broken down over such a long time scale that they effectively become permanent additions to the aquatic environment. They accumulate in organisms and some may biomagnify in food chains (p. 40). The major uptake route for many aquatic organisms is directly from the water so that, to a certain extent, tissue concentrations reflect concentrations in water. Carnivores at the top of the food chain, however, such as birds and mammals, including humans, obtain most of their pollutant burden from aquatic ecosystems by ingestion, especially of fish, so there exists the potential for considerable biomagnification.

Evidence has accumulated over the last few years to suggest that environmental contaminants can alter the embryonic and early postnatal development of animals. This may lead to early death but often the effects are not seen until adulthood, with a loss of fertility. Contaminants may interfere with the development of reproductive, endocrine, immune and nervous systems. Many compounds interfere with or mimic both male and female hormones, modifying development and reproduction. Chemicals producing these effects include some pesticides, such as DDT, certain PCB congeners, dioxins, some herbicides and some heavy metals (Colborn and Clement, 1992).

Falling sperm counts (a 50 per cent drop over the last 50 years) and increases in testicular cancer in men may be related to oestrogen-mimics released into the environment (Sharpe and Skakkebaek, 1993). Feminization of males has been recorded in several vertebrate species living in aquatic environments. Particular attention has been directed at alkylphenols (e.g. nonylphenol), widely used as antoxidants and in the formulation of detergents, paints, toiletries, etc. They are known to be oestrogenic to fish in the laboratory. Many other compounds, however, may exert similar effects.

Male rainbow trout were caged at various distances from sewage treatment works in the River Lea north of London. Vitellogenin levels in the fish were elevated immediately below the sewage works and an increase could still be detected up to 15 km below a discharge (Purdom *et al.*, 1994). Vitellogenin is normally produced in the liver of female fish in response to the hormone oestradiol and is incorporated into the yolk of developing eggs, providing food for larval fish before they feed freely. Clearly some chemical in the sewage effluent was behaving like an oestrogen.

Heavy metals

The heavy metals of most widespread concern to human health are mercury, cadmium and lead. Nriagu (1988) has suggested that over one billion (10^9) human beings are currently exposed to elevated concentrations of toxic metals and metalloids in the environment and several million people may be suffering from subclinical metal poisoning.

Heavy metals suppress the immune system, leading to increased susceptibility to disease in animals. They may also be carcinogenic (Peakall, 1992).

Mercury

Environmental pollution by heavy metals became widely recognized with the Minamata disaster in Japan in the early 1950s. In 1932 a factory producing acetaldehyde and vinyl chloride began using mercuric oxide as a catalyst and effluents discharged into Minamata Bay contained mercury. Seafood was a major part of the diet of the local population. In the early 1950s a nervous disease began to affect dogs, cats and pigs and many died. In 1956 there was the first recorded human case, a young girl suffering from speech disturbances, delirium and difficulties in walking. A number of similar cases were detected over the ensuing months and an investigation team was set up, which established in 1956 that the cause of the illness was the consumption of seafood from Minamata Bay. In 1958, when the number of victims had exceeded 50, 21 of whom had died, a ban was placed on the sale of fish from Minamata Bay, though there was no restriction on catching fish. No controls were placed on the factory discharge. Concentrations of mercury in mud in the drainage canal from the factory were as high as 2000 mg kg^{-1} wet weight and in sediments in Minamata Bay of the order of 10–100 mg kg^{-1} wet weight. Mercury levels in fish and shellfish ranged from 5 to 40 mg kg^{-1} wet weight. It appears that by the end of 1959 the company itself had good reason to believe that its waste water was the cause of Minamata Disease, but that information was withheld. The discharge of mercury was stopped only in 1968, when the production unit became uneconomic. By 1988 there were 2209 confirmed cases of Minamata Disease. The disease killed 730 people. There is still serious mercury pollution in Minamata Bay but no controls on fishing have been enacted by the government. A detailed summary is provided by Laws (1993).

There are two sources of natural inputs of mercury to the environment. The weathering of mercury-bearing rocks releases about 3500 t^{-1} yr^{-1} and 25 000–150 000 t yr^{-1} is released as gases from volcanic areas. The burning of fossil fuels releases a further 3000 t yr^{-1}. The world production of mercury is around 10 000 t yr^{-1} and it has a wide range of uses. Much goes into electrical apparatus and into the chlor-alkali industry, which produces chlorine and sodium hydroxide electrolytically using mercury as a cathode. This industry is responsible for much contamination of lakes and rivers. The paper and pulp industry previously used mercury as a slimicide, again resulting in much environmental contamination, but this has now largely ceased. Paints, pharmaceuticals, dental applications and precision instruments also use mercury. The use of mercury as a fungicide in agriculture and horticulture may also lead to contamination of the environment.

In the aquatic environment, inorganic mercury is converted by microorganisms to the highly toxic methyl mercury, which is more readily taken up by tissues. Some 95 per cent of methyl mercury is absorbed by the gut and most of it is retained in the body, less than 1 per cent being excreted. Over 90 per cent of the body burden in fish is in the form of methyl mercury. Mercury biomagnifies along the food chain, an example from a Finnish lake polluted by effluents from a paper factory being given in Table 6.1. There is a marked increase in concentration from plants and invertebrates to fish, with carnivorous species having the greater loads and the pike (*Esox lucius*) being most heavily

Table 6.1   Average mercury content (μg kg^{-1} fresh weight) at different levels in the food chain in Lake Päijänne, Finland (adapted from Särkkä *et al.*, 1978)

Material	Mercury concentration
Sediment	87–114
Phytoplankton	15
Higher plants	9
Zooplankton	13
Herbivorous zoobenthos	77
Carnivorous zoobenthos	83
Herbivorous fish	332–500
Carnivorous fish	604–1510
Insectivorous duck	240
Piscivorous birds	2512–13 685

contaminated. Of the fish-eating birds, the black-throated diver (*Gavia arctica*) has the greatest concentration of mercury.

There have been many measurements of the concentrations of mercury in fish and bioconcentration factors from water in excess of 10^5 for total mercury and 10^6 for methyl mercury have been recorded (Zillioux *et al.*, 1993). Different studies are often difficult to compare, however, because factors such as age, size and condition of the fish may have a great influence on the amount of mercury which has accumulated. Mercury, for example, is directly correlated with fish length (which is closely related to age) in both eels (*Anguilla anguilla*) and roach (*Rutilus rutilus*), although eels, which spend much of their lives in intimate contact with sediments, generally have higher concentrations (Barak and Mason, 1990a). A comparison of mercury concentrations in eel and roach is shown in Fig. 6.1. Apart from in eels, concentrations of mercury in fish species are related more to size than trophic position, and open water species generally have lower amounts than eels (Barak and Mason, 1990b), emphasizing the importance of sediments as a source of contamination. After adjusting for length, however, concentrations of mercury in pike may be considerably greater than in other species living in open water (e.g. Wren and

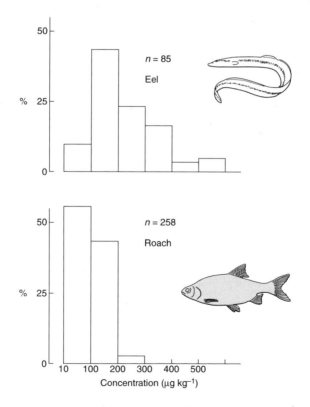

Fig. 6.1. Concentration (μg kg^{-1} fresh weight) of total mercury in eels and roach from freshwaters in Great Britain (from data in Mason, 1987).

MacCrimmon, 1986). The mercury concentration in fish is inversely related to the pH or alkalinity of lake water (Spry and Wiener, 1991). This is due, at least in part, to the microbial production of methyl mercury, itself inversely correlated with pH at the sediment–water interface.

Fish can take up heavy metals either in their diet or through their gills, though experimental studies have shown that the former is the predominant route, with around 60 per cent for pike and yellow perch (*Perca flavescens*) (Dallinger *et al.*, 1987). Although mercury is toxic to fish and accumulates to significant concentrations in wild populations, there is little evidence that fish themselves are disadvantaged by their burdens. Piscivorous vertebrates, however, feeding on fish from contaminated areas, may be at risk. For example great northern divers (common loons, *Gavia immer*) laid fewer eggs and more readily deserted nests when mercury in their fish prey averaged 0.3–0.4 mg kg^{-1} wet weight. If prey concentrations exceeded 0.4 mg kg^{-1} territory occupancy was severely reduced and almost no eggs were laid (Barr, 1986). In the Danube delta the eggs of vegetarian mallards (*Anas platyrhynchos*) had an average mercury concentration of 0.05 mg kg^{-1} wet weight, while those of ibises and herons, feeding on small fish and invertebrates, averaged 0.12 mg Hg kg^{-1}. Eggs of the piscivorous cormorant (*Phalacocorax carbo*) and white pelican (*Pelecanus onocrotalus*) had mercury concentrations averaging 0.74 mg kg^{-1} (Fossi *et al.*, 1984). The osprey (*Pandion haliaetus*) is a fish-eating bird of prey, taking its food by plunging, talons first, into the water. Figure 6.2 compares the concentrations of mercury in the feathers of nestling ospreys with mercury in the fish prey. Low concentrations of mercury in fish from two uncontaminated areas were reflected in low concentrations in feathers. At the three contaminated sites higher concentrations of mercury in feathers were found, but not in proportion to the increased amounts in the fish prey. There is therefore a limit to the amount of mercury which can be

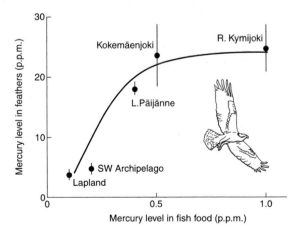

Fig. 6.2. Accumulation of mercury from fish eaten by osprey nestlings into feathers in five areas of Finland (adapted from Häkkinen and Häsänen, 1980).

acquired by feathers, though transferring the metal to the plumage can be seen as a method of removing it from the body.

Ospreys have declined in Finland and Sweden during this century, but there was no difference in breeding success between contaminated and uncontaminated areas and a mercury ban in Sweden, which was followed by a decrease in the contamination of nestlings, did not lead to an increase in breeding success. This indicates that mercury was not responsible for the decline and other pollutants, such as organochlorines, may have been more significant. Another fish-eating species in the region, the white-tailed eagle (*Haliaetus albicilla*) has also declined severely. Numbers in Finland decreased from about 35 in the late 1940s to 12 in 1964, and no nesting took place in 1966. Five dead birds which were analysed had sufficient levels of mercury to have killed birds in laboratory experiments (Henriksson *et al.*, 1966). However organochlorines were almost certainly also present so it is difficult to assess the role of mercury alone in the decline of eagles.

Even low concentrations of mercury may be considered potentially harmful to humans. Methyl mercury is believed to inhibit enzyme activity in the cerebellum, which is responsible for neuron growth in early developmental stages. Chronic exposure to mercury leads to mental retardation. Clinically observable effects on adults occur at blood levels of $0.2–0.5\,\mu g\,ml^{-1}$ and body concentrations of $0.5–0.8\,mg\,kg^{-1}$, or at concentrations in hair as low as $15–20\,mg\,kg^{-1}$ (Piotrowski and Inskip, 1981). The toxic effects of methyl mercury are dominated by neurological disturbances. A tolerable weekly intake for mercury is set at $0.005\,mg\,kg^{-1}$ body weight but people eating fish from contaminated lakes are liable to exceed this level significantly. In Finland, fish in some man-made lakes exceed the Finnish safety limits of $0.5–1\,mg\,kg^{-1}$ mercury. Increased concentrations of mercury were recorded in hair from people eating fish from the lakes, with the highest concentrations $(30\,mg\,kg^{-1})$ in middle-aged people who ate substantial amounts of fish (Lodenius *et al.*, 1983). In Sweden 250 lakes are blacklisted because the concentration of mercury in fish exceeds $1\,mg\,kg^{-1}$, but it is suggested that the true figure should be 9400 lakes (Håkanson *et al.*, 1988).

The European Community has suggested that a maximum average mercury content for fish taken for consumption should be $300\,\mu g\,kg^{-1}$. A reference back to Fig. 6.1 will show that 25 per cent of eels sampled in Britain exceeded this level. Nriagu (1988) suggests that, worldwide, some 40 000–80 000 individuals may be suffering from mercury intoxication.

Zillioux *et al.* (1993) provide a review of the impact of mercury on freshwater ecosystems.

Cadmium

Japan was also host to the first reported case of cadmium poisoning in the environment. In 1955 doctors reported a disease, which they later called itai-itai (ouch-ouch), characterized by severe back and joint pains, a duck-like gait, kidney lesions, protein and sugar in the urine and a decalcification of the bones, leading sometimes to multiple fractures. It was especially prevalent in women over 40. It was found that a mining company was releasing effluent into the River Jintsu, which downstream was used to irrigate paddy fields. Victims were confined to this region. There has been some controversy as to whether other factors (such as zinc or dietary deficiency) were involved, but a rapid decline in the incidence of itai-itai occurred when effluent controls were introduced. Some 200 people were afflicted, half of whom died.

World production of cadmium is about $21\,000\,t\,yr^{-1}$, for which the main uses are in electroplating, as pigments and as stabilizers for plastics. Mine drainage, sewage sludge applied to land and phosphate fertilizers are also significant sources.

Cadmium is highly toxic to some forms of life. The 48-hour LC_{50} values for invertebrates range from 7.0 to $34\,600\,\mu g\,l^{-1}$, with cladocerans being especially sensitive (Wren and Stephenson, 1991). In long-term experiments a 'no-toxic-effect' level of $0.37\,\mu g\,l^{-1}$ was determined for *Daphnia magna*, the most sensitive cladoceran tested (Canton and Slooff, 1982). Based on these data a water quality criterion of $0.1\,\mu g\,Cd\,l^{-1}$ was recommended for Dutch waters but it is often exceeded. In a study of the cadmium concentrations in the bottom-dwelling amphipod *Hyalella azteca*, 81 per cent of the variability could be accounted for by three parameters of lake water chemistry: concentration of calcium ions, total cadmium in water and dissolved organic carbon (Stephenson and Mackie, 1988).

Sublethal effects of cadmium toxicity have been reported from wild populations of fish. Perch (*Perca fluviatilis*) from a Swedish river contaminated with cadmium had a markedly enhanced lymphocyte count, slight anaemia and changes in the concentrations of potassium and magnesium in the blood (Larsson *et al.*, 1985). Residues of cadmium in the livers of exposed perch were six to eight times those of controls. Although cadmium bioaccumulates in tissues it does not appear to biomagnify along food chains (Spry and Wiener, 1991). Concentrations in fish flesh are generally less than $0.5\,mg\,kg^{-1}$ wet weight, though levels are higher in the liver and kidney.

More than 80 per cent of the body burden of cadmium in vertebrates is localized in the liver and kidney, where it binds to low molecular weight metallothionein. It accumulates with age and when cadmium concentrations in the kidney approach $200\,\mu g\,g^{-1}$ the ability of the metallothionein to protect cells begins to decline. This critical threshold in humans can be achieved on a rather low daily intake over an extended time and results in proteins being excreted into the urine and kidney damage. Nriagu (1988) estimates that more than 500 000 people may be at risk from cadmium-induced kidney damage worldwide, though freshwaters are only one of a number of sources of contamination.

Lead

Some 3.4 million t of lead were mined annually worldwide in the mid-1980s, a 22 per cent increase on the 1960s. The metal has a wide range of uses in pipes, battery cases, paints, etc. and tetra-ethyl lead is used as a petrol additive. It has been widely dispersed in the environment, with contamination increasing sharply from the 1950s, even in remote areas, with the growth in the number of automobiles.

Lead concentrates in organisms, bioconcentration factors in mosses, for example, being in the order of 3000–5000. There appears, however, to be no biomagnification along food chains (Spry and Wiener, 1991). A detailed study of the distribution of lead in macrophytes from Shoal Lake, Manitoba, has shown that different species vary considerably in their accumulation of the metal. The pondweeds *Potamogeton gramineus* and *P. richardsonii* contained concentrations of 45 mg kg^{-1} dry weight in shoot tissue, whereas four other species of *Potamogeton* contained less than 20 mg kg^{-1}, other genera of macrophytes having lower concentrations. Lead concentrations varied with depth but not season (Reimer, 1989). Invertebrates from lowland rivers with generally low levels of contamination had bioconcentration factors of between 32 and 360, accumulation being unrelated to trophic position (Barak and Mason, 1989).

The US state of Missouri has been a major producer of lead since the early 1800s and the residues (tailings) are rich in heavy metals, resulting in considerable contamination of freshwater ecosystems. In 1977 the dam of a tailings pond burst, releasing contaminated material into Big River. Fish were sampled from a number of localities upstream and downstream of the source of pollution for a period of 18 months from December 1979 to see if lead had accumulated in tissues. The results for three species, black redhorse (*Moxostoma duquesnei*), golden redhorse (*M. erythrurum*) and northern hogsucker (*Hypentelium nigricans*) are summarized in Fig. 6.3. Lead concentrations in fish flesh increased sharply at the source of pollution and remained above the recommended level for human consumption (0.3 mg kg^{-1}) for 125 km downstream. There was no correlation between lead concentrations in flesh and the length of fish in this study, as has also been found with other species, such as eel and roach (Barak and Mason, 1990a).

Waterfowl are the group most affected by the chronic contamination of aquatic ecosystems with lead. Those wildfowl which manage to avoid being killed or maimed by hunters may nevertheless ingest large quantities of spent shot while feeding. In the United States, where the problem has been most investigated, some 6000 t of spent shot each year are deposited and a minimum of 2.4 million waterfowl, out of a total North American population of 100 million, die each year of lead poisoning due to ingestion of shot. In Great Britain some 8000 mallard (*Anas platyrhynchos*) are estimated to die each year from the same cause (Mudge, 1983). Many other waterfowl suffer sublethal effects, making them vulnerable to disease and predation. The adverse physiological

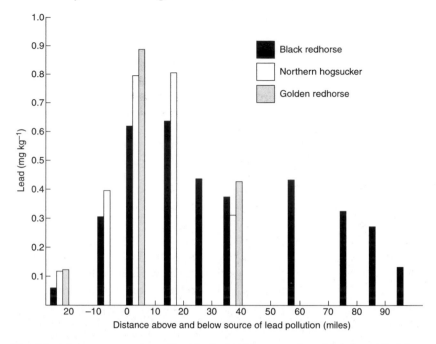

Fig. 6.3. Mean concentration of lead in flesh of three species of fish from Big River, Missouri (adapted from Czarnezki, 1985).

effects include muscular paralysis, damage to the nervous system, anaemia and liver and kidney damage.

The mute swan (*Cygnus olor*) is a familiar bird in Britain and many non-breeding flocks frequent urban stretches of river, where they are fed by the public. Lowland rivers are also used extensively by recreational anglers, who attach lead weights and shot to nylon lines to hold the bait beneath the water. Much of this is lost or discarded and swans ingest it while collecting food from the river bottom. The ingested lead is quickly ground down by the grit in the gizzard and is absorbed into the bloodstream. Of more than 1500 swans post-mortemed over the period 1981–84, 60 per cent had died of lead poisoning.

In the 1970s, up to 90 per cent of swans on some rivers were killed by lead poisoning and the annual death rate in England alone was estimated at 3370–4190 swans (Birkhead and Perrins, 1986). Lead levels in the blood of living swans on the River Thames are at their lowest during the close season for fishing and increase rapidly in June when angling recommences (Fig. 6.4). A blood lead level of 40 μg/100 ml indicates lead poisoning. The swans on the River Thames were below this level only during the angling close season. Swans suffering lead poisoning develop problems with the neuromuscular system. In particular they are unable to maintain the upright posture of their necks (they develop 'kinky necks') and cannot propel food through the gullet, eventually dying of starvation (Fig. 6.5).

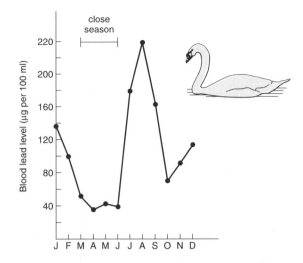

Fig. 6.4. Monthly median blood lead levels in 1981 from a flock of mute swans on the River Thames (adapted from Birkhead and Perrins, 1986).

Fig. 6.5. Young mute swan with a kinky neck caused by lead poisoning (photograph by Professor C. M. Perrins).

Many swans also die through collisions with power lines (Brown *et al.*, 1992). In Ireland such birds have been shown to have elevated concentrations of lead in their tissues, though lower than those dying of acute lead poisoning (O'Halloran *et al.*, 1989). Sublethal levels of lead may affect the co-ordination of swans, increasing their chances of collisions during flight. However, this is likely to occur only in the very early stages of lead poisoning for swans quickly lose condition, with wasting of the muscles, and are unable to take flight (Perrins and Sears, 1991).

A voluntary ban on the use of lead for angling was introduced in 1985 and the ban became law in 1987. In the early 1980s 84–89 per cent of swans on parts on the River Thames had abnormally high levels of blood lead. This percentage had fallen to 44 per cent by 1987 and 24 per cent by 1988 (Sears, 1989). The legislation banning lead weights has had a dramatic effect in reducing the incidence of lead poisoning in swans and the population has since steadily increased.

Humans have used lead for many purposes since antiquity and an increase in lead in the Greenland ice-cap reveals extensive smelting activity between 500 BC and AD 300: lead ore was smelted in open furnaces to extract silver. Its widespread use in Roman times has led to the suggestion that the fall of the Roman empire was the result of lead poisoning, especially as wine was stored in lead containers. An alternative view is that the mines became depleted, resulting in a shortage of silver coins.

Variable amounts of lead are found in food and a major aquatic source is lead pipes used to provide water to homes; lead is especially soluble in soft and slightly acid waters. Air pollution, particularly from vehicles, is a further major source. The adult gut absorbs 10–20 per cent of ingested lead, though children have a greater efficiency of uptake, and this provides the principal route into the body. Although less lead enters the body via the lungs, the percentage absorption is greater, around 40 per cent (Moore, 1986).

Exposure to relatively low levels of lead has been associated with metabolic and neuropsychological disorders, which include anaemia and lowered IQ. Children are especially at risk. The threshold for possible medical intervention in the United States has been set at a concentration of lead in blood of 250–300 µg/100 ml. It has been estimated that 590 000 young children in the country exceed that level and many may be suffering from lead poisoning. Worldwide as many as 130–200 million people may exceed this threshold and be at risk (Nriagu, 1988). Recent work, however, has suggested that levels of lead in blood lower than 25 µg/100 ml can affect children (Joyce, 1990). Furthermore, behavioural impairment in children associated with raised lead levels has recently been shown to be permanent (Needleman *et al.*, 1990), affecting the lifetime success of individuals. A detailed review of lead and health is provided by Smith *et al.* (1989).

Tap water is a significant source of lead and in the United States it has been suggested that a reduction in the allowable concentration of lead in potable water from 50 µg l^{-1} to 20 µg l^{-1} would benefit 42 million people whose supply is currently above the proposed new limit (Levin, 1987). The European Community currently also has a limit of 50 µg l^{-1} for potable water. In the Scottish city of Glasgow, before remedial action was taken at the end of the 1970s, some 50 per cent of random samples of tap water had lead concentrations exceeding 100 µg l^{-1}, many of them greatly exceeding this level.

Individual samples were higher than $1000\,\mu g\,l^{-1}$ (Richards and Moore, 1982). Nriagu (1988) considers that lead poisoning must be regarded as the most prevalent public health problem in many parts of the world.

Organochlorines

Organochlorine insecticides and polychlorinated biphenyls (PCBs) are hydrophobic, fat-soluble and biologically stable so that they accumulate in body fats. They also biomagnify along food chains and concentration factors from water to top predators, such as dolphins, may be as high as 10^7 (Tanabe *et al.*, 1984). Organochlorines have led to widespread declines in some of these top carnivores; the decline of the otter has been discussed on page 9. Dispersal by air and water leads to contamination of regions remote from sources of production or use (e.g. Lockhart *et al.*, 1992).

Because of their persistence, restrictions have been placed on the use of organochlorines and PCBs in the developed world over the past two decades. But concentrations from historical and current uses are still sufficiently high to pose problems to sensitive species (Klamer *et al.*, 1991). For example, 18 670 t of DDT, 43 t of dieldrin and 9994 t of lindane were used in national control programmes for disease vectors in 1978, out of a total insecticide use for this purpose of 43 235 t. The use of DDT and lindane in control programmes was predicted to double by 1984 (Chapman, 1987), DDT making up almost 50 per cent of total insecticide usage for vector control. The majority of these insecticides are applied to freshwater ecosystems.

Pesticides

As well as being applied directly to watercourses as described above, pesticides enter aquatic ecosystems in run-off from farmland. Sewage and industrial effluents are a further source of pesticides, while atmospheric transport, followed by precipitation in rainfall, is a major route of entry (Haines, 1983; Larsson, 1985). Aerosols produced during cropspraying can be dispersed over vast distances by winds. Unsuspected sources may have significant local impacts on the aquatic environment. In Northern Ireland, for example, low concentrations of lindane of the order of $10\text{--}23\,ng\,l^{-1}$ were found in rivers draining into Lough Neagh and there was a high correlation between the concentration in water and the urban population density within the catchment. The source of lindane was considered to be the use of wood preservatives in the home, not agricultural or industrial sources. The concentrations found in river water were sufficient to affect the survival of mayflies (Ephemeroptera) in laboratory toxicity tests (Harper *et al.*, 1977).

The toxicity of organochlorines to fish can be quite high. For example, the 96-hour LC_{50} of DDT to various species ranges from $1\text{--}30\,\mu g\,l^{-1}$; values for invertebrates are similar. The criterion, proposed by the US Environmental

Protection Agency, to protect the freshwater environment is a maximum of $0.001 \mu g \, l^{-1}$, taking account of the biomagnifying potential (Train, 1979). At sublethal concentrations, organochlorine pesticides result in impaired learning behaviour, slowed reflexes and a reduction in reproductive success.

The earliest example of biomagnification came from Clear Lake, California. Between 1949 and 1957 the lake was repeatedly sprayed with DDD, an insecticide related to DDT, to control populations of *Chaoborus astictopus*, a non-biting phantom midge whose swarms irritated bathers and anglers. An application of DDD to the lake water of $14 \mu g \, l^{-1}$ in 1949 resulted in a 99 per cent kill of midge larvae, but they had recovered by 1951. A second treatment in 1954 was at a higher concentration of $20 \mu g \, l^{-1}$, as was a third in 1957. Following the 1954 application 100 western grebes (*Aechmorphus occidentalis*) were found dead in the lake. The transfer of DDD from water to this top predator is illustrated in Fig. 6.6. The amount of DDD in the fat of the birds averaged $1600 \, mg \, kg^{-1}$, a bioconcentration factor of 80 000. The population of grebes fell from 3000 to only 30 pairs by the end of the 1950s, and most of these were sterile. An organophosphorus insecticide, which does not biomagnify and is non-persistent, was substituted for DDD and the grebe population gradually recovered during the 1960s.

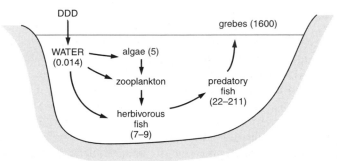

Fig. 6.6. Transfer of an organochlorine pesticide, DDD, through the food chain of Clear Lake. Numbers are concentrations in $mg \, kg^{-1}$ (from data in Hunt and Bischoff, 1960).

Organochlorine pesticides are widely dispersed in the environment and have had a significant impact on populations of aquatic wildlife, especially of birds. DDT affects the calcium metabolism of birds, resulting in the thinning of egg shells, which then break more easily. In studies of the night heron (*Nycticorax nycticorax*) concentrations of DDE (a metabolite of DDT) exceeding $8 \, mg \, kg^{-1}$ in eggs were associated with decreased clutch size, increased incidence of cracked eggs and a lower number of fledged young. In eight colonies in the western United States, from 10 per cent to 59 per cent of nests had some eggs with DDE concentrations in excess of $8 \, mg \, kg^{-1}$ and in one colony, Ruby Lake, the *mean* concentration averaged $8 \, mg \, kg^{-1}$.

Productivity in this colony was below the level necessary for population maintenance (Fleming *et al.*, 1983).

Eggs of red-necked grebes (*Podiceps grisegena*) from Manitoba, Canada, contained, on average, three times as much DDE as eggs from Saskatchewan and seven times as much as eggs from Alberta (PCBs were seven times higher and sixteen times higher than those respectively from Saskatchewan and Alberta). Thickness in egg shells from Manitoba declined significantly in the main period of DDE usage (1948–71) compared with earlier years and egg shells were still thinner in 1972–82. Declines in thickness in the other two areas were less and recovery was complete. The study sites were all remote from sources of industrial pollution and it was considered that contaminants were accumulated on the wintering grounds. Manitoba grebes winter on the Atlantic coast and in the Great Lakes, and Saskatchewan birds probably do also. Alberta grebes winter on the Pacific coast (Forsyth *et al.*, 1994). This shows that organochlorines can be widely dispersed by living organisms as well as by wind and water. In Europe internationally important wetlands for wildlife conservation are worryingly contaminated with organochlorine pesticides, areas such as the Coto Doñana in Spain (Hernandez *et al.*, 1985), the Danube delta (Fossi *et al.*, 1984) and Lake Mikri Prespa in Greece (Crivelli *et al.*, 1989). The egg shells of the endangered Dalmatian pelican (*Pelecanus crispus*) were thinned at this latter site.

A reduction in organochlorine usage has led to a reduced contamination of the aquatic environment by some compounds. An example, from Lac Léman (Lake Geneva), France, is given in Fig. 6.7. In the tissues of roach a rapid decline in hexachlorobenzene (HCB) and two hexachlorohexanes (alpha and gamma HCH) occurred. There was, however, no noticeable decrease in the more persistent DDT and PCBs. By contrast, in spottail shiners in the Great Lakes of North America DDT and PCB residues have declined steadily at most sites over the period 1975–1990 (Suns *et al.*, 1993). In Great Britain, DDT and dieldrin have been progressively restricted for agricultural uses since the mid-1960s, although they still have some industrial uses. Despite this the first national survey for organochlorines in freshwater fish, conducted in 1986-88, revealed some alarmingly high levels (Environmental Data Services, 1988). Of 62 sites surveyed, eels from 39 sites had mean dieldrin concentrations greater than $0.1\,\mathrm{mg\,kg^{-1}}$, a level at which eels should not be consumed. Mean dieldrin levels exceeded $0.5\,\mathrm{mg\,kg^{-1}}$ in eight rivers. In view of the high mercury levels also occurring in British eels (p. 167), they are a delicacy probably best avoided. Wildlife, unfortunately, cannot make the choice.

The human population, of course, is also widely contaminated with organochlorine pesticides. In samples of breast milk from Hong Kong, DDT and DDE concentrations ranged from 0.67–4.04 and 4.07–$22.96\,\mathrm{mg\,kg^{-1}}$ fat weight respectively (Ip and Phillips, 1989). These are some of the highest values reported in the literature and may be due to the importance of seafood in the diet of ethnic Chinese in the colony. In an earlier summary of data for 20 countries, mean concentrations of DDT in breast milk ($\mathrm{mg\,kg^{-1}}$ fat) ranged

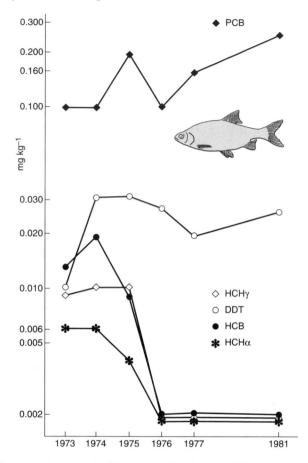

Fig. 6.7. Concentrations (mg kg^{-1} fresh weight) of organochlorines in tissues of roach (*Rutilus rutilus*) from Lac Léman, France, 1973–81 (after Venant and Cumont, 1987).

from 1.7 (Australia) to 19.5 (India), with means from seven countries exceeding 10 mg kg^{-1} (Fig. 6.8).

The maximum allowable concentration of DDT in human foodstuffs is 0.74 mg kg^{-1}, so even the lowest mean in breast milk was more than twice this standard. It may be that, in cases where organochlorine concentrations in breast milk are high, the benefits to infants of breast-feeding are outweighed by the transfer of substantial amounts of contaminants. Fortunately there is evidence that concentrations of DDE and PCBs in human adipose tissue are declining, for example in Japan and Canada (Loganathan, 1993; Mes, 1990).

PCBs

PCBs have a great chemical stability, a low flammability, good heat conducting properties, but a low electrical conductivity and a high di-electric constant.

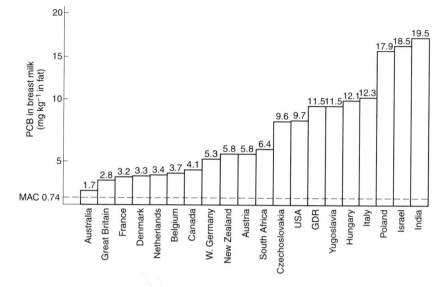

Fig. 6.8. Average concentration (mg kg^{-1} in fat) of DDT in breast milk. The maximum allowable concentration in human foodstuffs is 0.74 mg kg^{-1} (from Ramade, 1987, after Schüpbach, 1981).

These properties have ensured that they have been widely used in transformers and capacitors. They have also been used in heat exchangers, hydraulic systems, vacuum pumps and in lubricating oils and as plasticizers in paints and inks. They enter the general environment by leakage from supposedly closed systems, from landfill sites, from incineration of waste and in sewage effluents. They are dispersed widely in the atmosphere (Larsson, 1985). An analysis of sediment cores from Esthwaite Water in the English Lake District showed that PCB fluxes increased slowly from the late 1920s, then increased rapidly from the late 1940s, peaking during the late 1950s and early 1960s. Following restrictions on their manufacture and use, fluxes declined sharply (Sanders *et al.*, 1992). Despite restrictions on use, however, it is estimated that there are still 780 × 10^3 t of PCBs in electrical equipment, in landfills or in storage (Tanabe, 1988).

PCBs were commercially produced as mixtures and they also contained as impurities highly toxic dioxins and furans. Until recently analyses were conducted on total PCBs, comparing amounts in tissues with the commercial mixture (e.g. Aroclor 1260). Advances in the toxicology of PCBs have allowed the differentiation of individual isomers (congeners) in environmental samples. There are 209 PCB congeners, which differ in the number of chlorine atoms (1–10) and their position in the biphenyl ring. Toxicity varies largely with the number of chlorine atoms and their positions in the ring, coplanar (flat) PCBs being especially toxic (De Voogt *et al.*, 1990). The less toxic congeners may be slowly metabolized but this may mean that the PCBs eaten in food may be more toxic to vulnerable carnivores at the top of the aquatic food chain than are technical mixtures, because fish are less able to metabolize the more toxic congeners (Aulerich *et al.*, 1986).

The mechanism of action of the most toxic congeners is similar to that of dioxin (2,3,7,8-tetrachlorodibenzo-*p*-dioxin, or 2,3,7,8-TCDD), generally considered to be the most potent synthetic environmental contaminant. They bind to a receptor protein (arylhydrocarbon, Ah) in the cell which subsequently translocates to the nucleus. This PCB–receptor complex increases the production of mRNA that controls the production of mixed-function oxidase enzymes involved in the metabolism of fat-soluble xenobiotics. The metabolites produced may be harmless compounds which can be excreted, or they may be reactive intermediates, which may react with critical receptors, resulting in direct toxic effects or the initiation of cancer. The toxic effects observed in laboratory animals include liver damage, skin disorders, reproductive failure, growth of the thymus gland, loss of body weight, damage to the immune system and the production of deformities in young (teratogenesis).

Because PCBs occur as complex mixtures within the biota we need a method for assessing overall toxicity. One approach is to calculate toxic equivalency factors (TEFs) (Safe, 1987, 1990). The ability of each compound to induce particular hepatic enzymes, which is generally correlated with other toxic effects, is determined and compared with the induction caused by dioxin (TCDD). The toxic equivalency factor is:

$$TEF = \frac{EC_{50} \ (TCDD)}{EC_{50} \ (compound)}$$

EC_{50} is defined on p. 24. The TEF is then multiplied by the concentration in a tissue to determine the toxic equivalent:

TEQ = TEF × concentration

These can then be added together to find the total toxic equivalence of the congeners within a tissue and determine which individual congeners are likely to be toxicologically of most significance. Because they occur in much higher concentrations than the highly toxic dioxins and furans, the non-ortho and mono-ortho PCB congeners are the main contributors to dioxin-like toxicity in environmental samples.

The Hudson River rises in the Adirondack Mountains of New York State and discharges into the Atlantic Ocean via New York City. Two large generating stations, at Hudson Falls and Fort Edward, used PCBs in their capacitors, and a substantial amount found its way into the environment. Much of the PCB was retained by a dam, but this was removed in 1973, releasing large amounts of contaminated sediment into the river. Other sources of PCB were municipal waste tips and the burning of refuse. It is estimated that the Hudson River basin contains 1351 t of PCB (Clesceri, 1980). Less than 0.1 per cent is held within the biota. Nevertheless all but three species of fish have a PCB concentration exceeding $5 \, mg \, kg^{-1}$, the tolerance limit for edible fish, and angling is banned along long sections of the river. The shellfish industry, worth many millions of dollars annually, is completely closed.

Sediments are the greatest potential source of PCBs within aquatic ecosystems. In experimental systems PCBs originating from sediments are taken up by zooplankton and planktivorous fish. PCBs are released from sediment to water, from where they are taken up directly as a result of water/fat partitioning processes in animals (Larsson, 1986). Bottom-living fish, such as eels, accumulate greater amounts (Larsson, 1984a). Benthic invertebrates are also directly exposed to PCBs. Many of these eventually emerge as adults into terrestrial environments and it has been estimated that $20\,\mu g$ PCBs $m^{-2}\,yr^{-1}$ are transferred by chironomid midges from the aquatic to the terrestrial environment (Larsson, 1984b). Chironomids, of course, are extensively eaten by insectivorous birds.

Figure 6.9 illustrates the concentration of PCBs in flesh of roach and perch collected at various stations down the Lahn River in western Germany. The upper part of the river is rural. From station 5 the river flows through increasingly industrial areas and the PCB concentrations increase. At station 15 the river receives a large influx of waste water and there is a dramatic increase in PCBs. It was concluded that point sampling of fish such as roach and perch, which move relatively little along rivers, is sensitive enough to allow accurate localization of sources of contamination such as sewage and industrial waste.

The accumulation of PCBs in fish is a function of size, lipid concentration and trophic position. Figure 6.10 illustrates the concentrations of PCBs in a number of fish species from Lake Michigan in relation to mean weight. There is an apparent relationship between weight and concentration, except for

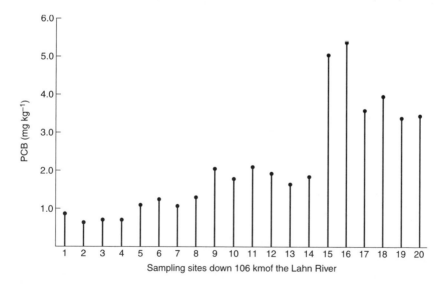

Fig. 6.9. PCBs ($mg\,kg^{-1}$ fresh weight) in fish at sampling points down the Lahn river (adapted from Schüler *et al.*, 1985).

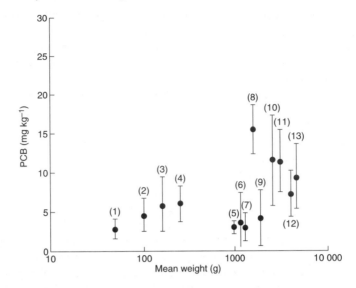

Fig. 6.10. PCB concentrations in whole fish (mg kg^{-1} wet weight) from Lake Michigan as a function of wet weight. 1, smelt; 2, alewife; 3, yellow perch; 4, bloater; 5, redhorse sucker; 6, white sucker; 7, whitefish; 8, lake trout; 9, carp; 10, Coho salmon; 11, chinook salmon; 12, brown trout; 13, rainbow trout (from data in Veith, 1975).

species such as carp and suckers which are detritus feeders. The relationship with weight is thought to reflect differing positions in the food chain, the lake trout having the highest concentrations.

Although fish may accumulate high concentrations of PCBs there is little evidence of any adverse effects on wild populations. In farmed Baltic salmon (*Salmo salar*) fed pellets made from Baltic herring (*Clupea harengus*) several physiological functions were disturbed, with effects on bone metabolism, steroid synthesis, the immune system and the induction of liver enzymes. As contaminant levels in farmed salmon were lower than in wild Baltic salmon, it was considered that the long-term survival of this stock should give cause for concern (Andersson *et al.*, 1993). In the Great Lakes there was a strong correlation between PCB concentration and the hatching success of lake trout (*Salvelinus namaycush*) (Mac *et al.*, 1993).

The Great Lakes region has also supplied much of the evidence for the effects of PCBs on wild populations of higher vertebrates. In Green Bay, Lake Michigan, a population of the fish-eating Forster's tern (*Sterna forsteri*) has impaired reproduction. The terns have an extended incubation period compared with birds from other localities. Fewer eggs hatch and the chicks have a lower body weight, with an increased liver size and the occurrence of oedema. They also have a high incidence of congenital deformities. The parent terns are inattentive in nesting, which further reduces reproductive output.

A detailed toxicological analysis and the calculation of toxic equivalents concluded that those PCB congeners that induce the enzyme aryl hydrocarbon hydroxylase (AHH) were the only contaminants present in sufficient amounts to cause the observed effects on eggs and chicks. Two pentachlorobiphenyls accounted for more than 90 per cent of the estimated TCDD equivalence. PCBs in general were the only contaminants causing the behavioural abnormalities in adults (Kubiak *et al.*, 1989).

Table 6.2 lists a number of effects and deformities reported in fish-eating birds from the Great Lakes. As well as Forster's tern the species include common tern (*Sterna hirundo*), Caspian tern (*Sterna caspia*), herring gull (*Larus argentatus*), bald eagle (*Haliaeetus leucocephalus*), night heron (*Nycticorax nycticorax*) and double-crested cormorant (*Phalacrocorax auritus*) (Fox *et al.*, 1991a; Giesy *et al.*, 1994). An example of a beak deformity in a cormorant is given in Fig. 6.11. The prevalence of malformed chicks was 52 per 10 000 in Green Bay, much higher than in other regions of the Great Lakes (Fox *et al.*, 1991b). The symptoms listed in Table 6.2 have only been observed over the past 20 years although the compounds have been in the environment for much longer. It seems likely that, previously, these symptoms were masked by the effects of DDE, which thinned eggs to such an extent that they did not survive long enough for symptoms to be expressed. It appears that all of the populations of fish-eating birds on the Great Lakes are currently displaying symptoms of exposure to PCBs, dioxins and furans at the biochemical level and these cause deaths and deformities in all populations examined. It is only in the more contaminated areas, such as Green Bay, however, that the effects are sufficient to cause population declines (Giesy *et al.*, 1994). A large number of compounds can cause these effects, which are expressed through a common mechanism of action via the Ah-receptor protein. There may be synergistic and antagonistic interactions between the various compounds, including individual PCB congeners. In the case of the deformed cormorants, 67 per cent of the total TEQ was composed of two pentachlorobiphenyl (coplanar) PCBs, those associated with reproductive effects in Forster's terns, emphasizing the

Table 6.2 Adverse effects in embryos and chicks of fish-eating birds of the Great Lakes (from Giesy *et al.*, 1994)

Effects	
Eggshell thinning	Cardiovascular haemorrhage
Deformities	Hormonal changes
Tumours	Enzyme induction
Behavioural changes	Metabolic changes, wasting syndrome
Immune suppression	Depletion of vitamin A
Oedema	Porphyria
Deformities	
Crossed bill	Eye
Clubbed foot	Brain
Hip dysplasia	Skull bones
Dwarf appendages	Gastroschisis
Ascites oedema	

Fig. 6.11. Cosmo, a double-crested cormorant from Lake Michigan. She was picked up by Dr James Ludwig and lived with him for two years. Her appearances on television in Washington and Japan, and in front of a Congressional Committee, did much to promote awareness of the environmental problems in the Great Lakes (photograph and information courtesy of M. Gilbertson).

threat posed by these two compounds to wildlife in parts of the Great Lakes (Yamashita *et al.*, 1993).

These effects are not confined to birds. Beluga whales (*Delphinapterus leucas*) in the St. Lawrence estuary downstream of the Great Lakes have failed to increase in numbers following a cessation of hunting. Necropsies carried out on 45 belugas revealed a high incidence of tumours (45 per cent), of lesions to the digestive tract (53 per cent) and to the mammary glands of females (45 per cent). There was tooth loss and evidence of immunosuppression, while one individual was a hermaphrodite (Béland *et al.*, 1993). No such symptoms were found in belugas from the Arctic. A wide range of contaminants at high concentrations were found in St. Lawrence belugas. Many of the pathological observations were similar to those found in seals from the Baltic Sea, which were related to the effects of PCBs on the reproductive, endocrine and immune systems (Bergman and Olsson, 1985). Not only has the seal population stopped breeding but immunosuppression has made it easier for infections that would otherwise have no adverse effects to become established.

In the Great Lakes region, mink (*Mustela vison*) and otter (*Lutra canadensis*) populations have also declined (Wren, 1991). Experiments with mink have shown that their reproduction is especially sensitive to non-ortho and mono-ortho PCBs, though the presence of di-ortho PCBs is also required to give expression to the effects (synergism) (Kihlström *et al.*, 1992). The relationship between declines in European otter populations and environmental contaminants was described on p. 9.

In laboratory experiments PCBs have a wide range of effects on mammalian reproduction (Fuller and Hobson, 1986). Are there any observable effects on humans? As usual it has been extremely difficult to prove and is controversial. Recent studies are, however, of concern. Mothers giving birth in hospitals close to Lake Michigan were asked about their history of eating fish from the lake. Blood was taken from the umbilical cords of their newborn babies and analysed for PCBs. A sample of 242 mothers ate moderate quantities of lake fish (2–3 salmon or trout meals per month) and 71 mothers ate no fish. Effects on health observed amongst those in the highly exposed categories included increased anaemia, oedema and susceptibility to infectious disease. Babies of those mothers who had eaten fish weighed 160 to 190 g less than controls and their head circumferences at birth were disproportionately small in relation to their age and weight. At age 4 years there was still a relationship between weight and PCB measured in the umbilical cord at birth: the most highly exposed children weighed, on average, 1.8 kg less.

Babies also displayed behavioural defects including increased startle reflexes and were classified by physicians within the 'worrisome' neonatal category. At 7 months there was more than a ten per cent decline in visual recognition memory amongst the most exposed babies, suggesting effects of PCBs upon the brain. At age 4 years there were still significant deficits in short-term memory and speed of information processing (Jacobson and Jacobson, 1993). Although the deficits were small they could have a significant impact on the ability of a child to master basic reading and arithmetic skills.

It was calculated that an infant nursed for a year by a mother equivalent to the group's 'average mother' eating Lake Michigan fish would have been exposed to 6.22 mg of PCB, of which 5.29 mg would be retained in the body. Infants nursed by mothers who had eaten fish from the lake exceeded the recommended consumption guidelines for *adults* of $1\,\mu g\,kg^{-1}$ body weight per day for their entire breast-feeding experience, some by as much as 25 times each day (Swain, 1988). Swain predicts that, if exposure to PCBs could be stopped *now*, the contaminants would be transmitted in measurable concentrations through transplacental passage of PCBs and postpartum exposure to breast milk for at least five generations beyond the original mother.

Behavioural changes have been detected in laboratory rats fed on a 30 per cent diet of Lake Ontario salmon over 20 days. The rats became hyperreactive to negative and positive events in tests compared with controls. Similar hyper-reactivity was also present in their offspring, who had never eaten lake fish, and persisted into adulthood. The experimental group grew normally and showed no signs of illness, so that it seems that behavioural changes may occur at doses lower than those needed to produce physical illness (Daly, 1993). These results may have relevance to observations on birds in the Great Lakes (inattentiveness to nesting, etc.) and to the children described above.

Restrictions on the use of PCBs have led to a gradual decline in environmental levels in some areas (e.g. Bignert *et al.*, 1993; Jones *et al.*, 1992). There is, however, such a large quantity of PCBs already in the environment and an even larger quantity in older electrical equipment, toxic dumps or storage, that, unless great efforts are made to prevent further discharge, significant pollution with PCBs is likely for the foreseeable future. Transfer across the generations and the persistence of PCBs within living tissues indicate that levels in top carnivores and humans are unlikely to decrease significantly for some years (Harrad *et al.*, 1994). In view of the apparent potential consequences for human health at very low levels of long-term exposure, this should be a matter of great concern.

Chapter 7

THERMAL POLLUTION

Most electricity generating plants, whether they use fossil or nuclear fuels, operate through the thermodynamic process known as the Rankine cycle, in which high-pressure steam is produced in boilers and then expanded through turbines, which convert thermal energy into mechanical energy. The basic plan is illustrated in Fig. 7.1 and a typical generating plant is shown in Fig. 7.2. In a fossil fuel plant, water from the condenser is pumped into feedwater heaters and pressurized, from where it passes to the boiler to be converted to saturated steam. This superheated steam is expanded through the turbine, creating mechanical energy to drive the turbine and hence the generator. The re-

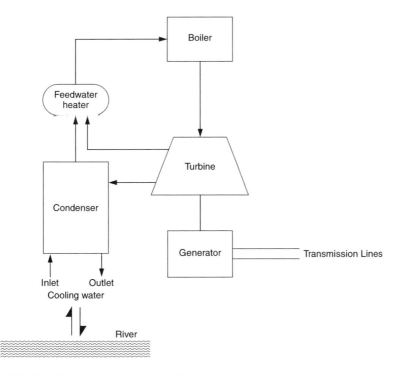

Fig. 7.1. Rankine power conversion cycle.

Fig. 7.2. A typical coal-fired electricity generating station with cooling towers, at Ironbridge on the River Severn (photograph by the author).

sulting expanded low-pressure steam is condensed, the heat being removed by cooling-water circulating through the condenser. Nuclear power plants generate at lower temperatures and pressures, so that less energy is added to the power cycle, but the Rankine cycle is less efficient, so that they still discharge a large amount of waste heat. The overall efficiency of power stations is less than 40 per cent.

The condensers require large amounts of cooling water, which is removed from the environment and returned at a higher temperature. Of a total of 35 259 Megalitres (ML) abstracted from groundwaters and surface waters for use in England and Wales in 1990, 12 612 ML (36 per cent) was used by the electricity generating board. In Wales, 74 per cent of water was used by the electricity industry (Department of the Environment, 1992). Most of this water, however, is returned to the environment and most of it is free of contamination other than heat, though there may be reduced levels of ammonia and organic nitrogen and increased levels of nitrate, chlorine and suspended solids compared with water in the power station intake. These chemical contaminants, though small in relation to the volume of cooling water, may in fact have a greater impact on the ecology of the receiving stream (Langford, 1983).

Effects on the environment

The effects of a power station at Ironbridge, in the English West Midlands, on the temperature of the River Severn have been studied in detail. The natural annual range of temperature was increased by up to 6°C (e.g. the maximum temperature in 1970 400 m above the power station intake was 22°C, while 2000 m below the outfall it was 28°C). The daily increments downstream ranged from 0.5°C during spates to 7.2°C in low flow conditions, and the diurnal variation in summer was increased by more than 100 per cent. In spring the rising mean temperature was advanced by 3–4 weeks, while the fall in temperature in autumn was delayed by 1–3 weeks (Langford, 1970).

An increase in temperature alters the physical environment in terms both of a reduction in the density of water and its oxygen concentration, which varies inversely with temperature. Nevertheless the oxygen concentration below a cooling water discharge may be substantially above that of the intake, because of turbulence and agitation produced within the cooling tower (Langford, 1983).

Effects on the biota

Possibly the most damaging environmental effect of a power station is that large numbers of organisms may be sucked in through the water intake. Larger organisms, such as fish, may be killed on the intake screens while smaller species pass through the plant. Even algae may be damaged, with permanent impairment of the photosynthetic mechanism (Nalewajko and Dunstall, 1994). At the Bay Shore power station, Lake Erie, over the period 1976–77, 284 million larval fish and 426 million fish eggs were destroyed by the intake, which was situated in shallow water where young fish congregate (Laws, 1993).

There are various possible effects of heated effluents on the biology of receiving waters. Those species intolerant of warm conditions may disappear, while other species, rare in unheated water, may thrive so that the structure of the community changes. Declines in species richness have been recorded in bacteria, benthic invertebrates and zooplankton living in thermal effluents. The standing crop and productivity of attached algae usually increases in heated effluents whereas species richness declines. Cyanobacteria are most tolerant and become dominant if temperatures remain above 32°C for any length of time. Exotic species, such as the tubificid worm *Branchiura sowerbyi*, are confined to heated waters, in which they build up large populations. Respiration and growth rates may be changed and these may alter the feeding rates of organisms. The reproductive period may be brought forward and development may be speeded up. Parasites and diseases may also be affected. Increased temperatures may render organisms more vulnerable to the effects of toxic pollutants present in the water. Any reduction in the oxygen concentration of the water, particularly when organic pollution is also present, may result in the loss of sensitive species.

As the temperature increases, the respiration and heart rate of a fish will increase in order to obtain oxygen for an increased metabolic rate, but at the same time the oxygen concentration of the water is decreased. At 1°C, for example, a carp (*Cyprinus carpio*) can survive in an oxygen concentration as low as 0.5 mg l^{-1}, whereas at 35°C the water must contain 1.5 mg l^{-1}.

Many species are able to acclimate to the normal range of temperatures occurring below thermal discharges, so that when exposed to increasing water temperatures, their upper lethal limit is raised. Temperature polygons can be constructed, enclosing the limiting temperatures which vary according to the history of previous exposure (Fig.7.3). In this example for the roach an increase in the acclimation temperature from A to B raises the upper tolerance from D to E. A further increase in the acclimation temperature, however, does not increase tolerance, so that the upper limit is reached (E to F). For fish acclimated to high temperatures (P), a sudden exposure to a low ambient temperature (4°C at Q in Fig. 7.3) may put the fish outside its tolerance polygon, so that it dies. Fish must, however, do more than survive, and different activities have different temperature tolerances. Figure 7.4 illustrates the temperature tolerance polygon for different activities of the brown trout. Whereas the fish can tolerate temperatures greater than 20°C, even after acclimation it will not feed or grow. At 19°C the swimming speed of trout declines, making them

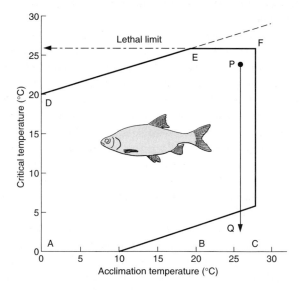

Fig. 7.3. Temperature polygon for the roach (*Rutilus rutilus*) (adapted from Hellawell, 1986).

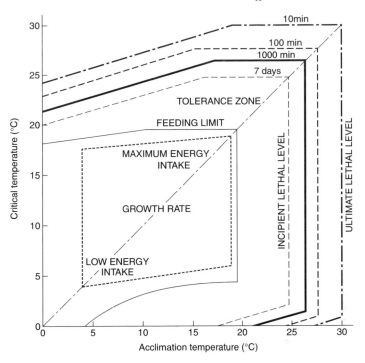

Fig. 7.4. Temperature tolerance polygon for different activities of the brown trout (*Salmo trutta*) (from Hellawell 1986, after Elliott, 1981).

less efficient predators. Prolonged exposure to such temperatures may lead to death by starvation.

Dheer (1988) maintained the fish *Channa punctatus* at temperatures of 23°C, 30°C and 35°C, with a control group at 14°C. At 30°C and 35°C fish lost weight and there was an increased mortality. Fish maintained at 14°C and 23°C showed normal growth. Within a week at 35°C there were significant falls in blood glucose level and depletion of glycogen reserves in liver and muscle, such stress symptoms becoming apparent after four weeks in those *Channa* kept at 30°C.

The general ability to acclimate means that large-scale mortalities of fish are infrequent. They occur mainly when populations are trapped in an effluent channel or when sudden discharges of hot water occur. Discharges of thermal effluents into water which is already polluted may also tip the balance for survival (Alabaster and Lloyd, 1982).

Some fish populations are unaffected by rapid temperature fluctuations of 12–15°C but these temperature changes may be too great for coldwater fish. Figure 7.5 illustrates the rates of survival of the yellowfin shiner (*Notropis lutipennis)* at different rates of temperature increase. A heating rate of 4°C per hour gave a 75 per cent survival at 32°C but a 100 per cent mortality in less than one hour at 34°C. With an increase in rate of 1°C per hour, 92 per cent

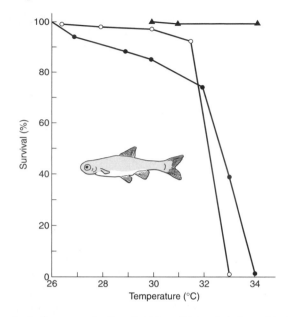

Fig. 7.5. Survival of groups of yellowfin shiner (*Notropis lutipennis*) at different rates of temperature increase. ●, 4°C h⁻¹; ○, 1°C h⁻¹; ▲, 0.125°C h⁻¹ (after McFarlane *et al.*, 1976).

survived at 32°C but the fish died very quickly at 33°C, whilst at a heating rate of 0.125°C per hour survival was excellent to 34°C, but at that temperature all fish died overnight.

With temperatures up to 26°C below thermal discharges fish tend to be attracted, whereas at temperatures above 30°C they tend to move away. Anglers certainly make use of the attraction of heated discharges to fishes.

The effects of a thermal effluent on the growth of an organism will depend on the relative apportionment of energy into respiration and the *scope for growth*. If the increase in temperature results in increased metabolism, with no increase in feeding rate, then less energy may be diverted to growth. However, the increased temperature may extend the period over which growth can occur, bringing forward the onset of growth in spring and extending the growing period into the autumn. Mobile species may avoid periods when discharges are resulting in higher temperatures than the species' optimum so that fishes living in thermally polluted areas will not show significantly different rates of growth from fish in neighbouring unpolluted waters. Other studies have shown an increase both in growth rate of fish and in the maximum size attained. Faster growth rates and larger maximum sizes have also been described for invertebrates (Langford, 1983).

The life-histories of organisms may be altered by thermal pollution, for instance by accelerating development time, and some fish spawn earlier in power station effluents than they do in neighbouring unheated waters, though other species, such as brown trout, may not spawn successfully (Langford, 1983). In an experimental stream maintained 10°C above a control stream at natural temperatures, peak macroinvertebrate density was 3–4 weeks earlier, while some species began reproduction 2–3 months earlier (Arthur *et al.*, 1982). The development of eggs was more rapid, while life cycles were shortened in some invertebrates living in a river below a heated discharge (Howells and Gammon, 1984). The emergence of insects from the River Severn, however, whose temperature regime was discussed above, was not noticeably altered by heated water and natural variability due to other environmental factors masked any effect due to temperature (Langford, 1975).

Temperature changes may increase the vulnerability of a species to predation and parasitism. Coutant *et al.* (1976) observed that rapid temperature decreases of about 6°C, which may occur when the quantity of thermal effluent is reduced, increased the susceptibility of juvenile channel catfish (*Ictalurus punctatus*) to predation by large-mouth bass (*Micropterus salmoides*). With reference to parasites Aho *et al.* (1976) found that, with the mosquito fish (*Gambusia affinis*), the metacercariae of the brain parasite *Ornithodiplostomum ptychocheilus* occurred at higher densities in fish from thermally polluted waters, whereas the metacercariae of the body-cavity digenean *Diplostomum scheuringi* were most numerous in fish from waters not receiving thermal pollution. These differences may have been due to the effects of temperature on the life cycles of the parasites or to effects on either the intermediate or definitive hosts.

Temperature changes are a feature of natural ecosystems so that organisms have the ability to adapt to the altered conditions provided by thermal effluents. Except for unusually severe thermally polluted conditions it appears that macroinvertebrate communities of rivers are relatively little affected by thermal discharges. The same is generally true of fish, which have the ability to vacate water which is temporarily inimical to them. The overall conclusion from an extensive body of research is that thermal pollution has not been as damaging to aquatic ecosystems as was originally feared and, within the usual range of temperature increases caused by power stations, the homeostatic mechanisms at work within the community minimize the damage.

Beneficial uses of thermal discharges

The production of electricity is an inefficient process and there have been a number of attempts to put the waste heat in cooling water to use. Power stations could, for example, be linked to sewage treatment works. Mixing warm power station effluents with raw sewage accelerates the treatment process, while the circulation of partially treated sewage effluent around the cooling tower accelerates the conversion of organic nitrogenous compounds and ammonia to nitrites and nitrates. Cooling water is also being used to heat glasshouses for the production of high-value crops such as tomatoes.

Most effort has been put into aquaculture, growing marine fish, such as sole (*Solea solea*) and turbot (*Scophthalamus maximus*), and freshwater fish such as eel (*Anguilla anguilla*), carp and channel catfish in cooling water, resulting in greatly increased growth rates. Figure 7.6 illustrates the growth of eels in power station condensers, tower ponds and tanks containing water pumped from the River Trent. The mean weight increased from 1 g to 25 g over the year in the condenser water, a markedly greater increase than in the pond and river water, though individual eels varied widely in size. It was considered that a marketable eel would take 2–2.5 years to produce in cooling water, compared with 10–14 years in the wild. Carp have shown similar good growth in condenser water, averaging 958 g in weight from a starting weight of 13 g after ten months, compared with a mean weight of 172 g in river water. Eels are highly prized in Continental Europe and many rivers have been overfished. Britain exports large numbers of eels to the Continent so that there is a potentially lucrative market in growing at least this species in cooling waters. Aquaculture units do, however, have to be sited close to, and preferably within, power station complexes, because the temperature advantage is quickly lost if water has to be transported. Rapid fluctuations in temperature and the frequent presence of chlorine, used to prevent fouling within the power station pipework, are also problems which have to be overcome.

Global warming

Global warming, or the greenhouse effect, is caused largely by an increase in carbon dioxide (and to a lesser extent methane and other gases) to the atmosphere, derived mainly from the burning of fossil fuels. Solar radiation passes largely unaffected through the carbon dioxide in the atmosphere but outgoing infra-red radiation is absorbed by carbon dioxide, preventing its escape to space and hence increasing surface temperatures. It has been estimated that temperatures may rise by between 0.2 and 0.5°C per decade over the next

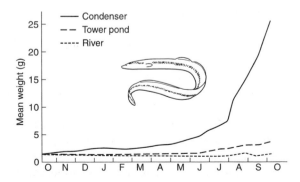

Fig. 7.6. Mean weights of eels in condenser, tower pond and river tanks over a 12-month period (adapted from Langford, 1983).

century (Houghton *et al.*, 1992) and this could have major effects on both the quantity and quality of freshwaters.

Hayes (1991) has listed the hydrological parameters likely to be affected by global warming: atmospheric humidity; evapotranspiration; soil moisture; surface run-off; groundwater recharge and base flow; seasonal snowfall; timing of snow melt; river flows and lake levels; incidence of droughts; hydrological variability and sensitivity; natural and artificial storage of aboveground and impounded freshwaters; position, depth and movement of saline intrusions upstream of rivers flowing into seas; and sea-level rise and associated coastal flooding. Clearly a potentially complex situation could develop, leading to great difficulties in prediction.

Global sea levels are likely to rise by 20 cm by the year 2030. Many coastal areas are protected by sea walls and the cost of upgrading all of these to cope with the increased tidal flow will be prohibitive. Large areas of semi-natural and agricultural land, claimed from the sea in the past, may therefore revert to saltmarsh and tidal lagoons, while increased coastal protection will be targeted at areas of high population density. The rise in sea level will mean that saltwater will penetrate further upriver than nowadays, resulting in changes in the fauna and flora.

Storms are likely to be more frequent and more severe in many areas, leading to a greater risk of flooding. In Australia, where global warming may result in increased rainfall, it has been predicted that even a relatively small rise in the frequency of flooding will lead to significant increases in financial loss (Smith, D.I., 1993).

Higher summer temperatures and lower rainfall will result in increased evapotranspiration and reduced soil water content. Many areas of intensive agriculture are already highly dependent on irrigation water and abstractions are likely to increase. Rises in domestic and industrial demand will lead to over-abstraction from both surface and groundwaters. In southeast England the headwaters of many rivers run dry in summer and, in drought years, there may be no surface flow even in winter (Fig. 7.7). In more southerly countries the impacts are severe, with dire consequences for the aquatic biota. Many wetlands may disappear.

As described earlier (Chapter 3) effluents already make up a substantial proportion of the low summer flows of many rivers. Effluents may become more concentrated, affecting all but the most resilient aquatic species. Higher concentrations in water of those metals and organics which bioaccumulate could have ramifications throughout the food chain, far beyond the boundaries of the river.

If surface water temperatures rise then the effects described above for waters receiving thermal discharges may become general. In the Experimental Lakes Area of Canada, water temperatures have risen by 2°C over 20 years and the length of the ice-free season has increased by three weeks (Schindler *et al.*, 1990). There has been a decrease in water renewal time and an increase in the concentrations of most chemicals. With a deepening of the thermocline

Fig. 7.7. Upper reaches of the River Stort, southeast England, dry because of over-abstraction (photograph by the author).

there has been a decline in cold-water species such as lake trout (*Salvelinus namaycush*) and the opossum shrimp (*Mysis relicta*).

Fisheries in general are highly responsive to fluctuations in temperature so that fish communities are likely to undergo marked changes with global warming. It has been predicted that whitefish (*Coregonus lavaretus*) in Lake Constance, Germany, will increase in numbers in the early stages of global warming but are likely to decline in the later stages (Trippel *et al.*, 1991). The growth of the larval stages of this fish is particularly responsive to temperature.

The sex of some animals, for example certain reptiles, is dependent on temperature during a critical period in the egg's incubation. In cool years more males hatch, in warm years more females. In a six-year study on the Mississippi River, Illinois, it was found that hatchling male painted turtles (*Chrysemys picta*) became rarer in years when July temperatures were higher than average. It is suggested that the predicted temperature rises for the next century would be sufficient to eliminate male turtles altogether (Janzen, 1994).

Because of natural, long-term fluctuations in climate it is difficult to show unambiguously that global warming is currently taking place and indeed, there are still some who deny its likelihood. However, because of the potentially profound effects on ecosystems and humankind it is prudent to take action now to conserve energy and reduce emissions of carbon dioxide.

Chapter 8

RADIOACTIVITY

The nucleus of an atom contains positively charged protons and electrically neutral neutrons and is surrounded by orbiting electrons, each of which carries a negative charge equal to the positive charge on a proton. Normally they are balanced. Atoms of the same chemical element may vary in the number of neutrons they have and are known as isotopes. Some isotopes (radioisotopes or radionuclides) are unstable and they seek stability by giving off particles or electromagnetic rays. There are four main types of radiation, with differing powers of penetration (Fig. 8.1):

1. **Alpha particles**, consisting of two protons and two neutrons, which have very little penetrating power and lose their energy in a short distance. If, however, they do penetrate living tissue, by inhalation or ingestion, they can do considerable damage because they are strongly ionizing.
2. **Beta particles** have greater penetrating power than alpha particles, but they can be stopped by layers of water, glass or metal, though like alpha particles they can be hazardous if taken into the body. Beta particles may be negatively charged electrons or positively charged positrons, depending on whether a neutron spontaneously changes into a proton, or vice versa, in an unstable nucleus.
3. **Gamma rays**, like X-rays, are short-wave electromagnetic radiation, having neither mass nor charge, but they are capable of penetrating thick material.

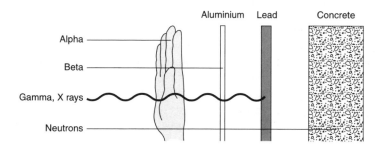

Fig. 8.1. The penetrating power of different types of radiation (source, United Kingdom Atomic Energy Authority).

4. **Neutron radiation** also occurs but it is generally found only in nuclear reactors. Some heavy, unstable elements have nuclei with a large excess of
   neutrons and the nucleus breaks into two fragments (spontaneous fission),
   with the production of free neutrons which are highly penetrating.

Apart from their radioactivity, radionuclides behave chemically like the
stable form of the element because the pattern of orbital electrons is the same.
An important feature of radioactive material is that radioactivity steadily declines with time. Each material has a fixed time for the radioactivity of a given
quantity of the material to fall by half, known as the half-life. In two half-lives
the radioactivity falls to a quarter of its original level, in ten half-lives to about
a thousandth. Radioactivity is inversely related to the half-life of a substance,
so that substances with short half-lives are intensely radioactive, while those
with long half-lives emit very little radiation, have generally low penetrating
power and are only hazardous if taken into the body.

There are several units used in the measurement of radioactivity:

1. **Becquerel** (Bq). Radioactivity is measured by the frequency with which
   radioactive disintegrations of the nuclei take place in a substance. The
   becquerel is one nuclear disintegration per second. The becquerel super-
   sedes an older unit, the curie (Ci) and 1 Bq = 27.03 pCi.
2. **Gray** (Gy), the measure of the amount of radioactivity absorbed by a tissue or organism. One gray is the amount of radiation causing 1 kg of tissue
   to absorb 1 joule of energy. The gray is the SI unit which replaces the rad
   (1 Gy = 100 rad).
3. **Sievert** (Sv), an arbitrary unit which accounts for the fact that different
   kinds of radiation do different amounts of damage to living tissue for the
   same energy. Neutrons and alpha particles have about ten times the effect
   of beta or gamma particles for the same number of grays. The sievert replaces the rem and 1 Sv is equal to 100 rem.

Sources of radioactivity

The environment is naturally radioactive, with radiation coming from outer
space, from the earth, from the atoms within the body and from normal activities such as burning fuel and cultivating soil. The average person in Britain
receives an annual dose of 2500 μSv of radiation. In general some 88 per cent
of radiation received is of natural origin, over which we have no control.
Natural sources include cosmic rays from outer space (10 per cent of total),
terrestrial gamma rays from rocks and soil (14 per cent, but very variable depending on geology), radon and thoron gas inside buildings (52 per cent but
variable according to local geological sources) and radiation accumulated in
tissues from food and drink (12 per cent). Levels of radionuclides in the
human food chain in the United Kingdom are discussed in MAFF (1994).
Despite our constant exposure to natural sources of radiation there is *no*

threshold limit for radiation damage and the amount of damage caused is roughly proportional to the total amount of radiation absorbed, so that it is essential to restrict man-made emissions of radiation to the environment.

The major controllable sources of radiation to the general environment are those from nuclear weapons testing and from the nuclear industry. Nuclear weapons testing began in the 1940s and fallout from this source reached a peak in the 1950s. With the signing of a test ban treaty in 1963 between the USSR, the United States and the United Kingdom inputs from this source have been greatly reduced, but unfortunately not entirely eliminated. Of course, while many nuclear weapons are retained for defensive purposes there remains the possibility of nuclear war, but in this event, the release of radioactivity would be only one of many catastrophes facing the world (Dotto, 1986), including the possibility of nuclear winter and major global climatic change.

A nuclear power station is in many ways similar to a conventional station in that steam is used to power generators which produce electricity (p. 189). The major difference is that the source of heat is the splitting of atoms rather than oil or coal. The steam is produced in the reactor building, in which the basic fuel is uranium, usually enriched so that the more reactive U-235 is concentrated relative to the more abundant U-238, making up 3.1 per cent of total uranium in the reactor compared with 0.7 per cent in natural uranium. The fuel is cast into ceramic pellets which are built into a stack and encased in a zirconium alloy to prevent direct contact with cooling water.

The uranium nucleus can be split into two roughly equal pieces (nuclear fission) if hit by a neutron. The total mass of the fragments is slightly less that that of the original nucleus and this reduction in mass (m) appears as energy (E) according to Einstein's equation:

$$E = mc^2$$

where c is the velocity of light. Even a small loss of mass releases a considerable amount of energy. The fission process releases two or three neutrons which can split other atoms, releasing more neutrons in a chain reaction. The energy of the fission reaction appears as the velocity of the fission fragments and is converted into heat, to produce steam for power generation, as the fragments are slowed down by collision with surrounding material. To increase the deceleration of neutrons, the fuel is embedded in a material known as a moderator and the neutrons bounce off atoms in this moderator, losing energy with each collision. In the Pressurized Water Reactor (PWR), the most common type of reactor in the world, the moderator is water, which is also used as a coolant to carry heat away. The power of the reactor is controlled by adjusting the number of available neutrons by using neutron absorbers, either cadmium or boron, which can be raised or lowered into the reactor core.

To carry away the heat, water is pumped through the fuel assembly and passed to a heat exchanger or steam generator, where it boils water which is in a separate circuit connected to the steam turbines. The water in the circuit passing through the fuel assembly must not boil, so that it is kept under pressure. The steam emerging from the turbine must be condensed using cooling water drawn in from the environment. In Great Britain most of the nuclear power stations are situated on the coast and use sea water for cooling (Fig. 8.2). In continental land masses nuclear power stations are sited by major rivers. Inevitably the cooling waters acquire some radioactivity and the discharge also contains liquid wastes, with low radioactivity, which come from processes involved with the handling of spent fuel rods from the reactor. The fuel rods in the reactor eventually lose their efficiency, but they still contain large amounts of U-235 which has not undergone fission. They can be reprocessed, and the facilities which undertake this release large amounts of waste water with a low content of radioactivity, but in overall quantities much greater than is released by nuclear power stations. They also produce small quantities of high-level waste which has to be stored for several decades for the radioactivity to decay.

The nuclear industry is strictly controlled and it is generally considered that the amounts of radioactivity normally released are not harmful either to the environment or to people. However, it is the *potential* problems with the nuclear industry that cause so much concern. The use of radioactive materials produces wastes which must be disposed of. High-level waste is produced only by the nuclear industry, but there is much intermediate- and low-level waste from nuclear power stations, defence establishments, the radioisotope industry, laboratories and hospitals. In Britain, some 2500 m³ of intermediate-level waste and 25 000 m³ of low-level wastes are produced each year, compared

Fig. 8.2. Sizewell B nuclear power station, Suffolk, eastern England, a pressurized water reactor (PWR) (photograph by the author).

with only 1100 m^3 of high-level waste over 30 years. Some low-level liquid and gaseous wastes from laboratories and hospitals are disposed of directly to the environment, but solid wastes, such as glassware, clothing, sludges, used reactor components, etc., and many radioactive liquid wastes need very careful management. Disposal presents a political if not a scientific problem. Much low-level waste has previously been dumped at sea, after being packed in steel drums and encased in concrete, but land disposal seems to be the current politically preferred option. The concern is that such disposal sites, whether deep or shallow repositories, may eventually leak, leading to the contamination of surface or ground waters. The potential pollution arising from the decommissioning of nuclear power plants and the disposal of their wastes is a problem for the future. Eisenbud (1987) discusses radioactive waste management and Fishlock (1994) describes the problems of managing a particularly nasty site in the United States.

Other major concerns are those of sabotage or accidents at nuclear installations. The nuclear power industry necessarily has a good safety record. Of several accidents, only that at Chernobyl (discussed below) has resulted in widespread and long-term environmental contamination. Nevertheless, that *any* accidents occur explains the unease of many members of the public over the nuclear industry. However well regulated and safety conscious it may be, accidents can happen, and eventually will, with dire consequences for ecosystems and human populations over large areas. Therefore, from the standpoint of public safety, the nuclear power industry should probably be considered unsafe. Openshaw (1992) provides further discussion.

Biological effects

The main concern over radionuclides is obviously the contamination of those food chains which lead to man. A small number of radionuclides tend to be of special importance. These are the ones which are incorporated into the food chain and biomagnify, and are either radioisotopes of essential nutrients or show chemical behaviour similar to essential nutrients. Some radionuclides, for example uranium and plutonium, are of concern because they may be inhaled with dust and concentrated in the lungs. Table 8.1 lists radionuclides of particular biological concern.

At high concentrations radionuclides cause acute toxicity. The radiosensitivity of a tissue is directly proportional to the mitotic activity of its cells (i.e. whether they are actively dividing), chromosome damage being the major cause of cell death. Different organs therefore cease to function at different times after exposure or at different levels of exposure. After a whole-body X-irradiation of 2–5 Gy, for example, a deficiency in lymphocytes becomes evident within two days, while a reduction in erythrocytes is delayed for 2–3 weeks. In mammals a dose of 6–10 Gy leads to failure of the tissues that generate the blood cells, 10–50 Gy causes intestinal failure, including vomiting

Table 8.1   Radionuclides of especial biological concern

Nuclide	Half-life	
^{3}H	12.4 yr	Assimilated by body in water
^{14}C	5730 yr	Passed up food chain
^{32}P	14.3 days	Concentrated in bones
^{40}K	1.3×10^9 yr	Found throughout body
^{90}Sr	28.9 yr	Concentrated in bones
^{131}I	8.1 days	Concentrated in thyroid
^{137}Cs	30.2 yr	Found throughout body
^{226}Ra	1622 yr	Concentrated in bones
^{238}U	4.5×10^9 yr	Concentrated in lungs and kidneys

and diarrhoea, and more than 50 Gy causes central nervous system syndrome, including tremor, convulsions and lethargy (Grosch, 1980). These effects are only likely in the event of nuclear energy accidents or warfare.

Particular concern is expressed over levels of exposure which may cause chronic effects, either somatic (affecting the body cells) or genetic (affecting the germ cells). Somatic effects are those which result in various types of cancer. Genetic effects are defects and abnormalities which occur in the next generation as a result of irradiation of the parent's reproductive organs. As such conditions normally appear long after the period of exposure to radiation they are very difficult to relate to particular exposure events.

Those aquatic organisms living below discharges from nuclear power stations or research installations have been most exposed to the long-term chronic effects of radiation. Fish in a stream adjacent to Oak Ridge National Laboratory in the United States were subjected to 0.6–3.6 mGy per day of radiation from sediments and produced larger than average broods but with a higher incidence of abnormal embryos. Chironomid midge larvae living in the same stream showed an increased frequency of chromosomal aberrations, though there was no effect on the abundance of the larvae (Laws, 1993). Salmon spawning below a nuclear installation on the Columbia River, in the United States, have apparently been unaffected by radiation doses of 1–2 mGy per week (Laws, 1993). In British waters no evidence of radiation damage, either acute or chronic, to aquatic organisms has ever been recorded (Langford, 1983). This does not mean that environmental effects have not occurred, but that investigations have not been sufficiently sophisticated to detect them.

Because of the concern over accumulation, the distribution of radionuclides in waters receiving effluents from nuclear installations is carefully monitored. It is however impracticable to monitor the whole system and usually it is found that only one or two pathways are important to humans, so-called *critical pathways*. If radionuclide contamination is controlled along critical pathways, then adequate control can usually be ensured along other pathways. A gener-alized critical pathway model is shown in Fig. 8.3. The critical pathways and

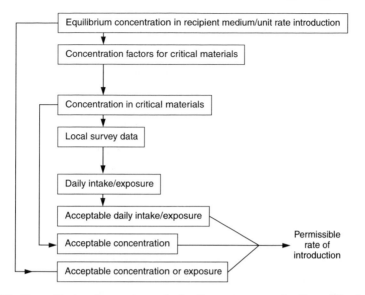

Fig. 8.3. The critical pathway approach for the assessment of radionuclides in the environment (adapted from Preston, 1974).

critical materials (those materials receiving or leading to the greatest degree of exposure) are identified after carrying out a study, usually a desk study, of the various pathways. The limiting environmental capacity is then set by comparison with primary exposure standards and on the basis of concentration of critical materials per unit rate of introduction. The critical material may be the receiving medium or some living component of the ecosystem that is affected by the radionuclide or is important in the human diet. Jackson (1992) describes the approach in more detail.

Below a nuclear power station on the Columbia River, it was considered that ^{32}P was the critical material, the human population being exposed through eating fish. The average concentration in river water below the plant was about 5.6 mBq ml^{-1} and the average concentration in the flesh of whitefish was 9 Bq g^{-1}. Concentration factors from water to fish were estimated at around 5000 in summer but <1 in winter. In those fish species eaten in greatest numbers, the average concentration of radiation was 1.5 Bq g^{-1}. A person who ate 200 fish meals a year (around 40 kg of fish) would receive a dose to the bone, the critical organ, of 3 mSv per year, which is 20 per cent of the annual limit (Laws, 1993).

At the Trawsfynedd Nuclear Power Station in Wales, cooling water is discharged into an enclosed, oligotrophic lake. Consumption of trout caught from the lake was higher than that of fish at marine sites but the dose level to consumers was estimated at only 7 per cent of the dose limit recommended by the International Commission on Radiological Protection (Langford, 1983). Concentrations of selected radionuclides in samples from Trawsfynedd Lake are shown in Table 8.2.

Table 8.2   Mean radioactivity concentration in samples from Trawsfynydd Lake, North Wales, 1986. Units are Bq kg^{-1} wet weight for biological materials, dry weight for mud and peat (from Hunt, 1987 © British Crown copyright, 1987)

Material	Total beta	^{134}Cs	^{137}Cs	^{238}Pu
Brown trout	380	36	340	0.00013
Rainbow trout	140	1.4	6.6	0.00013
Perch	1200	180	1200	0.00014
Eel	330	25	220	0.00011
Moss (*Fontinalis*)	330	6.6	16	–
Mud	3000	14	2000	3.9
Peat	3600	10	2000	13

From the evidence available, the discharge of radionuclides from the normal operation of the nuclear power industry appears to cause no measurable effects on the biota *in situ* and despite potentially large accumulation factors does not result in levels of radioactivity in food organisms which approach dose limits set for humans by the International Commission for Radiological Protection.

The Chernobyl accident

On 26 April 1986 there was an accident at the Chernobyl nuclear power station, some 60 miles north of Kiev in the Ukraine. A sequence of events occurred, involving a sudden surge of power in the reactor while it was at low power, followed by an explosion of steam and hydrogen and a fire. Initial reports of an accident came the following day from Sweden, where higher than normal background levels of radiation were recorded. The accident was reported in the Soviet Union on 28 April. It was estimated that 3.5 per cent of the radioactive materials in the reactor, some 2×10^{18} Bq, were released to the atmosphere.

Considerable environmental damage was done in the area surrounding the reactor, some 23 000 km^2 being designated as contaminated, and high levels of radioactivity have been found in terrestrial animals, aquatic insects, fish and waterbirds. Birth defects and cancers have been recorded in children born in the Kiev region since the accident and it is estimated that several thousand deaths can be directly related to the radiation released, with up to 15 000 persons affected by disease to a lesser extent.

The radioactive cloud initially flowed over Scandinavia, but then moved southwards across central Europe to Britain (ApSimon *et al.*, 1988). The Mediterranean region was also contaminated and subsequently evidence of the accident could be found over much of the world. Heavy rain led to substantial depositions of radioactivity over parts of Scandinavia, central Europe and western and northern Britain. Of the cocktail of radionuclides deposited, only caesium-137 has a significantly long half-life, of 30 years. Caesium behaves like potassium and hence becomes widely dispersed in ecosystems. Livestock in some northern and upland areas became severely contaminated and remain unfit for human consumption (Lundgren, 1993), but much of the deposited radioactivity will be leached or washed, sooner or later, into streams. The highest deposition of Chernobyl fallout occurred in areas with soft waters, where high concentrations from water to fish are likely.

In Sweden levels of radioactivity in freshwaters rose quickly. Fish species varied in the amount and rate of accumulation of radionuclides. For example Lake Tröske bream, a bottom-feeding fish, accumulated 1000 Bq kg^{-1} Cs-137 as early as May, 10 times greater than levels in pike and perch. By July, perch, which feed on zooplankton, had accumulated more than 8000 Bq kg^{-1}, whereas the fish-eating pike had accumulated only 4600 Bq kg^{-1}, a little higher than the bream at 3800 Bq kg^{-1} (Petersen *et al.*, 1986). Table 8.3 provides data on the maximum concentration of Cs-137 recorded in samples of aquatic animals in Sweden in the first three months following the Chernobyl

Table 8.3   Maximum amount of ^{137}Cs recorded in aquatic animals in Sweden in the first three months following the Chernobyl accident (from Petersen *et al.*, 1986)

	^{137}Cs $(Bq\,kg^{-1})$
Fish	
Trout (*Salmo trutta*)	18 700
Perch (*Perca fluviatilis*)	14 240
Grayling (*Thymallus thymallus*)	10 590
Char (*Salvelinus alpinus*)	9 890
Rainbow trout (*Oncorhynchus mykiss*)	6 280
Pike (*Esox lucius*)	4 690
Bream (*Abramis brama*)	3 840
Whitefish (*Coregonus lavaretus*)	3 130
Ide (*Leuciscus idus*)	2 840
Crucian carp (*Carassius carassius*)	1 870
Roach (*Rutilus rutilus*)	980
Vendace (*Coregonus albula*)	81
Herring (*Clupea harengus*)	23
Atlantic codfish (*Gadus morhua*)	2
Pike perch (*Stizostedion lucioperca*)	2
Aquatic birds	
Canada goose (*Branta canadensis*)	3 840
Mallard duck (*Anas platyrhynchos*)	1 290
Wigeon (*Anas penelope*)	1 190
Red-breasted Merganser (*Mergus serrator*)	110
Diver (*Gavia* sp.)	107
Greylag goose (*Anser anser*)	64
Shellfish	
Signal Crayfish (*Pacifastacus leniusculus*)	2 280
Crayfish (*Astacus astacus*)	1 180

accident. Over the two-year period 1986–1987 the average concentration of Cs-137 in perch dropped from 9800 Bq kg⁻¹ to 5040 Bq kg⁻¹, but levels were still high in 1989 and there was great individual variation in concentrations in fish from the same lake. It was estimated that between 4000 and 7000 Swedish lakes had perch with a Cs-137 load more than 1500 Bq kg⁻¹, the present guideline for consumption, and angling has virtually ceased (Håkanson *et al.*, 1989). Recovery is expected to take a very long time. Indeed Cs-137 activity levels in pike could exceed the guidelines well into the 21st century (Lundgren, 1993).

Because wild-caught freshwater fish feature little in the diet of much of the human population it was considered that the increase in radioactivity from this source due to the Chernobyl accident presented no cause for concern to human health. Indeed it had been calculated that even the critical group, that is those members of the population eating 100 kg annually of fish from the Baltic Sea, where contamination was greatest, would receive less than 0.08 mSv y⁻¹ (Camplin and Aarkrog, 1989). However, other species within the aquatic ecosystem are more exclusively fish-eating. Figure 8.4 shows the radioactivity in otter spraints (faeces) collected from river banks in central Wales and Galloway, southwest Scotland. Central Wales was outside the main deposition area for Chernobyl fallout. In the September following the accident, radioactivity in otter spraints was more than double that in a sample collected some 15 months before the accident and stored in a deep freeze. By the following January levels had returned to normal. By contrast, Galloway, which is an area

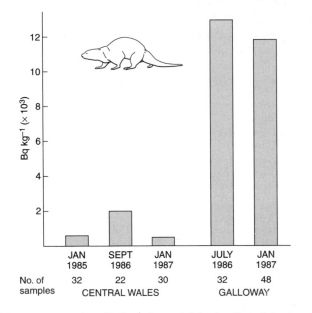

Fig. 8.4. Mean concentration (Bq kg⁻¹ dry weight) of radioactivity in otter spraints (from Mason, 1990).

of soft waters, received large quantities of Chernobyl fallout and average radioactivity was more than six times that in Wales (13 000 Bq kg^{-1}). Unfortunately, there was no pre-Chernobyl sample for Galloway but levels were still high in the following January, as has been found for other biological material from this area (Mason and Macdonald, 1988a). Whether or not such high levels of radioactivity could adversely affect otters is unknown and it would be impossible to disentangle any mortality caused by radiation from other mortality factors, a major problem with many ecotoxicological studies.

The Chernobyl accident stimulated considerable interest and research into radioactivity in the environment. Warner and Harrison (1993) and Savchenko (1995) provide detailed syntheses.

Conclusions

The normal wastes released from nuclear power plants have not been shown to cause harm either to the natural environment or to humans. Compared to generating plants burning fossil fuels and producing acidifying gases nuclear technology would seem to be clean. This does however ignore the problem of the long-term disposal of high- and intermediate-level wastes and the decommissioning of redundant plants. It is planned that reactors will be left for a long time before decommissioning so that they become less radioactive. However no full-size reactor has ever been fully decommissioned and the true cost of the process is unknown. It may even prove not to be technically feasible. And there remains the ever-present possibility of a major nuclear accident, the resulting release of radioactivity to the environment having 'the potential to be more insidious, more widespread, more prolonged, and less reversible than the effect of almost any other environmental disaster' (Depledge, 1986). The benefits of nuclear power need to be weighed extremely carefully against future and potential costs.

Chapter 9

OIL

Oil pollution incidents in freshwaters and their effects on living resources have received much less attention than those in marine ecosystems. The oiling of seabirds, for example, is commonplace and widely commented upon. Some accidents at sea have devastated marine life, as when the *Exxon Valdez* ran aground in Prince William Sound, Alaska, in March 1989 spilling 35 000 tonnes of crude oil. This spill killed some 4000 sea otters (*Enhydra lutris*) and hundreds of thousands of seabirds, as well as ruining the extraordinarily rich fisheries in this once pristine coastal region (Pain, 1993). The clean-up and compensation costs to the Exxon Oil corporation are likely to exceed $8 bil-lion (though the annual revenue of the company is $111 billion). Accidents in freshwater ecosystems are much more localized and hence attract much less public outcry.

The Monongahela River flows into the Ohio River at the Golden Triangle in the city of Pittsburgh, Pennsylvania, thence down through Cincinnati to the River Mississippi. At Florette, some 37 km above Pittsburgh, a riverside stor-age tank was being slowly filled with no. 2 diesel fuel in early January 1988. The tank had recently been moved and reconstructed and the company had not carried out the normal testing before filling it with oil. When almost full the tank collapsed, releasing some 3.5 million litres of oil, most of which flowed into the Monongahela River and downstream to Pittsburgh. Within a day the water supply of 23 000 people was cut off and a further 750 000 people had their water rationed. More than a thousand families were evacuated, dozens of factories using water from the river stopped production, schools were closed and commercial traffic on the river came to a halt. Numerous fish, ducks and geese were killed and, as the oil continued downriver, drinking-water intakes were closed. Access to the river was difficult because of steep banks, hinder-ing clean-up operations, while the swift flow resulted in oil being forced under booms positioned to contain it. Only 380 000 litres were recovered (Lemonick, 1988). Had the accident occurred in the summer, when fish and invertebrates were more active, biological damage would have been much more severe.

In the marine ecosystem most oil pollution derives from tanker operations (cleaning of compartments, etc.) and accidents, incidents at production instal-lations being of less significance. Some 28 per cent of the world input of pet-roleum hydrocarbons to the sea is by way of rivers, however, suggesting that chronic pollution of freshwaters is widespread. In England and Wales, over the period 1988–92, 23 per cent of all reported pollution incidents affecting rivers were caused by oils, while of those incidents threatening the closure of water intakes, 55 per cent were due to spills of hydrocarbons (Cole *et al.*, 1994). Much oil in freshwaters is derived from petrol and oil washed from roads or other hard surfaces and from illegal disposal of used engine oil. Other sources include irrigation pumps and boats, and accidents involving transporters, spillages from storage tanks and vandalism are also significant.

Detailed reviews of oil and its impact on freshwater ecosystems are provided by Vandermeulen and Hrudey (1987) and by Green and Trett (1989).

What is oil?

It is difficult to predict the potential toxic effects of oil because of its very complex chemical nature. Crude oil consists of thousands of different organic molecules, the majority of them hydrocarbons with between 4–26 atoms per molecule. There are also some sulphur and nitrogen compounds, and metals such as vanadium. Oils from different sources have very different compositions. The three major types of hydrocarbon are alkanes (e.g. ethane, propane, butane), cyclohexanes (naphthenes) and aromatics (e.g. benzene, toluene, naphthalene).

Crude oil is refined by a process essentially of distillation, which separates off different fractions at increasing boiling points (Fig. 9.1). At lower temperatures, the compounds used in the production of petroleum are separated off. At high temperatures naphtha, which forms the basis of the petrochemical industry, separates, and yet higher temperatures boil off diesel oil, bunker oil used to fuel ships and power generating stations, and tars. Further refinements are often necessary to produce commercial products. Oil products may also

Fig. 9.1. An oil refinery on the Thames estuary. Several accidental spills at the refinery complex in this area have caused ecological damage (photograph by the author).

contain toxic compounds such as polynuclear aromatic hydrocarbons (PAHs), PCBs and metals, especially lead (p. 173).

Effects on biota

Animals and plants may be affected by the physical properties of floating oil, which prevents respiration, photosynthesis or feeding. Higher vertebrates, whose coats get covered in oil, lose buoyancy and insulation, while the ingestion of oil, frequently the result of attempts to clean the fur or plumage, may prove toxic. Many water-soluble components of crude oil and refined products are toxic to organisms, their eggs and young stages being especially vulnerable. There are also a whole range of sublethal effects.

The principle effect of oil discharges on the microbial community is one of stimulation, especially of heterotrophic organisms which utilize hydrocarbons. Shales *et al.* (1989) concluded that algae are less sensitive to the direct effects of oil than many other organisms, but they are especially sensitive to secondary effects. These may include an increase in primary production caused by the death, decomposition and nutrient release of sensitive species, an increase in primary production by nitrogen-fixing species or, in the absence of these organisms, a decrease in primary production.

The effect of aqueous extracts of various oils has been investigated on three species of floating duckweed (*Lemna*). They proved generally tolerant of crude oils, but synthetic oils, such as raw coal distillate, were much more toxic, even a dilute 4 per cent aqueous solution resulting in a 50 per cent decrease in growth rate (Fig. 9.2). The toxicity was probably due to higher concentrations of aromatic hydrocarbons. Severe oiling of higher plants may reduce photosynthetic rates, either by interfering with the permeability of cell membranes or by absorbing the light required by the chloroplasts. Affected leaves may eventually turn yellow and die. The responses of plants to oil may be very species-specific, influenced especially by the thickness of the cuticle and hence the permeability of the leaves.

The toxicities of various oil derivatives to several algae, invertebrates and fish are shown in Fig. 9.3. The compounds tested were two phenols, two azaarenes from coal-derived oil and two polycyclic aromatic hydrocarbons. Snails were less sensitive than arthropods and fish were generally less sensitive than invertebrates. Within each pair of compounds the toxicity increased with increasing ring number of the molecule, for example naphthol (two rings) was 45 times more toxic to *Gammarus minus* than phenol (one ring).

A very wide range of differences have been revealed in the toxicity of oils to fish in short-term tests and these may depend on the type of oil, differences between the batches of the same oil, the changing nature of oil with time, the type of test system or the length of exposure (Hedtke and Puglisi, 1982). The aliphatic components of oils are relatively innocuous, while the monohydric aromatic compounds are generally toxic, increasing unsaturation of the mol-

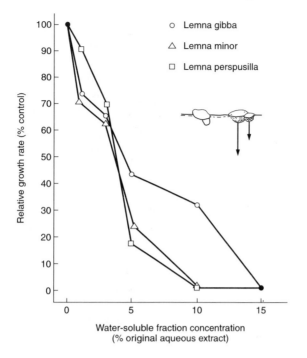

Fig. 9.2. Effects of aqueous extracts (water-soluble fraction) of raw coal distillate on frond multiplication rate of duckweeds (*Lemna gibba*, *L. minor* and *L. perpusilla*) (adapted from Green and Trett, 1989, after King and Coley, 1985).

ecules being associated with increased toxicity. Emulsifiers and dispersants, used to remove oil, contain surface active agents which render membranes more permeable, thereby increasing the penetration of toxic compounds. In general terms the uptake of hydrocarbons by tissues and the histological damage due to exposure to hydrocarbons are both increased when an oil dispersant is mixed with the test oil. Most dispersants are quite toxic to fish and mixtures of dispersants with oil are much more toxic than oil alone. Frequently the oil–dispersant mixture is also more toxic than the dispersant alone, though this is not always the case.

There has recently been considerable research directed at the toxicity of polycyclic aromatic hydrocarbons (PAHs). Early studies showed that anthracene was acutely toxic to bluegill sunfish (*Lepomis macrochirus*) under conditions of ultraviolet light and Oris and Giesy (1987) have investigated the phototoxicity of 12 PAHs to the larvae of the fathead minnow (*Pimephales promelas*). Six compounds exhibited acute toxicity, two slight toxicity and four showed no effect. Kazan *et al.* (1987) found that pyrene was not toxic even in ultraviolet light, but piperonyl butoxide, when present at non-toxic levels, synergized pyrene toxicity even in the dark. These experiments may be of considerable environmental significance for butoxide-like compounds are widespread.

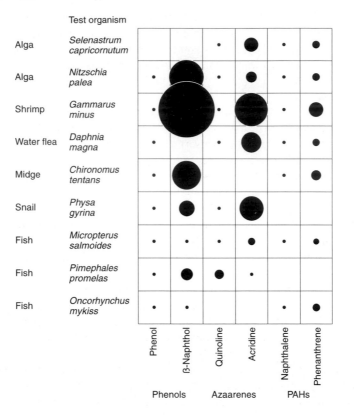

Fig. 9.3. Relative toxicities of single hydrocarbon compounds to freshwater organisms. The diameter of the circle is proportional to the toxicity. Comparisons can only be made between compounds within a class (i.e., between phenols, azaarenes or polycyclic aromatic hydrocarbons (PAHs)) and only within a species. Where boxes are empty, toxicities could not be compared (adapted from Milleman *et al.*, 1984).

Fish readily take up hydrocarbons, but, when they are placed in clean water, the compounds quickly disappear, indicating that fish have a metabolic capability for removing hydrocarbons. Mixed-function oxidase systems in fish liver respond to foreign chemicals (xenobiotics) and the microsomes in fish livers contain cytochrome P-450. Fish react to PAH inducers by showing an increase in associated enzyme activities, resulting in the excretion of the foreign compounds (see review by Müller, 1987). Rapid metabolism has been reported for quinoline, anthracene and benz(a)pyrene.

Amphibians are likely to be vulnerable to the effects of oil because of their permeable skins and aquatic larval stages. In experiments with different concentrations of used engine oil, hatching success of the eggs of the tree frog (*Hyla cinerea*) was unaffected by concentration. However the growth of tadpoles was severely reduced at higher concentrations and at the highest concentration (100 mg l^{-1}) no tadpoles metamorphosed (Mahaney, 1994).

Figure 9.4 illustrates the general effects of petroleum hydrocarbons on birds. Oil spills in areas where birds congregate in large numbers can be especially damaging to populations: for example, many ducks, geese and herons were killed following an oil spill on the St. Lawrence River (Alexander *et al.*, 1981). The thermal conduction of oiled ducks greatly increases, resulting in rapid heat loss.

Ingested oil inhibits the movement of water and sodium across the gut wall, so that contaminated birds may die of dehydration. Growth rates, behaviour and egg production may also be adversely affected (Shales *et al.*, 1989). Birds may be contaminated though not incapacitated by oil, but they may nevertheless fail to breed successfully because oil may be transferred to the eggs during incubation and prove toxic. Oil applied to the shells of mallard (*Anas platyrhynchos*) eggs drastically reduces hatching success, toxicity probably being due to the aromatic components of the oil. When 10 μl was spread over the egg shell, there was 100 per cent mortality. The dead duck embryos had a number of abnormalities, including deformed bills, reduction in liver size, incomplete ossification and incomplete feather formation (Hoffman, 1979). These laboratory experiments do have relevance to the field. Parnell *et al.* (1985) observed that, following a spill of no. 6 diesel oil in North Carolina, 24 of 98 nests of the brown pelican (*Pelecanus occidentalis*) were contaminated with oil transferred from the breast feathers of incubating adults. The hatching success of oiled eggs was significantly less than that of uncontaminated eggs.

The effects of the *Exxon Valdez* oil spill, two years after the accident, have been studied on a population of river otters (*Lutra canadensis*) which foraged intertidally (Duffy *et al.*, 1994). Otters abandoned ranges (measured in terms

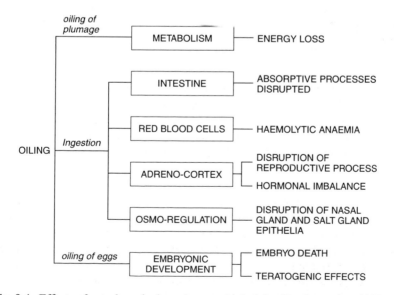

Fig. 9.4. Effects of petroleum hydrocarbons on birds (after Vandermeulen, 1987).

of usage of iatrine sites) more than three times as often in oiled than in non-oiled areas, indicating a continuing population decline. Blood samples taken from live-trapped otters in the oiled area showed elevated levels of hapto-globins, interleukin-6 ir and three enzymes. The high level of interleukin-6 ir indicates that the otters may have impaired immune systems, which could pre-dispose them to disease long after the oil has disappeared.

Because several components of oil and petroleum products, most notably PAHs, are known from laboratory studies to be mutagenic and carcinogenic, there has been much concern about their potential role in the environment, especially when they get into drinking water supplies. Vandermeulen *et al.* (1985) tested the mutagenicity of several oils and found only crankcase oil to be mutagenic. Proving a relationship between cancers and degree of pollution in the field is extremely difficult (p. 37). Given the high mutagenic and carcinogenic potential of a number of oil products, however, considerably more work is required to assess the true risk to freshwater communities and to humans.

Tainting

Fish that live in waters receiving oily wastes develop an objectionable taste, known as tainting, which may result in serious economic losses to commercial fisheries. Wastes from outboard motor engines can also taint fish flesh. Taints are due mainly to unsaturated aliphatic hydrocarbons and some aromatic hydrocarbons. It is often difficult to pinpoint sources of taint in rivers receiving wastes from many sources.

Field observations on spills

Although the accidental pollution of freshwaters with oil occurs frequently and may cause severe mortality to aquatic life, very few incidents have been adequately described and reported. Bury (1972) observed the effects of a spillage of some 9000 l of diesel oil, following a road tanker accident, into Hayfork Creek, California. Fuel entered the study area, 1.6–4 km below the site of the spillage, 36 h after the accident, the water turning brown and murky with floating droplets of oil. Thousands of aquatic invertebrates were killed, as were 2500 fishes, a very large number of amphibian larvae and 36 aquatic snakes. Pond turtles (*Clemmys marmorata*) were found with eyes and necks swollen, damaged epidermis and uncoordinated movements. Most of the oil had flushed out of the stretch within three weeks, but the animal community had by then been severely damaged.

Guiney *et al.* (1987) studied the effects of a leak of 1310 barrels of aviation kerosene into a stream in Pennsylvania. At stations below the spill the benthic invertebrates and fish were eliminated. Recovery began within six months and was almost complete six months later. Thus although the short-term effect of this incident on the fauna was catastrophic, the community recovered relatively quickly. The factors which accelerated the recolonization process were immediate and effective clean-up activities, plentiful unpolluted water upstream, invertebrate drift from upstream and immigration of fish from unpolluted sections of the watershed. Ecosystems that receive spills of crude oil, however, may take several years to recover, or indeed may never regain their former community of plants and animals.

Biodegradation and clean-up

Wherever possible, oil should be prevented from reaching the river in the first place. The River Rhine is a major shipping route and oil-contaminated bilge water is a potential source of pollution. Since 1965, Germany has established a bilge-draining facility. Currently eight bilge de-oiling boats pump out bilges while ships are in motion, at no financial cost or wasted time to the transporter. It is estimated that this keeps 8000–10 000 t of oil out of the river each year (Malle, 1994).

After an incident, oil that is deposited on the surface of water will be dispersed by physical and chemical processes, while some of the components will be evaporated, or lost by photochemical oxidation. As most of the remaining compounds are hydrocarbons they can be broken down by microorganisms. Brown (1989a) lists 21 genera of bacteria, 10 genera of fungi and 5 genera of yeasts that can degrade hydrocarbons. Complete degradation of a hydrocarbon may occur if a single strain of a microorganism uses that compound as its sole source of carbon and energy. Mixed populations of microorganisms may degrade up to 97 per cent of crude oil (Berwick, 1984). In general the substrate preferences of microorganisms for hydrocarbons are in the order aliphatics > heterocyclics > asphaltenes (Berwick, 1984; Brown, 1987).

Some of the reactions used by fungi and bacteria in oxidizing aromatic hydrocarbons are shown in Fig. 9.5. The filamentous fungus *Cunninghamella elegans* oxidizes aromatics initially to arene oxides that can then isomerize to phenols or undergo enzymic hydration to form *trans*-dihydrodiols (Gibson, D.T., 1983). Arene oxides have not been detected as intermediate products when bacteria oxidize aromatics and *cis*-dihydrodiols are formed. Gibson con-

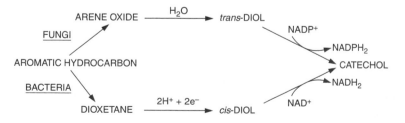

Fig. 9.5. Reactions used by fungi and bacteria for the oxidation of aromatic hydrocarbons (from Brown, 1989a).

siders that these differences in metabolism may represent a fundamental difference between procaryotic and eucaryotic organisms. The more complex aromatic molecules are more resistant to decay, probably because of their low solubility.

A number of factors influence the rate of biodegradation of oil, including oxygen availability, temperature, the type of oil spilled, the amount of emulsification and the availability of nutrients. Nitrogen and phosphorus are frequently limiting to microbial activity in freshwaters and the addition of these nutrients to lake water samples increased the amount of mineral oil and hexadecane degraded over three weeks (Cooney *et al.*, 1985). A review is provided by Prince (1992).

The microbial breakdown of spilled oil is a natural process which slowly removes the often substantial residue left after physical and chemical clean-up technologies have been applied. The microbial process can be speeded up by seeding the oil spills (bioaugmentation), and a number of companies have developed inocula of microorganisms to clean up spills. One product has been developed specifically to attack hydrocarbons in freshwaters and includes mutants which can degrade a variety of hydrocarbons. Once the substrate has been degraded the seeded microorganisms quickly lose their competitive ability and die out (Brown, 1989b). Nutrients are usually added to enhance the rate of breakdown.

In areas where oil spills are likely to occur (e.g. oil loading and handling bays, garage forecourts, etc.), interceptors can be built into surface drains to separate oil from water. Interceptor design and operation is described by Ellis (1989). When oil does pollute freshwaters, the clean-up consists of two stages – confinement followed by removal. Confinement involves the use of floating booms, or in very shallow waters the construction of temporary dams (Brown, 1989b; Ellis, 1989). The removal of the contained oil takes place in two stages: surface oil is skimmed off and the remaining traces removed using oil-absorbent devices. A variety of skimmers are in use and their operation is described by Ellis (1989). Sorbents include vegetable fibres and synthetic felted organic fibres, available in the form of cushions, blankets or loose chips. Some materials may hold up to 150 times their own weight of thick fuel oil. The chemical dispersal of oil is generally an undesirable option in the majority of spills in fresh water because of the highly toxic nature of dispersants (see above).

Conclusions

Oil-related industries have the potential to cause considerable environmental damage at the exploration, drilling, transporting, refining and storage stages of operation. Countries in the developed world have therefore placed strict regulations on oil and related industries and the larger multinationals employ teams of scientists, including ecologists, to assess and reduce the impacts of their industry, often at great expense. Most of the problems in the developed world are caused by accidents at the facilities of both producers and users of oil products. Such accidents, unfortunately, occur with considerable frequency and very many could be avoided if adequate safeguards were implemented. It is surprising, in view of the many millions of dollars payed out for clean-up operations, compensation and fines in accidents such as the grounding of the *Exxon Valdez*, and also the very poor public image that such accidents project, that safety standards and accident procedures are often so lax within the industry.

Away from the developed West, environmental protection appears to be a very low priority of the oil industry in its exploration and exploitation activities. Pearce (1993) reports how the vast wilderness of Russian Siberia is being turned into an ecological disaster area by oil and gas exploitation. The road infrastructure, the primitive drilling techniques, production of oil wastes and numerous pipeline leaks are resulting in widespread, but unmonitored, contamination of swamps, lakes, rivers and groundwaters. A spill in the Pechora river catchment in 1994 is estimated by some to have released six times as much oil as the *Exxon Valdez*. The impact on the tundra and its wildlife, and indeed on the Arctic environment as a whole if the oil reaches the Barents Sea, is as yet unknown.

Similarly, multinational companies appear to take a buccaneer attitude when exploiting oil reserves of developing countries, if the report of Read (1989) is typical. He describes the exploration for oil by more than 20 foreign oil companies in the forests of the upper Amazon basin in Ecuador. Some 630 000 ha of forest are currently being exploited. Initial exploration includes building roads and trails, the use of dynamite, crop destruction, erosion, attendant massive fish kills, and uncontrolled hunting by employees, with no concern for the indigenous population. At drilling sites, the surrounding land and waterways are polluted with drilling muds and oil wastes, while gas from the wells is burnt, rather than piped to cities where it could be used – Ecuador imports 50 per cent of its gas for domestic consumption. Spills are commonplace and are neither contained nor cleaned up. The main pipeline has already lost twice as much oil as was spilled by the *Exxon Valdez*, polluting hundreds of kilometres of the headwaters of the Amazon. Crops from flooded fields and cattle drinking from the rivers have died, while the indigenous peoples must use polluted water for drinking and cooking. The network of roads allows access to speculators and colonists who destroy the forest and its wildlife and drive out the native peoples.

Chapter 10

BIOLOGICAL ASSESSMENT OF
WATER QUALITY IN THE FIELD

To manage effectively a water resource receiving polluting substances we need information concerning:

1. The pollutants entering the aquatic environment and their sources, quantities and distribution.
2. The effects of these pollutants within the aquatic environment.
3. Trends in concentrations and effects, and the causes of these changes.
4. How far these inputs, concentrations, effects and trends can be modified and by what means at what cost.

The first stage in this management is to carry out a *survey*, which is a programme of measurements that defines a pattern of variation of a parameter in space. As an example, we may be concerned about the output of zinc into a river in an effluent from a rubber-processing factory. Our initial survey may involve measuring the concentrations of zinc in the river sediment at a number of stations downstream from the factory, together with sampling the fauna and flora at these stations. The survey will only inform us of the situation at one point in time.

The next stage will be *surveillance* and *research*, which will enable us to learn more about a problem before any policy decisions are made. Surveillance is defined as the repeated measurement of a variable in order that a trend may be detected. In our example we may measure the concentration of zinc in the sediments at three-monthly intervals to see how sedimentary loadings vary. Similarly, the animals and plants will be sampled to see if the original observations are repeated. The research function will be to examine the pollution process in more detail, using experimental and analytical techniques. For instance, the survival of fish in concentrations of zinc flowing from the rubber-processing factory could be studied in the laboratory, or experimental streams might be used to study whole communities. Furthermore, the tolerance of organisms to concentrations of zinc lower than that in the effluent might be studied as a guide to fixing a standard for the zinc level.

From the surveillance and research programme, and taking into account economic considerations, a policy for managing the pollution might be decided. For example, the level of zinc in the effluent might be reduced by 75 per cent by installing a treatment plant in the factory. It will be necessary to see that this reduction has the desired effect – an improvement in the state of the receiving river – and also to ensure that the effluent's quality is maintained. These observations on performance in relation to standards are known as *monitoring*.

It should be noted that these terms are used rather loosely and the boundaries between them are indistinct. Furthermore, there is not necessarily an ordered sequence in the procedure. The monitoring programme might produce useful research data which can be used to redefine the policy, which may then require modifications in the monitoring strategy.

The strategy outlined above for examining pollution involves both the experimental and observational approach (Fig. 10.1). The experimental approach involves the simulation of the pollutant's behaviour and its action on resources in experimental systems, whereas the observational approach examines the distribution of pollutants and their pattern of effects on natural resources. The integration of these two approaches is essential if a management policy for a particular pollutant or pollutional source is to be effective. Both approaches have their disadvantages. The experimental simulation does not take into account the complexity of the normal pollutional situation in which a variety of factors influence the way a pollutant affects its target. This very complexity, however, makes the interpretation of observational data exceedingly difficult.

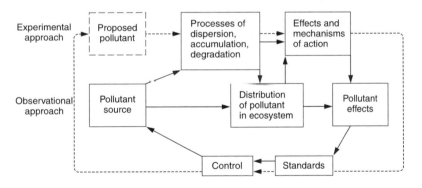

Fig. 10.1. Flow diagram of pollution and its control (from Edwards, 1975).

Why biological surveillance?

When carrying out a pollution surveillance programme, why do we not merely measure the levels of pollution at key points in the transfer pathway, especially if, for economic reasons, we are only interested in man and his food supplies? Pollutant levels can be measured at the sites of discharge and the sites of abstraction from the watercourse for potable supply, irrigation, etc., and we need not concern ourselves with problems in the river. Moreover, the concentrations of chemicals can be measured accurately and repeatedly, using standard methods. Biological material is notoriously difficult to sample and is inherently variable. There are, however, considerable advantages in biological surveillance.

Animal and plant communities respond to intermittent pollution which may be missed in a chemical sampling programme. For example, a river may be sampled at station A every Tuesday and analysed, back in the laboratory, for 10 chemical determinands, one of which is zinc. Such regular programmes may be necessary for logistical reasons where a large number of sites are under surveillance. On Wednesday morning each fortnight a factory immediately upstream may be discharging an effluent containing zinc. By the following Tuesday this will have disappeared downstream and the chemist will not detect zinc. A biologist, however, sampling at monthly intervals alongside the chemist, will record an unexpected depression in the diversity of the biological community, as some species may be eliminated and many individuals killed by the zinc. Some of the species missing may be known to be especially sensitive to zinc; they are acting as *indicators*. As replacement of organisms at station A will have to be by immigration and reproduction, the polluting incident will be apparent for several weeks or even months after the event.

The chemist may get round the problem of periodic sampling by installing an automatic analyser at the station in permanent contact, via telemetry, with a central control where alarms can sound if significant changes in water quality occur. These are expensive, however, they deal with only a few determinands and they are liable to failure under the often rigorous and unpredictable conditions in the field.

This brings us to the next advantage of biological surveillance: biological communities may respond to new or unsuspected pollutants in the environment. It would obviously be uneconomic and impracticable regularly to determine concentrations of the 1500 or so known polluants. The water industry at present regularly tests for about 30 determinands. If, however, a change in a biological community, or in members of that community, is detected and gives cause for concern, then a detailed screening for pollutants, and indeed of chemicals hitherto not considered as pollutants, can be made. The decline in western grebes at Clear Lake, California (p. 178) gave the first evidence that organochlorines were biomagnified in food chains to have detrimental environmental effects.

The example of the grebes demonstrates another advantage of biological monitoring, namely that some chemicals are accumulated in the bodies of some organisms and these levels can reflect the environmental pollution levels. Chemicals may accumulate in the body over long periods of time, whereas at any particular point in time the pollutant may be present at too low a level in the water to be detected without the concentration of large volumes of water. In this way many organisms may be used in bioassays in surveillance programmes. Bioassays may not only measure concentrating pollutants, but also growth responses (e.g. of algae in nutrient-rich water) or mortality (e.g. of fishes placed in effluents). This is the subject of the next chapter.

Biological survey

The objectives of a particular study must be clearly defined before the work programme is begun. A biologist working within the water industry will mainly be involved in large scale survey, surveillance or monitoring studies, necessitating visits to many, often widely scattered, localities during the course of a year, frequently for different purposes. Severe constraints will be placed on the amount of sampling that can be undertaken at any particular site and a compromise will have to be reached between the level of sampling effort and the amount of data required to produce meaningful results. It will not be feasible to sample the entire biota and a decision must be made as to which group of organisms will provide the most information for solving a particular problem.

Ideally a sample should be compared with detailed information from past samples for that site, but frequently the biologist is involved only after a pollution incident has occurred. It may be possible to find an acceptable control site, for example upstream of the incident or a local tributary, but the assumption then has to be made that the biological community was identical in the unpolluted control and the polluted site before the incident took place. As an example, 28 shallow lakes in the Broadland area of eastern England were surveyed and many of them were found to have a impoverished flora and fauna and were suffering severe eutrophication (p. 102) (Mason and Bryant, 1975). Despite the fact that the area had been renowned for decades for its biological diversity and importance for conservation, no previous survey had been carried out, so that the extent of the ecological degradation had to be pieced together from published natural history notes, unpublished reports and the personal recollections of local naturalists. In no cases were the survey methods of previous observations recorded. Of these 28 sites, three were still biologically rich and could be considered as possible controls, but it was known that in two of these lakes marked changes had occurred (severe pH fluctuations in one and changes in the dominance of macrophytes in the other). It was not known whether these changes were due to natural factors or human influences. The biologist investigating pollution is often faced with these dilemmas.

The ecology and sensitivity to pollution of many of the organisms used in water quality assessments are still very inadequately known, even in those countries with a long tradition of research in freshwater biology. The key organisms in the ecology of one particular watershed may be insignificant in a neighbouring watershed of differing geology or flow regime. Pollutants affect organisms differentially at different life stages or different times of year, while natural changes in population size may underlie any of the effects of pollution. Disentangling the effects of pollution from natural events is often very difficult.

Biologists in the water industry have also to explain their findings to managers, who are usually not biologists and often not scientists, or to the general public. Biologists therefore frequently resort to simple indices of water quality, single numbers which may help in communication but which result in a considerable loss of biological information. The danger is that the biologists themselves may come to rely too heavily on interpreting change in terms of indices, such that more subtle analyses of the data, giving increased biological understanding, will not be attempted. Indeed biologists in the water industry may not be allowed time by their managers to undertake more penetrating approaches to the understanding of pollution problems.

Despite these many difficulties, biological assessments of water quality have proved very successful and even mild, intermittent pollution, frequently missed by routine chemical sampling, has been detected. This chapter will deal chiefly with survey and monitoring techniques, rather than with the undertaking of specific research projects.

Sampling strategy

The definition of objectives is the essential first stage in the design of a sampling programme. There are three prime objectives in surveillance and monitoring studies for water quality.

1. **Environmental surveillance**, where the objective is to detect and measure adverse environmental changes, such as the effects of unknown or intermittent pollutants, or to follow the improvement of conditions once a pollutant has been removed.
2. **Establishing water-quality criteria** in which causal relationships between ecological changes and physico–chemical parameters are determined.
3. **Appraisal of resources** which may involve a large-scale survey to assess general water quality.

Alternatively, particular water resource problems, involving nuisance species or the impact of new developments, may be investigated. Such problems will also include the management of fisheries or the conservation of threatened species.

In designing a sampling programme it is essential to include randomly allocated replicates and controls. A preliminary sampling should be undertaken to evaluate the sampling design and examine options for subsequent statistical analysis; this often saves considerable time later in the programme by emphasizing design faults at the beginning of the study. The efficiency of the sampling device should be determined at the beginning, as should the size and number of samples in relation to the size, density and spatial distribution of the organisms being sampled (Green, 1979). Resh and McElravy (1993) discuss sampling design.

Surveys may be either extensive or intensive. Extensive surveys aim to discover what species are present in an area, usually with a measure of relative abundance, and are especially used where the water quality over many sites is being monitored or compared. Such surveys have been criticized, or even considered valueless, because they are too superficial to detect or interpret subtle environmental changes such as alterations in species dominance due to biological interactions, so that it is impossible to disentangle natural changes from those caused by pollution. Such a view is undoubtedly too pessimistic and there is ample evidence of the ability of faunal surveys to detect pollution without a detailed foreknowledge of the ecology of a site. It is probably true, however, that sampling design is frequently given inadequate consideration in extensive surveys.

Intensive surveys usually aim to determine population densities. The main considerations in designing a quantitative survey are the dimensions of the sampling unit, the number of sampling units in each sample and the location of sampling units within the sampling area (Elliott, 1977). Populations of organisms are usually highly aggregated so that a large number of samples are frequently required to obtain a population estimate that is statistically meaningful (Peckarsky, 1984).

Ninety-eight samples were required to obtain an estimate of mean density (± 40 per cent) with 95 per cent confidence limits for the limpet *Ancylus fluviatilis*, for example (Edwards *et al.*, 1975). Such sampling intensity would clearly be impossible in an extensive survey and even in an intensive survey the rarer species will be inadequately sampled. Edwards *et al.* (1975) have shown that, sampling a riffle, only 44 per cent of the species taken would be common to both of any two random samples. For studies of pollution it is the difference between means over time, or the difference between sites, which is important (Resh and McElravy, 1993). Therefore accuracy in determining populations of individual species is not required and six samples may be sufficient to provide estimates of the total number of individuals in a community with confidence limits ± 40 per cent of the mean (Canton and Chadwick, 1988).

The number of samples required for a specified degree of precision can be readily calculated if an estimate of the mean abundance is made from a pilot survey:

$$D = \frac{1}{x} \sqrt{\frac{s^2}{n}}$$

where D is the index of precision, x is the mean, s^2 the variance and n the number of samples. A standard error of 20 per cent of the mean is usually acceptable for ecological studies so that the number of samples required to obtain this level of precision ($D = 0.2$) is:

$$n = \frac{s^2}{0.2x^2}$$

If a series of samples is taken over time it must be remembered that the number of samples required to maintain this level of precision will change as the population size and degree of aggregation change. For example, on collecting 30 random benthic samples of the tubificid worm *Potamothrix hammoniensis* in a shallow lake, the standard error with a population density in July of $10\,129\,m^{-2}$ was 16 per cent, but after the death of adult worms during the late summer the standard error in October increased to 45 per cent of the population mean of $660\,m^{-2}$ (Mason, 1977a).

A small sampling unit is generally preferable to a large unit because more samples can be handled for the same amount of effort and the statistical error in estimating the mean is reduced. Many small units cover a greater range of habitats than a few large units so that the population estimate will be more representative of the sample area.

Sampling units must be selected at random from within the sampling area for the sample to be representative of the whole population. To prevent all of the sampling units falling randomly in one part of the sampling area, or if the location to be sampled shows environmental pattern, stratified random sampling is often used. The sampling area is divided into smaller areas (strata), usually of equal size, and sampling units, divided equally among the strata, are selected at random from within each stratum on each sampling occasion. Sampling stategies are discussed by Resh and McElravy (1993).

The choice of organisms for surveillance

It is usually impossible to study the entire biota present in a sampling area because of the constraints of time and of the wide variety of sampling methods required for different groups of organisms. A survey or monitoring programme must therefore be based on those organisms that are most likely to provide the right information to answer the questions being posed.

The use of a single species as a water quality indicator is generally avoided because individual species show a high degree of temporal and spatial variation due to habitat and biotic factors, and these confuse any attempt to relate presence, absence or population level to water quality. Similar constraints may limit the value of ratios of species or groups, such as the *Gammarus/Asellus* ratio (Hawkes and Davies, 1971). Nevertheless the *Gammarus/Asellus* ratio has been shown to perform well when compared with biotic indices (see below) in assessing organic pollution and it is extremely simple to use (Whitehurst, 1991). Considerable care is also needed in identification, as similar species may show very different reactions to pollution. For indicator species to be worthwhile, they must be able to register subtle, rather than gross and obvious, effects of pollution.

The use of communities of organisms allows this more subtle approach. To be suitable for a broad survey or monitoring programme a biological system requires the following features:

1. The presence or absence of an organism must be related to water quality rather than other ecological factors.
2. Water quality must be reliably assessed and expressed in a simplified form,

but the system must be sufficiently quantifiable so that comparisons can be made.

3. The assessment should indicate water quality conditions over an extended period of time, not just at the time of sampling.
4. The assessment should relate to the point of sampling, not the watercourse as a whole, to locate sources of pollution rather than describe the general water quality of the catchment.
5. A minimum of time and manpower should be required for sampling, sorting, identification and data processing.

Numerical abundance at some sites, widespread distribution and a well-documented ecology are also important factors to take into account in selecting a group of organisms for water quality assessment.

Despite the ubiquity of bacteria in aquatic ecosystems and the large populations developed, little attention has been given to the use of indigenous bacteria in the assessment of pollution, with the exception of the determination of biological oxygen demand. It has been suggested that, because of their morphological, physiological and genetic characteristics, microbial communities could act as an excellent early-warning system for pollution; by the time a change has been detected in the communities of larger organisms it may be too late to reverse (Bianchi and Colwell, 1986). Bacteria that metabolize petroleum hydrocarbons, for example, are ubiquitous in aquatic ecosystems but are present in much higher numbers in environments exposed to oil (Bartha and Atlas, 1977). Similarly, the number of bacteria resistant to a heavy metal increases even in the presence of very low concentrations. Genes for resistance to heavy metals are often carried on independent pieces of bacterial DNA, called plasmids, which can be transmitted from one bacterium to another. Because the acquisition of specific plasmids may render bacteria tolerant to several metals, bacteria could prove useful indicators of metal pollution (Bianchi and Colwell, 1986). The further development of techniques using responses of indigenous bacteria to pollution could prove invaluable for pollution evaluation (Pipes, 1982).

Algae have been popular organisms for the assessment of water quality and are especially valuable in examining the eutrophication of lakes. A Compound Quotient has been developed, where the number of species of Chlorococcales, Cyanobacteria, centric diatoms and Euglenophyta are divided by the number of Desmidaceae in a water sample. Values of the quotient less than 1.0 (rich in desmid species) indicate oligotrophy, while values of 5.0 or more are indicative of hypertrophic conditions. In Loch Leven, Scotland, Compound Quotients of the order of 1.6–1.9 were recorded in the early part of this century but 50 years later, following cultural eutrophication, the quotient had increased to 7.2 (Brook, 1994). The Compound Quotient may be useful for synoptic surveys but the development of a more subtle index, suitable for the routine management and monitoring of lakes and reservoirs, is proving elusive.

For the monitoring of river water quality epilithic diatoms, attached to stones or to artificial substrates such as glass slides (see p. 78), have much support. A diatom community index has been produced which is calculated from the relative frequencies of tolerant and indifferent/sensitive species, each taxon being given a value depending on its sensitivity (Watanabe *et al.*, 1988; Coste *et al.*, 1991). Considerable time and taxonomic skill (and hence cost) are required to identify and enumerate diatoms, though indices involving identification only to the genus level are being tested. A review of diatoms in water quality assessment is provided by Round (1991), while Whitton *et al.* (1991) provide a series of examples.

Macrophytes are conspicuous and relatively easy to identify in the field. In Great Britain a large-scale survey has led to macrophyte vegetation being classified into four broad groups, with 16 main subtypes (Holmes, 1989). Altitude, gradient and catchment and soil type are the major determinants of the groups and these will, of course, influence the use of vegetation in pollution assessment. Macrophytes are also seasonal and are tolerant of intermittent pollution, as well as being strongly influenced by management practices (e.g. for flood prevention). Haslam (1987) describes in some detail the relationship between water quality and plant distribution.

Fish tend to be too mobile and hence can avoid intermittent pollution incidents, while their capture requires considerable manpower. Fish populations are widely manipulated by managers. They are, however, easy to identify, their ecology and physiology are relatively well known and, as some are at the top of the food chain, they may reflect changes in the community as a whole.

In running waters most workers nowadays prefer macroinvertebrates for water quality assessments and the remainder of this chapter will concentrate on these.

Macroinvertebrates and water quality assessment

There are a number of reasons for preferring benthic macroinvertebrates to other groups (Metcalfe, 1989). The sampling procedures are relatively well developed and can be operated by someone working alone, and there are identification keys for most groups. Macroinvertebrates are reasonably sedentary, with comparatively long lives, so that they can be used to assess water quality at a single site over a long period of time. The group is heterogeneous and so a single sampling technique may catch a considerable number of species from a range of phyla. It is likely, therefore, that at least some species or groups will respond to a particular environmental change. Macroinvertebrates are also generally abundant.

They have some disadvantages. Their aggregated distribution means that, to obtain a representative sample of a site, many samples must be taken. The muddy, depositing substrata of the lowland reaches of rivers, or of lakes, are often dominated by chironomids and tubificid worms, which are difficult to identify, while the water in these situations is frequently deep, making sampling difficult. The insect members of the community may be absent for part of the year, so that care needs to be taken in interpreting the results of monitoring.

Sampling for macroinvertebrates

The simplest method of sampling, suitable for shallow waters over eroding substrata, is the kick sample. The operators face downstream and hold a standard pond net vertically in front of them, with the bottom against the substratum (Fig. 10.2). The substratum upstream of the net is then vigorously disturbed with the feet and the dislodged invertebrates flow into the net. In

Fig. 10.2. A kick sample being taken with a standard pond-net (photograph by the author).

shallow waters stones can also be turned over by hand in front of the net. By attempting to disturb a known area, or by kicking for a fixed period of time, this method can be made semi-quantitative for relative abundance estimates. The technique is rapid and inexpensive and is particularly suited to faunal surveys or extensive surveillance programmes. There are obvious potential problems in comparing results between sites of different flow regimes, substratum types and so on, and between individual operators, which will be inevitable in extensive survey programmes. Furse *et al.* (1981) have examined this variation. Using a standard pond net (a frame of 230 mm × 250 mm, with a 900 μm mesh net of depth 275 mm, on a 1.5 m handle) they found that a 3-minute kick sample of all available habitats at a site collected 62 per cent of families and 50 per cent of species that could be attained by sampling for 18 minutes. They considered that their results justified the use of a 3-minute kick sample with a standard pond net for extensive surveys.

A variety of samplers have been designed for the quantitative collection of invertebrates (Hellawell, 1986). The Surber sampler, combining a quadrat with a net (Fig. 10.3a), and cylinder samplers (Fig. 10.3b) are widely used. The Surber sampler consists of a net with a hinged frame, attached to its lower margin, which can be pushed onto the substratum and locked into place. The frame quadrat encloses 0.09 m² (1 sq ft) and stones and gravel within this area are lifted and stirred so that invertebrates are dislodged into the net. Side-wings on the net help to reduce the loss of animals around it.

In shallow waters, open-ended cylinders can be pushed into the substratum to enclose a known area, a typical example being the Neil sampler, which has two openings near the base. The upstream opening is covered with a coarse metal mesh which allows a through-flow of water but excludes drifting invertebrates. The downstream opening has an attached net for collecting animals. The openings of the Neil sampler have sliding doors which are opened after vigorously stirring the enclosed substratum. The sampler is traditionally made of stainless steel or aluminium. A considerably cheaper version can be made by using a 50-cm length of 25-cm diameter plastic sewer piping. The moveable doors can be dispensed with and a standard pond net can be held over the downstream aperture as the substratum is stirred (Fig. 10.3b).

For shallow, still waters, such as ponds or the muddy edges of reservoirs, a cylinder sampler without inflow and outflow apertures can be used. The sediment may be removed with a plastic beaker or a small, fine-mesh hand net and I have found this method highly satisfactory for sampling the benthos of shallow, coastal lagoons. In water less than 2 m deep sediment can be collected from a boat using a pond net and this can be calibrated against a standard sampler (Mason, 1977a), which results in a considerable saving of time.

For sampling deeper waters of lakes and rivers a variety of grab and core samplers have been devised (Downing, 1984). The jaws of grabs (Fig. 10.3c) close beneath a known area of bottom, the mechanism being operated using a messenger released from the surface. For very fine sediments, corers (Fig. 10.3d) are usually favoured. These have a smaller diameter than grabs (less

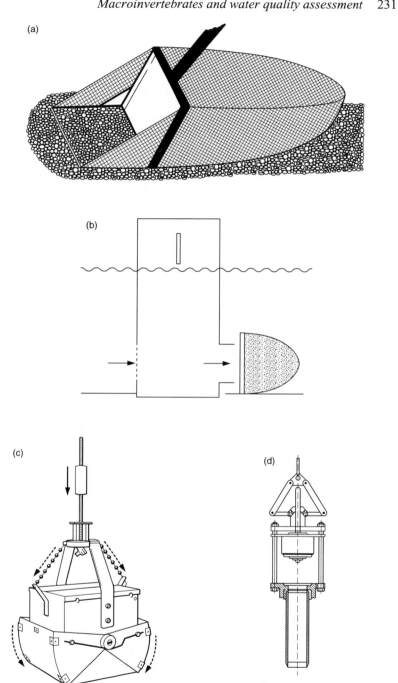

Fig. 10.3. Types of samplers for benthic invertebrates. (a) Surber sampler; (b) cylinder sampler; (c) Ekman grab; (d) corer.

than 15 cm) and consist of an open-ended perspex tube. The top of the tube has a valve which allows the free passage of water during the descent of the corer but which is closed, either automatically or with a messenger, when the sampler is lifted, to prevent the loss of sediment. Multiple units, which take several samples at once, have been developed.

Artificial substrata

Considerable interest has been shown in the use of artificial substrata for the collection of macroinvertebrates. A variety of substrata has been tried (Flannagan and Rosenberg, 1982; Rosenberg and Resh, 1982), including wire mesh trays filled with stones and placed on the river bed, or wire mesh baskets filled with crushed limestone (Fig. 10.4a), or bark chippings from coniferous trees, placed on the river bed or suspended at various depths in the water. Artificial substrata may be mimics of plants, consisting of nylon ropes attached to a plastic or wire base. Units of multiple hardboard plates (Fig. 10.4b), suspended in the water, have also been used.

Artificial substrata are, of course, initially devoid of invertebrates, so that it is important for a representative invertebrate community to have developed before they are retrieved. Using modified Hester-Dendy multiplate samplers exposed for 60 days in a stream, it was found that the number of individuals present peaked at 39 days and then declined, whereas new taxa were still colonizing at the end of the exposure period (Meier *et al.*, 1979). Using a similar collector in a canal, it was found that the number of species present increased very little after three weeks (Cover and Harrel, 1978), and using artificial macrophytes in a shallow lake, an equilibrium community was reached after 35 days (Mason, 1978). In these cases, however, species disappeared and new ones were added in substrata exposed for long periods, while overall populations increased, affecting diversity. It is generally agreed that an exposure period of six weeks will allow a reasonably representative community to develop.

There are several advantages in using artificial substrata for the assessment of water quality. The samples can be readily processed because there is usually little extraneous material, such as silt. The effect of natural substrata is reduced and the samples can be used in sites where more conventional techniques are difficult to apply, for instance in deep, swiftly flowing rivers. A higher level of precision is obtained and comparison between sites is made easier. Disadvantages are that the sample obtained may not be representative of the community present in the natural substratum of that site, for example the fauna of stony riffles may colonize artificial substrata positioned in mud and substrata may be selective. They only collect invertebrates colonizing during the period of exposure and this makes comparisons with other techniques difficult. They are subject to vandalism and to loss during spate conditions.

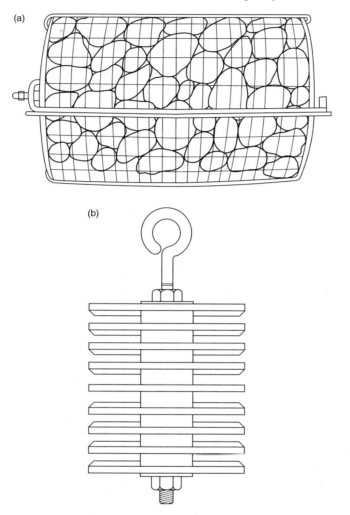

Fig. 10.4. Artificial substrata for sampling invertebrates. (a) Rock-filled basket; (b) Hester-Dendy multiplate sampler.

Drift and emergence samplers

Drift and emergence samplers are reviewed by Peckarsky (1984) and Davies (1984). Emergence traps collect only a proportion of the community (i.e. insects) at certain times of the year, while the emergence period for different species varies, so the technique is of little use in the assessment of water quality. The collection of drifting invertebrates is subject to similar problems in that different species have different predilections to drift, and daily and longer-term variations and other factors make it difficult to relate drift to the macro-invertebrate community in a river bed. One component of drift, however, the exuviae (pupal skins) left behind by hatching chironomid pupae, has been shown to have considerable potential in the assessment of water quality (p. 85). Exuviae are collected with a pond net from streams or lakes, especially where concentrations of scum or flotsam occur. As surface samples are taken, the type of water body does not affect collection. The exuviae are easily separated from detritus by sieving. Identification requires considerable expertise. The technique can provide a rapid biological assessment of an entire catchment from a single set of field collections taken on a single day at a good time of year (late May to September). The method also reveals points of change in the fauna due to changes in water quality. It is most effective during the summer and will have little value for the examination of sporadic pollution incidents occurring at other times of year (Wilson and McGill, 1977; Wilson, 1994).

A comparison of sampling methods

Few studies have been made of the efficiencies and relative merits of different samplers. Hughes (1975) compared a Surber sampler, a modified Neil sampler, which was also used in conjunction with an electric shock pulser, and an artificial substratum. The Surber and Neil samplers gave similar results, but the electric shock method was highly selective, particularly for mayflies, and was the least consistent. The artificial substratum collected the most species and numbers and was the most consistent, but Hughes considered that it failed to represent the fauna in the surrounding river bed and so preferred the Surber and Neil samplers.

When sampling sediments beneath deeper waters, the accuracy of the sampling device will depend on the substrate to be sampled. The efficiency of seven grabs in sampling small plastic pellets (model invertebrates) from different substrata was tested in a large tank. Three grabs had low efficiencies in all substrata. The remainder were adequate in a fine gravel substratum, if the minimum acceptable efficiency was 50 per cent, but only two grabs sampled efficiently when small stones were present and none of the grabs were adequate for sampling when stones greater than 16 mm diameter were present (Elliott and Drake, 1981).

Flannagan (1970) compared a range of core samplers and grabs for sampling sediments beneath deep waters, core samples also being taken by a diver. The Ekman grab and multiple corer gave good quantitative estimates of total biomass when compared with the diver. The Ekman grab collected fewer oligochaete worms and the corer fewer chironomids. Overall the multiple corer seemed to perform better and was the preferred sampler in Hellawell's (1986) review.

Several studies have compared different types of artificial substrata. In general, all samplers that have been tested have developed a diverse fauna but marked differences between the collections of various substrata at the same site may occur. The sample variability between artificial substrata at a site is relatively low compared with samples collected by traditional methods, so that a small number of replicates will give a representative sample of the macrofauna. If the aim of a study is to assess water quality, rather than to examine the local macrofauna, then artificial substrata have much to recommend them because, even if the sample collection is richer than in the surrounding river bed, as it frequently will be if artificial substrata are placed on depositing beds, it shows that a diverse fauna can live in those water quality conditions. As Green (1979) has succinctly remarked, 'the health of canaries can be a good indicator of the safety of coal mines, even though canaries are not natural inhabitants of coal mines'. For extensive comparative surveys, however, it is necessary to standardize the artificial substrata and their placement within a waterbody, but this is easier than with traditional sampling methods.

Sorting samples

Experienced personnel may sort samples in the field, especially if the taxonomic precision required is low, for example to the family level only, and if estimates of population size are not needed. If data are entered into a portable computer, rather than on to a checksheet (Whiten and Barton, 1988), instant assessments of water quality can be obtained!

Sorting samples in the laboratory is tedious and time-consuming especially when it involves material from depositing substrata. Coarse detritus can be removed using a 4-mm sieve, but a smaller sieve to retain invertebrates has to be a compromise between retention efficiency and the time available for sorting, bearing in mind that hand-sorting will be inefficient if large amounts of detritus remain with the sample. Some 87 per cent of chironomids and 82 per cent of tubificid worms were retained by a 500-μm aperture sieve, and using a 250-μm sieve, the retention efficiency was 98 per cent and 99 per cent respectively (Mason, 1977a). The smaller mesh sieves readily become clogged with clay particles so that wet sieving becomes very time-consuming. For routine surveillance programmes a 500-μm aperture sieve should be adequate.

A variety of flotation techniques has been tested to separate invertebrates from the substratum using solutions such as calcium chloride, carbon tetrachloride or sucrose. These are likely to add another source of loss to the sorting programme and can, in themselves, be time-consuming. The most effective method is probably to transfer the contents of a 500-μm sieve, using a gentle jet of water, to a white tray which has been divided into squares, and then sort each square systematically under a good light. A stain, such as rose bengal, can be added to help sorting, but where possible the sorting of a live sample is best because the movement of animals aids detection.

Data processing

Surveillance data may be analysed by biotic or diversity indices or by multivariate techniques, the former two methods being the most widely used in the water industry. Biologists in the water industry, as already stated, must communicate their findings to managers and the public, who are usually unfamiliar with ecological techniques. A single figure describing the biological impact of water quality at a site is therefore a tempting way to present data and is generally preferred by water managers. Water quality indices, consisting of a single figure representing a large number of chemical determinands, are being developed (e.g. House and Newsome, 1989), to produce a precise and unambiguous assessment of water quality which will replace subjective and much criticized classification systems. The use of a biotic or diversity index to describe the biological impact of water quality reduces the amount of information extracted from the data, however. Ready calculation of indices also tends to replace more sophisticated analyses which would enable the relationships between organisms and the measured physico–chemical parameters of water quality to be better understood. This would eventually place the biological management of freshwaters on a sounder footing. Green (1979), who is decidedly unenthusiastic about derived indices, has stated, 'the literature has been filled with wrangling over which diversity index should be used of the countless ones proposed while there has been too little emphasis placed on the critical importance of proper sampling design, or on appropriate multivariate statistical analysis methods that can efficiently test exactly the hypotheses one wishes to test in environmental studies'. Biotic and diversity indices can be used for overall assessments of water quality and their ability to detect changes in quality will be illustrated later in this chapter, but it must be stressed that information, and indeed the expertise of biologists, is tragically wasted if indices become the be-all and end-all of biological surveillance programmes, as they so often are in the water industry.

Biotic indices

Several biotic indices have been developed to assess water quality (see reviews in De Pauw and Vanhooren, 1983; Washington, 1984; Hellawell, 1986; Metcalfe, 1989). A biotic index takes account of the sensitivity or tolerance of individual species or groups to pollution and assigns them a value, the sum of which gives an index of pollution for a site. The data may be qualitative (presence–absence) or quantitative (relative abundance or absolute density). They have been designed mainly to assess organic pollution.

The earliest biotic index, devised in the early years of this century, was the Saprobien system, which recognizes four stages in the oxidation of organic matter: polysaprobic, α-mesosaprobic, β-mesosaprobic and oligosaprobic. The presence or absence of indicator species in the zones is recorded. The Saprobien index and derivatives are widely used in continental Europe but the method has received little support in Britain or North America.

In Britain widely used biotic indices, such as the Trent Biotic Index and Chandler Biotic Score, have been largely superseded by the Biological

Monitoring Working Party (BMWP) Score. This was designed to give a broad indication of the biological condition of rivers throughout the United Kingdom. Sites are sampled, where possible, by a 3-minute kick/sweep with a standard pond net, all major habitats (bottom substrata, vegetation, margins, etc.) within the site being included. Where it is not possible to use nets, artificial substrata can be substituted. Identification of taxa is to the family level only and no account is taken of abundance. Each family is given a score, between 1 and 10, depending on its perceived susceptibility to pollution. Those taxa least tolerant, such as families of mayflies and stoneflies, are given the highest scores (Table 10.1). The BMWP score is the sum of the scores of each family present in a sample. This total score can then be divided by the number of taxa to produce the Average Score Per Taxon (ASPT), which is independent of sample size (a larger sample is likely to include more families, thus inflating the BMWP score if not standardized).

The performance of the BMWP score using invertebrate samples from 268 sites in 41 river systems in England and Wales, collected using the standard method, has been examined (Armitage *et al.*, 1983). The ASPT was less influenced by season than the BMWP score, so that samples taken in any season will provide consistent estimates of ASPT. ASPT was also less influenced by sample size than the BMWP score, so that more information can be obtained for less effort. Physical and chemical data, and physical data alone, were used to predict BMWP scores and ASPT using multiple regression techniques. On average a higher proportion (65 per cent) of the variance was explained in equations used to predict ASPT than in those used to predict BMWP scores (22 per cent). The authors suggested that the ratio of observed to predicted ASPT could be used to indicate the possible influence of pollution on the macroinvertebrates. Somewhat different conclusions were drawn from a Spanish study, where ASPT, but not BMWP, was found to be dependent on temperature and hence season. Water quality explained 54 per cent of the variation in BMWP, better than for the English study, possibly because a wider range of water qualities were investigated. ASPT and BMWP were highly correlated (Zamora-Muñoz *et al.*, 1995).

Various biotic indices have been compared during a biological surveillance programme of a chalk stream in southern England (Pinder *et al.*, 1987; Pinder and Farr, 1987a, b; Pinder, 1989). It was concluded that the ASPT was the best indicator of water quality over the range of conditions encountered. At one site, below a small discharge of partially treated sewage, the ASPT was depressed, while the BMWP increased compared to a site further upstream. It was concluded that the discharge was resulting in mild enrichment of the stream, rather than causing pollution, and that there may be some advantage in evaluating both ASPT and BMWP scores.

The BMWP score has been used to assess the effect of a spillage of the insecticide Dursban, the active ingredient of which is the organophosphorus compound chlorpyrifos, on the macroinvertebrate fauna of the River Roding, which flows from the northeast of London into the Thames estuary. A road

Table 10.1   The Biological Monitoring Working Party (BMWP) score

	Families	*Score*
Mayflies	Siphlonuridae, Heptageniidae, Leptophlebiidae, Ephemerellidae, Potamanthidae, Ephemeridae	
Stoneflies	Taeniopterygidae, Leuctridae, Capniidae, Perlodidae, Perlidae, Chloroperlidae	
River bug	Aphelocheiridae	10
Caddisflies	Phryganeidae, Molannidae, Beraeidae, Odontoceridae, Leptoceridae, Goeridae, Lepidostomatidae, Brachycentridae, Sericostomatidae	
Crayfish	Astacidae	
Dragonflies	Lestidae, Agriidae, Gomphidae, Cordulegasteridae, Aeshnidea, Corduliidae, Libellulidae	8
Caddisflies	Psychomyidae, Philopotamiidae	
Mayflies	Caenidae	
Stoneflies	Nemouridae	7
Caddisflies	Rhyacophilidae, Polycentropidae, Limnephilidae	
Snails	Neritidae, Viviparidae, Ancylidae	
Caddisflies	Hydroptilidae	
Mussels	Unionidae	6
Shrimps	Corophiidae, Gammaridae	
Dragonflies	Platycnemididae, Coenagriidae	
Water Bugs	Mesoveliidae, Hydrometridae, Gerridae, Nepidae, Naucoridae, Notonectidae, Pleidae, Corixidae	
Water Beetles	Haliplidae, Hygrobiidae, Dytiscidae, Gyrinidae, Hydrophilidae, Clambidae, Helodidae, Dryopidae, Elminthidae, Chrysomelidae, Curculionidae	5
Caddisflies	Hydropsychidae	
Craneflies	Tipulidae	
Blackflies	Simuliidae	
Flatworms	Planariidae, Dendrocoelidae	
Mayflies	Baetidae	4
Alderflies	Sialidae	
Leeches	Piscicolidae	
Snails	Valvatidae, Hydrobiidae, Lymnaeidae, Physidae, Planorbidae	
Cockles	Sphaeriidae	3
Leeches	Glossiphoniidae, Hirudidae, Erpobdellidae	
Hoglouse	Asellidae	
Midges	Chironomidae	2
Worms	Oligochaeta (whole class)	1

accident resulted in the release of about 500 litres of Dursban into a tributary and within two days the insecticide had reached the tidal reaches of the river, 26 km downstream. Some 90 per cent of the fish biomass and all of the aquatic arthropods were destroyed over a 23-km stretch of river. There was little effect on molluscs and annelids. Figure 10.5 illustrates the BMWP score for a control site and affected sites 4 km and 23 km downstream of the spillage. Soon after the spill, in April 1985, the BMWP score fell sharply at both sites as sensitive arthropod species, including those species which contributed most to the score, were eliminated. The effect was most severe at the upstream site (B), which had better water quality before the accident and received the pulse of pesticide in a more concentrated form. The BMWP then gradually recovered as recolonization occurred, largely by drift from unaffected sites above the spillage. Chironomids began to recolonize 13 weeks after the spillage and *Asellus aquaticus* also recovered quickly, probably because of the greatly reduced predation by fish. Most other arthropods had recolonized within 79 weeks, but the beetle *Oulimnius tuberculatus* and the mayfly *Caenis moesta* were still absent 108 weeks after the spillage.

Studies over the past decade have therefore shown that the BMWP score,

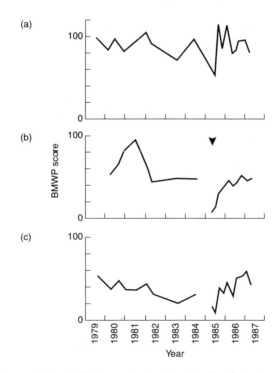

Fig. 10.5. Changes in the BMWP score in the River Roding following a spill of the insecticide chlorpyrifos. (a) Control site above the spillage; (b) 4 km downstream; (c) 23 km downstream of the spillage. The arrow marks the spillage (adapted from Raven and George, 1989).

and especially its derivative, the ASPT, are reliable indicators of water quality. How these can be used to develop a water quality index is described on p. 243.

Diversity indices

Biotic indices have been developed largely to measure responses to organic pollution and may be unsuitable for detecting other forms of pollution. Stoneflies, for example, have a prominent role in most biotic indices owing to their great sensitivity to a decrease in oxygen in the water, but they are markedly more tolerant of metals and may be abundant in rivers receiving substantial quantities of metal wastes from mine tips. Diversity indices are used to measure stress in the environment. It is considered that unpolluted environments are characterized by a large number of species, with no single species making up the majority of the community. Maximum diversity is obtained when a large number of species occur in relatively low numbers in a community. When an environment becomes stressed, species sensitive to that particular stress will be eliminated, thus reducing the richness of the community. In addition, certain species may be favoured (e.g. as a result of the reduction of competition or predation) so that they become abundant compared with other members of the community. Most species diversity indices take account of both the number of species in a sample and their relative abundances, but the sensitivity of individual species to particular pollutants is not allowed for. Diversity indices are widely used by North American workers and they are frequently calculated, along with biotic indices, by British workers.

Diversity indices and their application to aquatic studies are described by Washington (1984) and Metcalfe (1989). The most widely used indices of diversity are those based on information theory, the most frequent measure being the Shannon index, which assumes that individuals are randomly sampled from an indefinitely large population:

$$H' = -\sum_{i=1}^{s} P_i \log P_i$$

where P_i is estimated from N_i/N as the proportion of the total population of N individuals belonging to the ith species (n_i), i.e. the number of individuals in a given taxon is divided by the total number of organisms in the collection, the resulting ratio being multiplied by the logarithm of that ratio. The results of this calculation for each taxon are added together to provide the diversity index. Wilhm and Dorris (1968), after examining diversity in a range of polluted and unpolluted streams, concluded that a value of H' greater than 3 indicated clean water, values in the range 1–3 were characteristic of moderately polluted conditions and values of less than 1 indicated heavily polluted conditions. Different base logs are used in diversity indices so care is needed in comparisons.

The abundance of organisms is obviously important in assessing the effects of pollution, but it can make the interpretation of diversity indices very difficult, especially when water quality over time at a particular site is being examined or when diversity indices from a wider geographical area are being compared from samples taken at different times of year. There may be marked changes in the seasonal abundance of animals. Over two years, Mason (1977b) examined the diversity of monthly samples of macroinvertebrates collected from a hypertrophic lake, devoid of submerged plants, and a clear-water, eutrophic lake with rich macrophyte growth. Diversity was generally lower at the hypertrophic site, but in June of both years the diversity index was lower at the unpolluted site, due to the presence of a very high population of the chironomid larva *Tanytarsus holochlorus*, which developed rapidly and then emerged from the lake.

When sampling is infrequent, as with most surveillance programmes, the appearance of seasonally abundant species could result in the misinterpretation of water quality conditions using diversity indices. Sampling method, the area sampled, the time of year and the level of identification all influenced the diversity index (Hughes, 1978), while the seasonal variations in the index at a site were found to be greater than differences between sites along a river (Murphy, 1978; Pinder and Farr, 1987a). None of the four diversity indices compared by Pinder *et al.* (1987) was found to produce values which were independent of time of year of sampling, size of sample and level of identification. Such independence is a necessary requirement to facilitate comparisons taken at different times, by different operators at different sites. Extreme care is obviously needed in the interpretation of diversity indices. Mason (1977b) concluded that the number of species (*S*) alone gave a more consistent indication of the difference in eutrophic status of two lakes and Winner *et al.* (1975) drew the same conclusions in a study of streams polluted with copper. Green (1979) quoted a number of studies which suggest that *S* is a better and more realistic indicator of diversity than information statistics.

Pinder and Farr (1987b) compared the performance of two diversity indices (the Shannon and Simpson) with biotic scores. Contrary to theory, as water quality deteriorated in their study stream, diversity tended to increase. The ASPT, in contrast, decreased in relation to water quality and was recommended as the best method of analysis over both diversity indices and other biotic indices, for it detected relatively small changes in water quality.

Biotic indices, including ASPT, have been designed chiefly to detect organic pollution and there may be some merit in continuing to use diversity indices where other pollutants, such as metal discharges from mines or acidification, are significant. The most frequently used index, the Shannon Index (equation 10.4) has been much criticized and Magurran (1988) recommends that it should be replaced by the log series index, which has not been used routinely in biological surveillance studies of freshwaters, but has been extensively evaluated by Taylor (1978). The log series index is obtained from the equation:

$$\alpha = \frac{N(1-x)}{x}$$

where N is the total number of individuals in the sample. The parameter x must be obtained by an iterative procedure, using the following term:

$$S/N = (1 - x)/x \, [-\ln(1 - x)]$$

where S is the total number of species.

Magurran (1988) gives a worked example.

Multivariate analysis

While diversity indices and biotic indices reduce information from samples into a single value, multivariate techniques retain information on the taxa within a sample, which may be of great importance in determining the real effects of changes in water quality on the community. Multivariate analyses can be carried out on both presence–absence and quantitative data but it has been argued that, as abundance is easily influenced by extraneous factors, presence–absence data give a less ambiguous measure of association. Multivariate techniques can identify discontinuities present within communities, which can be related to environmental change. They can be used to generate hypotheses about the causality of distribution but the relationship of distribution to environmental features must then be studied using experimental techniques. Good general introductions to multivariate techniques are those of Gauch (1982) and Manly (1986).

Coefficients of similarity can be calculated before a cluster analysis is carried out (Hellawell, 1986). The Jaccard coefficient was used as the basis of a cluster analysis of macroinvertebrates collected from the polluted River Cynon in South Wales. Three major groups of invertebrates were identified. One group was abundant mainly in some tributaries. Another group was abundant in upstream stretches of the river but declined downstream, probably due to the intermittent release of coal washings. The third group increased downstream in response to organic enrichment (Edwards *et al.*, 1975).

To some extent the species groups which are detected by the use of similarity measures and clustering techniques are influenced by the particular method used. Learner *et al.* (1983) compared several methods and found that some species were grouped together by all or most of the methods, so-called robust groups, which the authors considered had ecological validity. They recommend the use of two measures together – product–moment correlation, based on absolute abundance data, and Kendall's *tau*, which uses relative abundance data – for determining robust groups.

Two powerful techniques for analysing community data are detrended correspondence analysis (DECORANA) and two-way indicator species analysis

(TWINSPAN), devised by Hill (1979a, b). DECORANA is an ordination technique which arranges sites into an objective order, those sites with similar taxonomic composition being placed closest together. An axis score is produced which can be used to relate the ordination to environmental factors. TWINSPAN classification arranges site groups into a hierarchy on the basis of their taxonomic composition, while species are classified simultaneously on the basis of their occurrence in site groups. The technique also identifies indicator species that show the greatest difference between site-groups in their frequency of occurrence.

TWINSPAN has been used to classify the running waters of Great Britain on the basis of their macroinvertebrate fauna. Some 438 sites have been sampled using standard kick-sampling methods (p. 229). The DECORANA ordination showed that the highest correlations (Axis 1) were with substratum type and alkalinity, probably representing variations between different river types. Axis 2 correlated with variables related to distance downstream and discharge category, a measure of the long-term average flow of the river. Many rivers showed a progressive change in TWINSPAN group downstream so that it was possible to assign TWINSPAN groupings to particular stretches, provided they had no atypical physical or chemical features (Wright *et al.*, 1989).

It is possible to predict the probability with which a given species or family will be captured at a particular site using environmental data. A software package, River Invertebrate Prediction and Classification System (RIVPACS) has been developed (Wright *et al.*, 1993, 1994). The main use is the provision of a target macroinvertebrate assemblage for a site against which the effects of changes in water quality on the assemblage can be assessed. Eight environmental variables are used in all predictions: distance from source (km), mean substrate type (phi), altitude (m), discharge category (9 groups in $m^3 s^{-1}$), mean water width (m) and depth (cm), latitude (°N) and longitude (°W) Further variables are added, of which there are six options. The preferred option requires the addition of alkalinity (mg $CaCO_3 l^{-1}$), slope (m km^{-1}), mean air temperature (°C) and air temperature range (°C) in the prediction of the invertebrate assemblage. Predictions can be made for three seasons (spring, summer, autumn) separately or combined (winter is excluded because of the difficulties of sampling). The predicted target assemblage of macroinvertebrates can be used to generate expected BMWP or ASPT scores against which to assess the results of field surveys.

This procedure was used in the 1990 River Quality Survey of the United Kingdom. A system of banding was used to place the biological class into four categories, class A being of the highest quality. Bands were derived from the ratio of observed to expected ASPT, number of taxa and BMWP score (Table 10.2). The overall classification of biological quality is the median of the three individual classes, except when the individual class for ASPT is lower, when the overall classification is taken as the band given by the ASPT (the environmental variables in RIVPACS generally explain more of the variation in the ASPT score than in the BMWP and total taxa).

Table 10.2   Biological banding of ASPT, number of taxa and BMWP score based on sampling at three seasons (from Wright *et al.*, 1993)

Biological class	Obs/exp ASPT	Obs/exp no. taxa	Obs/exp BMWP score
A	≥0.89	≥0.79	≥0.75
B	0.77–0.88	0.58–0.78	0.50–0.74
C	0.66–0.76	0.37–0.57	0.25–0.49
D	<0.66	<0.37	<0.25

The use of RIVPACS requires considerable expertise and generally the sorting and identification of macroinvertebrates must be done in the laboratory. In some circumstances a more rapid appraisal may be preferred. In western Britain there is a widespread problem with organic pollution from livestock farms (p. 50), there being many point sources, often episodic in nature, affecting fisheries and the quality of water used for drinking and recreation. In west Wales TWINSPAN has been used to generate indicator species of organic pollution which have been adopted into a simple flow chart (Fig. 10.6). Invertebrates are collected in a 1-minute kick sample and the assessment is made on site. The procedure is simple enough to be used by non-biologists

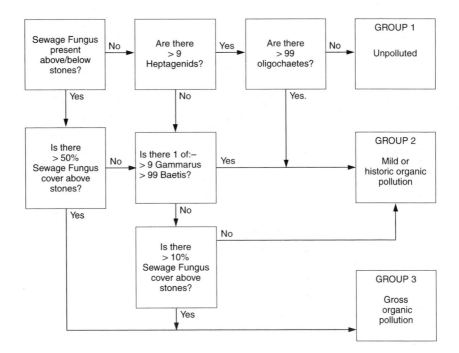

Fig. 10.6. Flow chart, derived from TWINSPAN indicator species, for assessing organic pollution from farms in west Wales (from Rutt *et al.*, 1993).

after a short period of training and it allows the rapid pinpointing of sources of pollution with the minimum of resources (Rutt *et al.*, 1993).

Conclusions

Only a few of the many approaches to processing data in routine surveillance programmes have been presented here and new developments and modifications of old methods are continually being tried. Multivariate techniques are potentially powerful tools, for they can be used to explore the relationships between biological communities and water quality characteristics and to generate hypotheses which can be tested experimentally. Nevertheless, as Edwards (1989) has emphasized, such techniques require taxonomic penetration to species level for a wide spectrum of organisms, which is the reverse of current trends in the water industry towards simple biotic indices and which will require an increased deployment of resources for biological surveillance. Biotic indices are likely, therefore to remain the major tool in everyday management of water quality. In conclusion, such an index for day-to-day use should have the following characteristics (Extence *et al.*, 1987):

1. The system should be based on established methods and it should be possible to calculate results retrospectively for historical data.
2. The method should be as simple as possible to use, both in the field and in the laboratory.
3. Non-specialists should be able to easily appreciate the meaning of any grading or index rating.
4. The index should use as much information as practically possible from the sample, as it is the whole community and not just 'key groups' which respond to variations in water quality.
5. The index should be applicable to all river types, whether they be fast or slow flowing, habitat-rich or habitat-poor.
6. It should be possible to associate index values with water quality classes, and existing or potential river stretch uses, and thus check for compliance with targets.
7. The index should be cost effective.

Chapter 11

INTRUSIVE MICROORGANISMS AND BIOASSAYS

The previous chapter was concerned mainly with methods of surveillance in the field, especially the sampling and enumeration of invertebrates, which appear on balance to be the most practical group to use. This chapter considers those methods in which most of the effort is in the laboratory, that is, the identification and monitoring of intrusive microorganisms (i.e. not natural inhabitants of freshwater) and bioassay tests, and those methods in which test organisms are introduced into the environment.

Intrusive microorganisms

Bacteria

The monitoring of intrusive bacteria to assess faecal pollution of freshwaters has received very wide attention. Various pathogenic bacteria, viruses and invertebrate parasites that are transmitted through faecal contamination present a potential public health hazard (p. 51). Clearly water used for drinking must be free of pathogens, but the microbial contamination of bathing waters sited close to discharges of poorly treated sewage is also responsible for widespread outbreaks of gastrointestinal diseases, as well as infections of the skin, ears and respiratory tract. The pathogens themselves may occur only in very low numbers, or intermittently, so other microbiological indicators of faecal pollution must be used. Indicators of faecal pollution must satisfy certain criteria (Prescott *et al.*, 1993).

1. The indicator bacterium should be suitable for the analysis of all types of water: tap, river, ground, impounded, recreational, estuary, sea and waste.
2. The indicator bacterium should be present whenever enteric pathogens are present.
3. The indicator bacterium should survive longer than the hardiest enteric pathogen.
4. The indicator bacterium should not reproduce in the contaminated water and hence produce an inflated value.
5. The assay procedure for the indicator should have great specificity; in other

words, other bacteria should not give positive results. In addition, the procedure should have high sensitivity and detect low levels of the indicator.

6. The testing method should be easy to perform.
7. The indicator should be harmless to humans.
8. The level of the indicator bacterium in contaminated water should have some direct relationship to the degree of faecal pollution.

The most frequently used indicators of faecal pollution are the coliform bacteria (e.g. *Escherichia coli*) faecal streptococci and *Clostridium perfringens*.

Coliform bacteria are Gram-negative, oxidase-negative, non-sporing rods, which are able to ferment lactose at 35–37°C in 48 hours with the production of acid and gas. In the past the total coliform count has been used as an indicator of faecal pollution, but it fails to satisfy the criteria for indicators of potential pathogens given above, because several types of coliform bacteria are non-faecal in origin. *Citrobacter* and *Klebsiella*, for example, are found in unpolluted soil and *Enterobacter aerogenes* and *Enterobacter cloacae* may be found on vegetation. Furthermore, pathogens have been found in water when coliforms were absent, and conversely, some coliforms can multiply in natural waters.

The presence in water of faecal coliforms, especially *E. coli*, is a better indicator of sewage contamination than total coliforms. Human faeces always contain high numbers of *E. coli* and the organisms can be readily distinguished from other coliforms. Direct relationships have been found between *E. coli* counts and the numbers of pathogenic organisms, such as *Salmonella*. There are also direct relationships between gastrointestinal illnesses (vomiting, diarrhoea, stomach-ache and associated fever) in swimmers and the density of indicator bacteria in bathing waters (Fig. 11.1).

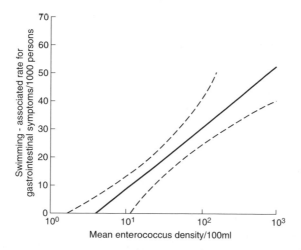

Fig. 11.1. Relationship between swimming-related gastrointestinal symptoms and the density of indicator bacteria in bathing waters. Dotted lines illustrate the 95% confidence limits of the regression line (adapted from Cabelli *et al.*, 1983).

Caution should be exercised in using *E. coli* as a specific indicator of faecal pollution in hot climates for it can multiply in warm waters. Certain non-faecal bacteria can also grow at 44°C and produce indole from tryptone, criteria used to identify *E. coli*.

Faecal streptococci are also used as indicators of faecal pollution and it is possible to distinguish between streptococci from human and animal sources. *Streptococcus faecalis* is the predominant streptococcus in the human gut, but is relatively rare in the gut of other animals. However its use as an indicator organism is limited, for a biotype (var. *liquifaciens*) which is largely non-faecal in origin and which cannot be separated during routine determination, may actually predominate at low concentrations of streptococci (Feacham *et al.*,1983).

The anaerobic spore-forming bacterium *Clostridium perfringens* has been suggested as an alternative or accompaniment to *E.coli,* particularly where re-mote or old pollution is being examined. *Cl. perfringens* is more resistant to toxic pollution than *E. coli*. Sorensen *et al.* (1989) showed that the spores of *Cl. perfringens* could be detected in decreasing concentrations for distances greater than 10 km below the discharge of effluent from a municipal waste-water treatment plant. Coliforms and faecal streptococci, in contrast, varied widely because of the presence of non-point sources, particularly livestock and animal feeding facilities in the catchment. It was therefore suggested that *Cl. perfringens* is a sensitive indicator of municipal waste water discharges even when agricultural non-point sources of faecal indicator bacteria are present.

Several other bacteria, such as *Pseudomonas aeruginosa* and *Salmonella* spp. have also been considered as potential indicators. None of them fully satisfies the criteria listed above.

Bacterial populations can be estimated either as the total count, or as the number of living bacteria – the viable count. Two counting techniques are in general use: multiple tube method and membrane filtration. Both give counts lower than the actual number of bacteria present.

The multiple tube method is now used mainly for clostridia. A series of di-lutions is incubated in a fluid medium and the appearance or non-appearance of growth is observed. The true bacterial count is estimated, using probability tables, as the most probable number (MPN).

Membrane filtration is used for total and faecal coliforms and faecal streptococci. Known volumes of water are passed through cellulose-ester membranes, with a fine pore size of 45μm, which retain the bacteria. The membrane is then placed on a suitable medium and incubated, individual bac-teria multiplying into visible colonies. Total coliforms are grown on Teepol medium, which prevents the growth of non-intestinal bacteria. Incubation is at 30°C for 4 hours, followed by 14 hours at 37°C for total coliforms and 44°C for *E. coli*. *E. coli* colonies are bright yellow. Faecal streptococci are incubated on membrane enterococcus agar medium for 4 hours at 37°C, followed by 44 hours at 45°C. Colonies are maroon or red in colour. The membrane technique produces highly reproducible results in a shorter time than the multiple tube method.

The presence of faecal bacteria in a sample of water gives information on the degree of contamination of the water by humans or animals but, as Ellis (1989) stresses, it is only an indication. The relationship between indicator bacteria and pathogenic organisms is often not direct, so careful interpretation of results is essential.

Viruses

More than 100 types of enteric virus are excreted in human faeces, where as many as 10^6 may be present in a single gram. They represent a considerable potential health hazard (see p. 54) in water used for potable supply and recreation, especially as viruses survive the sewage treatment process well. Many infections caught while swimming in poor quality water may be caused by viruses rather than bacteria. Viruses are present in water in much lower numbers than bacteria and large volumes of water must be examined to detect them, but they also cause infections at lower levels of contamination than bacteria. Virus levels in river water are in general dependent upon virus input through domestic sewage, and a close relationship between a sewage index for a stretch of river and the concentration of viruses in the water has been described (Walter *et al.*, 1989).

Arguments in favour of using human enteric viruses as indicators are that they appear more tolerant of the aquatic environment than bacterial indicators and survive sewage treatment better. Because viruses can initiate infection in humans at low environmental levels, a reliable indication of their presence may not be provided by bacteria. A routine test procedure, however, must be easy to perform and current techniques for enumerating viruses are both technically more complex and more expensive than routine tests for bacteria. It has been suggested that coliphages and bacteriophages (viruses which are obligate intracellular parasites of bacteria) may be reliable indicators of recreational water quality, for numbers in samples are directly correlated with numbers of enteric viruses and faecal coliforms (Palmateer *et al.*, 1991). A discussion of viruses in water is provided by Gerba *et al.* (1991).

Bioaccumulators

The concentration of poisons in the tissues of organisms has been discussed on pp. 40–42. Where pollutant levels in water are near the limit of detection, tissue analysis can bring detection within the scope of most instruments and operators. This is particularly advantageous for surveillance programmes involving a number of laboratories with widely differing facilities and expertise. Whereas the analysis of a water sample will record the level of a pollutant at a particular time, the bioaccumulator will reflect the level of pollution over much longer periods. Furthermore, dried tissues can be stored for long periods before analysis, whereas water samples need almost immediate analysis, particularly as transformations of material due to microbial activity may occur. Organisms are often monitored to assess the risk to the public of exposure to material with which they are likely to come into contact, *critical material* such as food. Fish are of particular importance in freshwaters.

Several criteria need to be fulfilled before an organism can be considered a satisfactory biomonitor. Organisms should be relatively sedentary so that they reflect only local pollutant levels. They need to be readily identifiable and in sufficiently large numbers to ensure genetic stability. They also need to be large enough so that low concentrations of pollutants can be detected within individuals. The life cycles should be long enough to ensure that there is a good balance of age groups throughout the population during the monitoring period. The degree to which organisms concentrate pollutants will vary with both the pollutant and the species so, for large-scale surveys, a single, widespread species is needed. Although the use of bioaccumulators has considerable appeal, the results are often difficult to interpret. The total pollutant content and concentration may vary with the age of an organism, its size, weight and sex as well as the time of year, the sampling position and the relative levels of other pollutants in the tissues (Phillips, 1993; Phillips and Rainbow, 1993).

In addition to measuring pollutant levels in organisms *in situ*, bioaccumulators can also be placed into the environment in cages and uptake rates can be measured over defined periods of time. This approach can be very valuable in comparing pollution levels between sites and at different times of year.

Bioaccumulators have so far mostly been used to monitor the coastal marine environment because of the economically important fisheries and shell-fisheries in these regions. Intertidal seaweeds, such as *Fucus*, and bivalves, especially mussels (*Mytilus* spp.) have received particular study because they satisfy the criteria given above (e.g. Tripp *et al.*, 1992).

Bryophytes

For freshwater, bryophytes (mosses and liverworts) have been particularly studied. They have a number of advantages for monitoring heavy metal pollution (Mouvet, 1985). They are easily sampled and identified and are abundant and widespread. They are generally tolerant of high levels of pollutants and concentrations of metals in moss tissues correlate strongly with concentrations in water and sediment (Goncalves *et al.,* 1992). Their populations are stable over many years, are homogeneous and, once dried, they can be stored indefinitely. The different moss species are distributed along a well-defined gradient of immersion. Methods for processing mosses are considered by Wehr *et al.* (1983) and the use of moss bags, in which mosses from unpolluted sites are transferred to sites of interest, is described by Kelly *et al.* (1987) and Lopez *et al.* (1994).

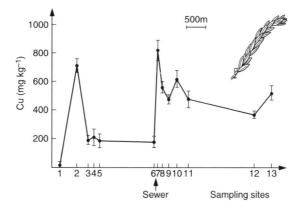

Fig. 11.2. Concentrations of copper (mg kg⁻¹ dry weight ± standard error of the mean) in the moss *Cinclidotus* down the Bienne River, France (after Mouvet, 1985).

Figure 11.2 shows the concentration of copper in the moss *Cinclidotus danubicus* down the Bienne River in northern France. Two sharp peaks revealed points of discharge of copper that were hitherto unknown. Further investigations led to the discovery of copper in effluents between sites 1 and 2, while a sewerage discharge between sites 6 and 7 included wastes contaminated with metals from factories upstream.

In Wales, a number of rivers are polluted with water from disused metal mines, some of them dating back to prehistoric times. The River Mawddach, flowing south from the mountains of Snowdonia, has very few people in the catchment and appears pristine but supports few fish and invertebrates. The catchment, however, receives water from a disused copper mine, an old gold mine which was re-opened in 1981 and which caused extensive pollution downstream in 1984, and an abandoned ammunition dump. One tributary, the Eden, appears to receive no contamination, supports a healthy trout population and can be considered a control. Below the ammunition dump, iron in samples of the moss *Fontinalis squamosa* was 4 times higher than the Eden controls, while aluminium was 9 times higher. Below the gold mine there were large increases in zinc (3 times) and lead (8 times), while below the copper mine, there was an increase of copper of 39-fold compared with the control. The distribution of a number of metals in moss samples throughout the catchment indicated that there were several unknown, old mines which were contaminating the river (Mason and Macdonald, 1988b).

Several other studies have used bryophytes to indicate the presence of heavy metals in rivers (e.g. Whitton *et al.*,1982; Say and Whitton, 1984; Jones *et al.*, 1985), and it has even been suggested that arsenic levels in mosses could be used as an aid in prospecting for gold deposits (Jones, 1985). Metals not detected in water can be detected in bryophytes, suggesting that these plants can be used to monitor episodic pollution. Metals and organochlorines have been shown to accumulate very rapidly following an accident so there is effectively no time lag, while high levels are present up to 13 days after an accident, when the pollutants have effectively disappeared from the water (Mouvet *et al.*, 1993). The analysis of metals (and possibly also accumulating organics) in bryophytes should become routine where discharges of these contaminants are likely, especially as the current methods of assessing water pollution biologically have been developed specifically for organic effluents (p. 240).

Mosses are more or less passive samplers. There has been recent interest in using non-biological accumulators for organic compounds. One such is the 'passive, *in situ* concentration–extraction sampler' (PISCES) which has been used to locate point sources of PCBs in the Black River, New York (Litten *et al.*, 1993). The sampler consists of standard plumbing units, sealed with Teflon and filled with hexane. Samplers were suspended in the river at various points and hydrophobic compounds, such as PCBs, were taken up by diffusion, simulating uptake across the gills of a fish. The point source of PCB pollution was the discharge of a paper mill.

The large filamentous alga *Cladophora glomerata* has been suggested as a useful monitor of heavy metals (Whitton *et al.*, 1989) while higher plants have also been used to monitor heavy metals (Nasu *et al.*, 1984; Reimer and Duthie, 1993) and organochlorines (Lovett Doust *et al.*, 1993, 1994). Higher plants, however, are less ubiquitous than bryophytes, being scarce, for example, in oligotrophic waters or in shade. Having a more complex structure they also exhibit different levels of accumulation between organs as well as marked interspecific differences (Reimer and Duthie, 1993).

Bivalves

Bivalve molluscs have been extensively used for biomonitoring in coastal environments, and to a more limited extent in freshwaters. The small bivalve *Sphaerium corneum* has been utilized to detect the organochlorine pesticide dieldrin, which was used as a moth-proofing agent for textiles in northern England and which entered streams in treated sewage effluent. Most of the dieldrin was obtained by direct partitioning of residues from water into tissue fats, rather than via particulate food, and a steady state was achieved in laboratory studies after a 20-hour exposure at 10°C, a bioaccumulation factor of 1000 being recorded. The time to attain a steady state decreased as water temperature increased (Boryslawskyj *et al.*, 1987).

The freshwater mussel *Quadrula quadrula* has been used to monitor copper wastes from an electroplating works. Animals were placed in cages at several distances below the point of discharge (Fig. 11.3). At 0.1 km below the discharge the mussels accumulated $20.64\,\mu g\,Cu\,g^{-1}$ in 14 days and died. The amount of copper accumulated with time decreased at greater distances from the outfall. At no time did the copper in the effluent exceed the legal limits, so the bioaccumulators were indicating how very low levels of the metal were having adverse effects on the stream fauna. The native mussel fauna was

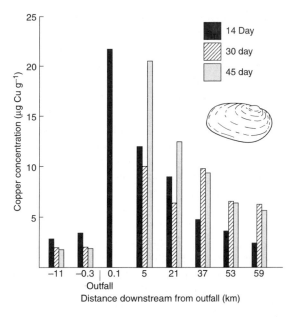

Fig. 11.3. Accumulation of copper in the tissues of caged mussels, *Quadrula quadrula*, exposed in the Muskingum River, United States (from Foster and Bates, 1978).

absent for 21 km below the outfall because of levels of copper. As mussels are very slow to recolonize, the stability of the native fauna had been seriously altered and recovery would be slow.

The freshwater mussel *Dreissena polymorpha* has been used since 1976 to monitor heavy metals in the Rivers Rhine and Meuse in the Netherlands. Mussels were both collected on site and transplanted from clean sites to the rivers. Since 1976 cadmium concentrations have markedly decreased in mussels exposed in the Rhine but increased in those in the Meuse. Copper concentrations have not changed (Kraak *et al.*, 1991).

The crustaceans *Asellus aquaticus* and *Gammarus pulex* have also been used in cages in streams to determine accumulation and heavy metal burdens. Zinc was the most rapidly bioaccumulated metal, concentrations stabilizing at 4–5 times the background level after 5–6 weeks (Shutes *et al.*, 1992).

Fish

Because fish are the main critical material in freshwaters, i.e. they are the major component of the food chain leading to humans, there is clearly much merit in using them as bioindicators. The main problem with fish is that they are highly mobile, so that levels of contamination cannot easily be related to specific sources of pollution. Notable exceptions are eels (*Anguilla* spp.), which are generally highly sedentary during their extended period of development in freshwaters, sometimes remaining up to 20 years before beginning their long journey to oceanic spawning grounds. Eels are carnivores and hence high in the food chain, and they spend much of their time in contact with the sediment, from which they may absorb contaminants. These features make eels potentially useful indicator organisms for both metals and organic contaminants (Beumer and Bacher, 1982; Mason, 1987; Brusle, 1991; Pieters and Hagel, 1992). Mercury in eels is generally highly correlated with length, so that standardization of the size of fish is desirable, though there is less correlation between eel length and cadmium and lead (Barak and Mason, 1990a). Mercury in the flesh represents long-term accumulation and if liver concentrations are higher than those in the flesh, it is likely that the water is severely contaminated (Barak and Mason, 1990b). Cadmium and lead levels in eel livers, but not flesh, also reflect environmental levels of these metals (Barak and Mason, 1990c).

Eels are the only freshwater fish species exploited commercially in significant numbers in eastern England. Most are exported to continental Europe, but in the East End of London and the neighbouring seaside town of Southend, 'jellied' eels are a local delicacy, eaten in considerable quantities by those who have acquired the taste. Eels were collected from more than 60 sites in rivers draining into the Colne/Blackwater estuary, eastern England, and analysed for heavy metals (Mason and Barak, 1990). Routine analysis of water had shown no problems with metals. In Fig. 11.4 are shown those sites where mean mercury concentrations in eel flesh or liver exceeded $0.3\,\mathrm{mg\,kg^{-1}}$, the level above which fish should be considered unfit to eat, according to a Directive of the European Community. Seventeen sites (33 per cent) had mean levels of mercury in eel flesh greater than $0.3\,\mathrm{mg\,kg^{-1}}$. The River Brain was the most contaminated, with concentrations as high as $0.64\,\mathrm{mg\,kg^{-1}}$ in flesh, and high concentrations also in liver, indicating continued chronic pollution. Increased mercury in eels tended to occur below point discharges of effluents from sewage treatment works, some of them serving quite small villages. Eels with high mercury concentrations also tended to have raised concentrations of lead and cadmium. Clearly, the analysis of eels detected a heavy metal problem which has been overlooked by routine chemical surveillance.

The eel is a bottom-living fish in contact with sediments. Pike (*Esox lucius*) have also been used as biomonitors; as top predators they have the potential to accumulate large amounts of contaminants. Mercury has been analysed in pike from 220 Swedish lakes and it was found that concentrations were related to local sources, atmospheric deposition and acidification (Björklund *et al.*, 1984). An analysis, over time, of pike from Swedish waters has shown declines in both DDT and PCBs as the use of these compounds has been restricted (Olsson and Reutergårdh, 1986). The bream (*Abramis brama*) has also been suggested as a good indicator of the contamination status of freshwaters (Scharenberg *et al.*, 1994).

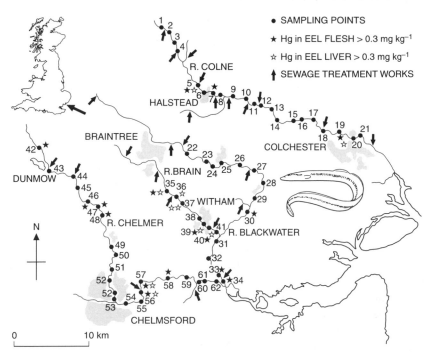

Fig. 11.4. Rivers in Essex, eastern England, showing sampling sites for eels and the location of sewage treatment works. Sites are indicated where the mean mercury concentration in eels was greater than 0.3 mg kg^{-1} in flesh (filled stars) and liver (open stars).

Biomarkers

In chapters 2 and 6 it was shown how environmental contaminants may exert effects at various levels of biological organization, down to the biochemical and sub-cellular. These effects may be quite subtle and can be measured long before any outward toxic effects become apparent. They may be used as *biomarkers* to assess the health of an organism in its environment. A biomarker can be defined as 'an xenobiotically-induced variation in cellular or biochemical components or processes, structures, or functions that is measurable in a biological system or sample' (National Research Council, 1987).

The measurement of biomarker responses can show that organisms have been exposed to pollutants at levels which exceed the normal detoxification and repair capabilities; this cannot be done by measuring only the body burden of a contaminant. Biomarkers may provide evidence that an organism has been exposed to a compound which does not bioaccumulate or is rapidly metabolized (e.g. polynuclear aromatic hydrocarbons, PAHs). They provide a biologically relevant measure of toxicant interactions in target tissues, expressing the cumulative effects of these at the molecular or cellular level. Biomarkers may also provide a more sensitive and precise indicator of ecological effects (such as population change) that are themselves difficult to measure because of the variability in monitoring in the field (McCarthy and Shugart, 1990).

Biomarkers have, however, a number of limitations. They may be specific to a species or phyletic group, there may be differences between species, they may show a slow response time or lack of consistent response, show seasonal variation or have low levels of precision or they may lack ecological relevance. It is difficult to relate biochemical analyses in a specific organ to the LC_{50} (Forbes and Forbes, 1994).

The biomarker must be part of a continuum of events which occur between exposure to the contaminant and the resultant disease or irreversible effect. Biomarkers may be indicators of exposure effects (impairment) or adverse effects (disability) depending on the events they relate to (Fox, 1993). Biomarkers of impairment include:

- induction of the stress response
- induction of detoxification systems
- inhibition of specific enzymes
- metabolic impairments which alter processes of synthesis or breakdown, or which deplete energy, vitamin or substrate stores
- impaired growth
- genetic damage or impaired repair
- impairment of the immune system
- impaired or altered reproductive function
- impaired tissue or organ function.

In using biomarkers it should be possible to detect effects of contaminants at an early stage, i.e., we can ask 'what levels of contaminants produce effects that *are likely to lead* to biological damage?' rather than 'what levels of contaminants *do lead* to biological damage'? (Depledge, 1989).

Porphyrin can be taken as an example of the use of biomarkers. Porphyrins are involved in the synthesis of haem and a number of xenobiotics result in excessive production of porphyrins. Concentrations of porphyrins in the livers of herring gulls (*Larus argentatus*) from the Great Lakes have been determined and compared to samples of porphyrins in gull livers from the Atlantic coast of North America (Fox *et al.*, 1988). The highest porphyrin levels were found in samples from Hamilton Bay, Lake Ontario (38 times levels in Atlantic samples) and Green Bay, Lake Michigan (28 times) (see p. 184). It was not possible, however, to determine which compound was responsible for the increased porphyrin levels.

Detailed discussions of biomarkers can be found in McCarthy and Shugart (1990), Peakall (1992) and Fox (1993).

Bioassay

Because pollutants often occur in complex mixtures and interact, the chemical determination of pollutant levels alone will frequently give little indication of their potential biological effects. In a bioassay, environmental conditions are carefully controlled so that the response of a test organism to particular pollutants can be defined, but the extrapolation from bioassay to field situations may be dubious. Bioassay techniques must go hand in hand with field observations and field experiments in order to fully understand a pollution problem.

The selection of a suitable organism for routine bioassays will depend on a number of factors:

1. the organism must be sensitive to the material or environmental factors under consideration;
2. it must be widely distributed and readily available in good numbers throughout the year;
3. it should have economic, recreational or ecological importance both locally and nationally;
4. it should be easily cultured in the laboratory;
5. it should be in good condition, free from parasites or disease;
6. it should be suitable for bioassay testing.

Small organisms with short generation times are generally preferred for bioassays, though fish are also popular because of their physiology and their recreational and economic importance. Bioassay tests can be divided into biostimulation and toxicity tests.

Biostimulation

Biostimulation tests are mostly carried out using algae and are suitable for evaluating the nutrient status of a waterbody, for distinguishing between total and biologically available nutrients and for determining the potential effects of changing water quality on algal growth. Algal bioassays can be done in the field but most involve laboratory studies under defined conditions.

The Standard Bottle Test (American Public Health Association, 1989) has been widely used and is a measure of either maximum specific growth rate or maximum standing crop. The specific growth rate (μ), defined as the number of logarithmic units of increase in cell number per day during the exponential phase of growth, is:

$$\mu = \frac{\log_e N_2 - \log_e N_1}{t_2 - t_1}$$

where N_1 is the initial cell density at time t_1, and N_2 is the final cell density at time t_2. The maximum standing crop is the maximum algal biomass achieved during incubation. Absorbance, chlorophyll measurements or total cell carbon can be substituted for cell counts or biomass determinations. Standard test algae are normally used, namely *Selenastrum capricornutum*, *Asterionella formosa* and the cyanobacteria *Microcystis aeruginosa* or *Anabaena flos-aquae*. *Selenastrum* is easiest to culture and use. The algae initially present in the test water must be removed, by membrane filtration or autoclaving, which may alter the quality of the water. Nutrient depletion and the build-up of metabolic wastes also occur during the test period in the closed bottle.

Fig. 11.5 shows the effect of adding primary and secondary treated sewage effluents to water from the oligotrophic Lake Tahoe and the eutrophic Clear Lake in California. Both effluents resulted in a large increase in the populations of *Selenastrum* in Tahoe water, but there was no effect in Clear Lake water. The effect of tertiary treatment was examined in bioassays by using water from Lawrence Lake, Michigan (Fig. 11.6), where spiking with tertiary treated sewage (from which nutrients had been removed) produced no significant growth, compared with the large growth when the lake water was spiked with secondary treated sewage.

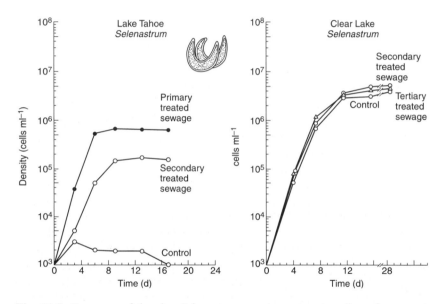

Fig. 11.5. Response of the alga *Selenastrum capricornutum* to spikes of sewage in oligotrophic Lake Tahoe and eutrophic Clear Lake waters (from Payne, 1975).

An alternative to bottle tests is the continuous flow or chemostat method, in which nutrient medium is added continuously so that its constituents are maintained at the required concentrations. A steady state is established, at which the test organism maintains a constant specific growth rate and cell concentration in constant substrate concentration.

Toxicity tests

The terminology used in toxicity testing and the responses of organisms to acute and chronic levels of toxic pollution have been described in Chapter 2.

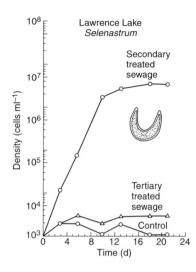

Fig. 11.6. Effect of secondary and tertiary sewage spikes on the growth of *Selenastrum capricornutum* in water from Lawrence Lake, Michigan (from Payne, 1975).

The chief uses of toxicity tests are for preliminary screening of chemicals, for monitoring effluents to determine the risk to aquatic organisms and, for those effluents which are toxic, to determine which component is causing death so that it can receive especial treatment.

Studies on toxicity can be conducted in the field, using caged organisms. Examples are given for fish on p. 44 and invertebrates on p. 159. The use of caged animals in the field, however, makes heavy demands on labour, especially with fish which require feeding. Vandalism is a recurring problem.

The simplest type of laboratory toxicity test is the static test, in which the organism is placed for 48–96 hours in a standard tank of the water under examination. There is normally a series of tanks with test water of different dilutions, usually in a logarithmic series. The organisms are removed at the end of the test period and mortality is recorded. In these tests, the poison may evaporate, degrade or adsorb onto the surface of the tank, so that toxicity may be underestimated. These tests are nevertheless useful when an effluent needs rapid evaluation. A more sophisticated method involves the periodic replacement of test water. Continuous flow systems are also used, but these require large volumes of clean water for dilution.

Fish are normally used in standard toxicity tests. In the United States, the fathead minnow and the bluegill sunfish, together with goldfish (*Carassius auratus*) and the guppy (*Poecilia reticulata*) are the main test species, though there is a tendency now to use a range of fishes. Early work in Britain concentrated on brown and rainbow trout but the tropical harlequin fish (*Rasbora heteromorpha*) is also now used extensively because it is small and has a similar sensitivity to trout.

Toxicity tests need to be replicated and this requires the maintenance of large numbers of fish, needing considerable volumes of clean water. The response time of fish to low concentrations of pollutant may also be slow. Furthermore, there are growing ethical objections to using vertebrates in routine toxicological assessments and, in Britain, such experiments on fish must be licensed under the Animals (Scientific Procedures) Act 1986 and are regularly inspected. There may therefore be considerable benefits in using other organisms for routine work. Eventually *in vitro* toxicity tests may be developed which could obviate the need for vertebrates in routine toxicity tests. Freshly isolated rainbow trout hepatocytes (liver cells), for example, have been used, though they were found to be, in general, less sensitive to a range of chemicals than *Daphnia magna* (Lilius *et al.*, 1994). Solbé (1993) provides a review.

The role of algae in biostimulation studies has been described above and they can also be used in toxicity tests, methods being critically evaluated by Nyholm and Källqvist (1989). The test alga *Selenastrum capricornutum* was used to investigate the possible influence of trade effluent from a nickel alloy plant on the ecology of the River Lugg, a tributary of the River Wye in Wales. Primary production in the river was low and algal bioassays indicated that River Lugg water was much less fertile than water from the River Wye. However, enrichment of River Lugg water with nitrogen and phosphorus did not significantly stimulate algal growth, suggesting that a deficiency of these nutrients was not the cause of the infertility and providing strong circumstantial evidence that toxic waste from the nickel alloy plant was contaminating the river (Bowker and Muir, 1981).

Vascular plants, in particular duckweeds (*Lemna*), have also been used in toxicity tests. Duckweeds are small, they have a simple structure of leaf and root, they reproduce rapidly and are easy to culture. They may be more effective than algae as test organisms (Wang, 1990). Of more complex plants, clones of *Vallisneria americana* are being used to identify point sources of pollution in the Detroit River (Lovett Doust *et al.*, 1994). A review of the use of primary producers in toxicity tests is provided by Lewis (1993).

Much recent work has considered *Daphnia*, especially the large *D. magna*, as a test organism. The genus is ubiquitous, has a pivotal role in many lake food webs and its ecology has been well studied. *Daphnia* is sensitive to pollutants, easily cultured and has a high reproductive rate. The advantages and drawbacks of using *Daphnia* in toxicological tests are considered by Baudo (1987). The amphipod *Gammarus pulex* has been proposed as a test organism because it is easy to culture and is sensitive to pollution (McCahon and Pascoe, 1988), and this species has also been used in 'scope for growth' studies, both in the laboratory and in the field. A wide range of other invertebrates have been used in toxicity tests but only those involving *Daphnia* are formally endorsed by international organizations (Persoone and Janssen, 1993).

For management purposes, the results of acute toxicity tests need to be related to field conditions in terms of safe concentrations of effluents and this has led to the introduction of application (or uncertainty) factors, most of which are purely arbitrary. For every poison it is assumed that there must be a concentration which is so low as to have a negligible toxic effect and this concentration has an approximate relation to the lethal threshold concentration. Multiplying the lethal threshold by an application factor should give an approximate indication of the acceptable concentration of the poison (Lloyd, 1992). Application factors vary from 0.01 for persistent or accumulative compounds to 0.3 for compounds of low toxicity that rapidly break down.

Despite their widespread use, toxicity tests using a single species are unlikely to yield reliable information on the potential effects of pollutants on the aquatic ecosystem, because there is no universally sensitive indicator species and the relative sensitivities of species will vary according to the pollutant being tested for (Cairns, 1984; Edwards, 1989). The loss of a sensitive species from the aquatic ecosystem could markedly alter the processes occurring within it. The use of artificial channels or streams allows the effects of pollutants on entire communities to be studied (Muirhead-Thomson,1987; Hill *et al.*, 1994).

Automated biomonitors

Organisms may show distinct physiological and behavioural responses to low levels of pollutants and attempts have been made recently to harness these in devising automatic alarm systems. Automatic monitors should provide a rapid indication that water quality has deteriorated and they have potential use in monitoring river waters and raw waters which are abstracted for potable supply, and for monitoring effluents from sewage treatment works and industrial plants. Such monitors must be on line to the operations control centre so that immediate action can be taken if an alarm is sounded.

A number of requirements must be met for an automatic biomonitor to be fully successful (Diamond *et al.*, 1988).

1. The operation of the system should be continuous, being always alerted when it is likely to be needed.
2. The alarm must be given soon enough so that any action required can be taken before toxic effects occur.
3. The system should detect a wide range of toxic substances.
4. The detector must respond unfailingly to the selected signal at the relevant level, always giving an alarm when the selected level is exceeded.
5. The number of false alarms owing to non-toxic variations in water quality should be minimal.
6. The monitoring systems should be easy to operate and the alarm signal should be clear, characteristic and easy to interpret.
7. The organism used as the sensor in the system should be fairly inexpensive and easy to acquire and culture.
8. The apparatus should require as little maintenance as possible.

Much of the early work on biomonitors made use of fish and these have developed the furthest, commercial models being available. Fish alarm systems have monitored movement and respiratory activity in relation to levels of pollution. A typical tank, in which increasing opercular rhythms and movement of fish in relation to sublethal levels of pollutants are monitored, is illustrated in Fig. 11.7. To remove environmental effects, the experimental tanks are usually enclosed in a chamber, which reduces sound and vibrations and allows the photoperiod to be controlled, while water temperature is maintained constant. Food is given from an automatic feeder. Water is provided on a continuous flow basis. One fish is held per tank. The opercular movements of the fish result in a change of potential between electrodes. The movement of a fish can be recorded as it swims across a light beam being detected by two photocells. Infrared emitters can be used so that the natural rhythms of fish are not altered, while ultrasonic emitters are ideal for turbid waters (Morgan and Kuhn, 1988).

Fig. 11.8 shows some typical results, the ventilatory profiles of bluegill sunfish exposed to sublethal concentrations of zinc and trichlorethylene. Note the changes in amplitude of the signals. In the case of zinc the amplitude is very much decreased, whereas with trichlorethylene it is increased, and the signal is irregular. With further work it may be possible to identify the nature of the pollutant as well as the existence of chronically toxic conditions (Diamond *et al.*, 1988).

To be effective a biomonitor should respond quickly to the presence of a

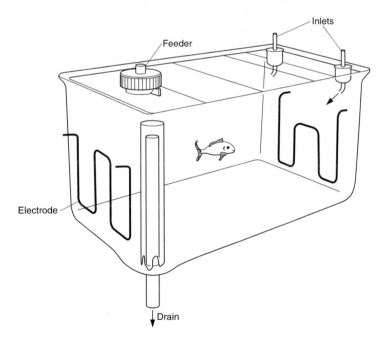

Fig. 11.7. An automatic fish monitor tank (from Westlake and Van der Schalie, 1977).

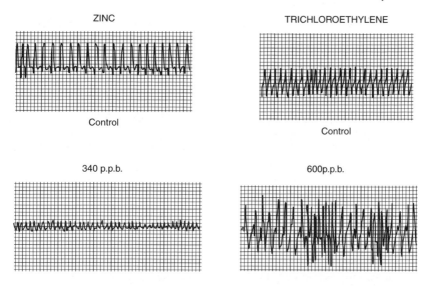

Fig. 11.8. Ventilatory profiles for bluegill sunfish exposed to sublethal concentrations of zinc and trichlorethylene (after Diamond *et al.*, 1988).

pollutant. Table 11.1 shows the detection limits and detection times for several pollutants tested in the Water Research Centre Fish Monitor, which measures the ventilation frequency of rainbow trout. In this case one hour was considered a reasonable time for the detection of a pollutant. If a longer response time is acceptable, then limits of detection may be lower than in Table 11.1. The sensitivity of the biomonitor was much greater than standard toxicity tests. For example the 96-hr LC_{50} for phenol is 9.3 mg l^{-1} and for lindane 0.06 mg l^{-1}, showing that the monitor responds more quickly to lower concentrations. The Water Research Centre biomonitor is used in particular to continuously monitor the quality of water in lowland river intakes, which, after treatment, is used for drinking water supply (Evans and Wallwork, 1988). It is considered to provide a degree of protection which no other currently available instrument can match, although it responds rather poorly to com-

Table 11.1 Limit of detection and detection time of a biomonitor, measuring ventilation rate of rainbow trout, to various pollutants (after Evans *et al.*, 1986)

Pollutant	Detection limit (mg l^{-1})	Detection time (min)
Ammonia	0.11	17
Phenol	2.00	28
Paraquat	0.80	26
Lindane	0.0003	21
Pentachlorophenol	0.14	33
Formaldehyde	100	38

pounds producing toxic symptoms which are likely to be chronic and long-term in nature (Baldwin *et al.*, 1994).

The signals produced by a biomonitor are usually passed on to a computer which, from pre-determined criteria, will assess whether the signals indicate toxic conditions. If they do, an alarm system will be activated or corrective action, such as the closure of a water intake, will be taken. Remote biomonitors may be linked to a control station by telemetry or earth-satellite.

There have been a number of recent developments with automated biomonitors. Several attributes of bacterial enzyme systems have been exploited. The most widely used bacterial test, however, is the commercially available Microtox test, which utilizes the bioluminescence of the marine *Photobacterium phosphoreum*. The reduction in light output is a measure of toxicity and the test compares well with toxicity tests using fish and other animals. The test is sensitive, precise and reproducible. It has been used, for example, to examine the toxicity of organic compounds in the River Meuse and showed that, over much of the length of the river, the toxicity of the water was below environmentally acceptable standards (Polman and de Zwart, 1994).

The locomotion and activity of *Daphnia* have been investigated, while the valve movements of freshwater mussels (*Dreissena polymorpha*) are also being exploited commercially, mussels rapidly closing their shells in response to environmental stress (Kramer *et al.*, 1989). Multiple species monitors, with invertebrates as well as fish, are also being developed, on the assumption that two or more species are more likely to respond to a wider range of pollutants than a single species (Morgan *et al.*, 1988; Cairns and Cherry, 1993).

Reviews of automated biomonitors are provided by Kramer and Botterweg (1991) and Gruber *et al.* (1994).

Chapter 12

MANAGEMENT OF WATER RESOURCES

The prime task of the water industry, considered here in the broad sense, is to manage the hydrological cycle for the benefit of users. The main sources of water are from reservoirs, and from natural lakes, river flows and groundwaters. Other sources, such as desalination of sea water or rainmaking by cloud-seeding, may be important in some parts of the world. The water industry develops these resources and manages them to provide water, wherever possible, in the quantities required by domestic, industrial and agricultural users. It must be of an acceptable quality for these purposes, requiring varying degrees of treatment.

Water is used extensively for recreation and amenity and the water industry may be responsible for the development and regulation of these demands. Included here are fisheries, many of which are of economic value. In underdeveloped countries fish may be the chief source of protein to the local community. Conservation of wildlife resources and landscape also fall within the general area of amenity.

Society produces a vast array of waste products and water provides an effective means of disposing of many of these. The water industry is responsible for ensuring that the disposal of wastes causes the minimum of damage to resources. Water can also be the source of electric power, using dams and turbines. The water industry is also responsible for reducing the damage caused by flooding, which can be highly destructive of life, industry and agriculture. Navigation must be maintained on many watercourses.

Water sources are often linked by management. The flow of rivers, for example, may be regulated by the controlled addition of water from reservoir or groundwater sources and water may be withdrawn from rivers at peak flows for storage in reservoirs or for refilling aquifers. Many of the outputs of management may be complementary but others may be in conflict. Land drainage and flood prevention schemes may, for instance, involve the destruction of prime wildlife habitats, a topic to be discussed later.

The water industry has little control over some key points in the water cycle, for instance precipitation and heavy floods. This is not to say that human influence is not profound on these parts of the hydrological cycle. The destruction of forests, particularly those of the tropics, is a case in point. Forests generate much of the rainfall, up to 50 per cent in the case of Amazonia, and much of this may be lost with forest clearance, a process occurring at a rate of some 164 000 to 204 000 km^2 per annum in the tropics. Forests also act as sponges, absorbing water and regulating its release to rivers. Felling results in rapid run-off, followed by drought, so that agricultural areas lower in the catchment alternately suffer devastating floods, followed by periods of water shortage, resulting in tremendous loss of life and agricultural production. Floods erode away valuable soil, depositing it in rivers and estuaries to damage fisheries and interrupt navigation (Fig. 12.1). In Nepal, for example, some 37–75 t of soil per hectare are stripped by monsoon rains from deforested mountainsides each year. Around 3 billion t of soil are carried annually by the rivers Ganges and Brahmaputra and deposited in the Bay of Bengal (Lean *et al.*, 1990).

The large-scale destruction of rainforests in the developing world, fre-

Fig. 12.1. Deforestation and overgrazing have denuded these hills in the Middle East. Rain has washed away the top-soil, leaving bare rock. The dark areas are isolated shrubs, obtaining subterranean water from the wadi using deep roots. The wadi runs only briefly during storms, resulting in frequent, severe landslips (photograph by the author).

quently by multinational organizations from the developed world seeking short-term profits, may also have major repercussions for world climate. More solar heat may be reflected from land cleared of forests, altering the global patterns of air circulation and wind currents and possibly decreasing rainfall in equatorial and temperate lands. Forests also act as sinks for carbon dioxide but, with large-scale burning, they become sources. An increase in atmospheric carbon dioxide will, through the greenhouse effect, increase the temperature of the earth; a rise of only 1°C could severely reduce the output of the productive grain belts of the world. Obviously such global effects, were they to occur, would be catastrophic for the carefully managed water resources of the developed world, thus emphasizing that resource conservation must, in the long term, be viewed on a global scale. Some 50 per cent of the world's population is directly affected by the way watersheds are managed.

Figure 12.2 (shaded area) illustrates the main points of intervention in the water cycle. Intervention involves regulating the passage of water through different storages, diverting between storages and altering the retention time so that water is available when required. Availability is thus, to some extent, emancipated from the pattern of precipitation. Water is abstracted into supply from groundwater, rivers and reservoirs and these sources are used in conjunction with one another, water frequently being moved between sources and between regions to where it is required.

The most efficient way to manage aquatic resources is at the level of the catchment or river basin, that is management of the river all the way from its various sources until it discharges into the sea. In this way resources are better used, for they are considered together in relation to regional needs. Water supplies are rationalized over a larger area, allowing greater integration, while effluent disposal can be considered in relation to water supply. This reduces the likelihood of conflicts between those providing water and those disposing of effluents and it allows the setting of environmental standards which reflect the characteristics of the particular basin.

Organization of the water industry

Before 1989, the management of the water cycle, apart from small private water companies, was in public ownership in England and Wales. In Scotland it still is but the tasks of supply and monitoring have always been separate. Water supply and sewage treatment are two functions of the Scottish regional councils, while river purification boards have been responsible for monitoring fresh and coastal waters.

Since the end of 1989, the provision of water and the treatment of sewage in England and Wales have become the responsibility of 10 privatized water companies, while a number of smaller companies provide only water. The companies are effectively regional monopolies for the customer has no choice of supply.

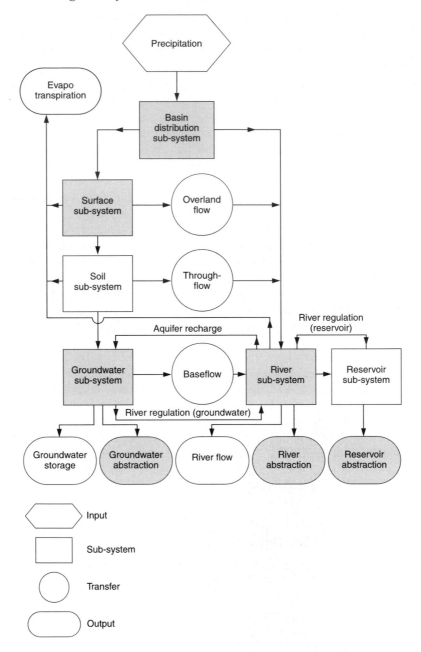

Fig. 12.2. Points of intervention by the water industry in the river basin system (from Porter, 1978).

The task of monitoring river quality and setting standards is undertaken by the National Rivers Authority (NRA) under the direction of central government through the Department of the Environment. The main functions of the NRA are:

1. Water resources: the planning of resources to meet the water needs of the country; licencing companies, organizations and individuals to abstract water; and monitoring the licences.
2. Environmental quality and pollution control: maintaining and improving water quality in rivers, estuaries and coastal seas; granting consents for discharges to the water environment; monitoring water quality; pollution control.
3. Flood defence: general supervision of flood defences; carrying out works on main rivers; sea defences.
4. Fisheries: maintenance, improvement and development of fisheries in inland waters including licencing, restocking and enforcement functions.
5. Conservation: furthering the conservation of the water environment and protecting its amenity.
6. Navigation and recreation: navigation responsibilities in three regions – Anglian, Southern and Thames – and provision and maintenance of recreational facilities on rivers and waters under its control (Fig. 12.3).

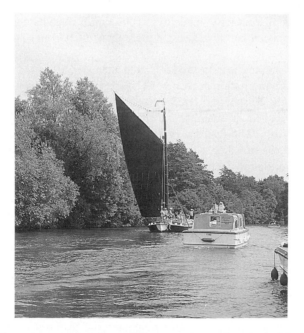

Fig. 12.3. Many rivers are heavily used for recreation, as here in the Norfolk Broads. Traditional wherries (left) cause little environmental damage, but motor cruisers (right) are noisy, leave oil films on the water and cause severe bankside erosion, adding to eutrophication problems (photograph by the author).

The history of water management in Great Britain is described by Kinnersley (1994).

At the time of writing (January 1995) water management in England and Wales is again set to change. A new Environmental Agency is planned to integrate the control of pollution, combining the functions of the NRA, Her Majesty's Inspectorate of Pollution and various agencies involved in the regulation of wastes. Current concerns are that the new agency will have much weaker powers than the NRA. The agency will have to regard costs and benefits in exercising its powers and this may thwart the development of new technology. It appears less committed to conserving and enhancing the environment. The NRA is involved in wider issues than pollution (see above) and it is unclear how these functions will develop within the new agency. It remains a possibility that much of the routine work of the NRA could eventually be contracted out to private companies. It will probably not be until the year 2000 that the benefits, or otherwise, of the new agency will become apparent (see page 295).

Water resource management in the United States is much more complicated and basin management is rarely possible because the majority of large rivers cross state boundaries. One exception is the Tennessee Valley Authority, established in 1933 to raise an underdeveloped region to a level at which it could compete with and contribute to other, more prosperous regions of the country. The authority has built 30 dams, producing 15 billion kilowatt hours per year of electricity, made over 1000 km of river navigable, reduced flooding and erosion and re-afforested more than 200 000 ha of the catchment (Black, 1987).

The use of surface waters in the United States is allocated according to Riparian and Appropriation doctrines. The Riparian doctrine, which applies to states east of the Mississippi River, gives the right to the reasonable use of water, undiminished in quantity and quality, to the owners of land which borders the stream. The doctrine protects owners against unreasonable withdrawals or the use of water which may reduce their rights to quality or quantity. The Appropriation doctrine applies in those states west of the Mississippi, where water is generally scarcer, and is based on a first-come, first-served principle, though the water taken must be put to beneficial use. In theory, therefore, upstream users could take so much water that rivers run dry downstream, an all too familiar occurrence in many drier parts of the world where dams are built for hydro-electric power or water supply. There are a very large number of organizations in the United States involved in various aspects of water resources at the national, regional and local levels, leading to many conflicts and inefficiencies in overall resource management. Nevertheless, over 90 per cent of water supply units are under direct municipal management and there are hardly any private sewage treatment and disposal companies. The Environmental Protection Agency (EPA) was established in 1970. The EPA is responsible for the administration of the Clean Water Act (1977) and it has the responsibility to do research, set standards, monitor emissions and enforce the

law on emissions for water, air and solid waste. There are increasingly re-
strictive standards and stricter enforcement, leading to a greater control over
water quality. This is seen by Black (1987) as the only way, in an
entrepreneurial society which dislikes legislation and long-term planning, of
maintaining a healthy aquatic environment.

In developing countries, the most serious threat from water pollution is that
to health (p. 1). Typically, uncontrolled industrial discharges are the first evidence of a deterioration in water quality (Ellis, 1989). Developing countries
are keen to attract foreign investment, while manufacturers from the developed world may be lured by the low environmental standards, in addition to
cheap labour. A vicious circle therefore develops. Even where quality standards are set, they may not be enforced because the government pollution officers are frequently poorly trained and poorly paid and hence have a very low
morale. Similar problems may beset the efficient running of sewage treatment
facilities, where these have been built into the municipal sewerage system.
Unfortunately they are often lacking and many large cities pump untreated
sewage into rivers. The provision of wholesome water for drinking and washing is an essential first step in the improvement of the quality of life for many
communities in the developing world.

Biology and the water industry

The construction of reservoirs, the drainage of land, the building of sewage
works and so on, are largely the province of engineers, and engineers dominated the water industry as it developed, initially to the exclusion of other
disciplines. Many large-scale works have been executed in the past without
advice on ecological consequences, which have frequently been severe. Large
reservoirs today are usually designed with multiple uses in mind, that is, as
well as the generation of electricity or the storage of water, reservoirs can also
be used for irrigation, fisheries, controlling floods and recreation. Multiple use
can help mitigate some of the adverse factors consequent on the building of a
reservoir.

The domination of the water industry by engineers meant that few biologists
were employed initially. The realization of the importance of biology to the
management of water resources has resulted in a marked increase in their em-
ployment over the past two decades. There are also, of course, many biologists
outside the mainstream water industry whose brief, wholly or in part, is concerned with aquatic resources, such as wildlife or landscape conservation, or
fisheries.

There are a number of objectives in the management of water resources in
which the biologist has an important role.

1. **The classification of water resources**. Water resources are increasingly
   classified according to the uses to which they are put, and the quality
   necessary for a particular use must be defined. For example, certain indus-

trial processes, raw water for potable supply and game fisheries require water of the highest quality, whereas low quality water is acceptable in a waterway used primarily as a receiving body for waste effluents. An assessment of the biological resources is necessary in an initial classification.

2. **The collection of baseline data**. These data will allow any changes in water quality caused by the development of a resource to be detected, together with changes which may interfere with the present and planned usage of water.

3. **Water quality surveillance**. This is routine work carried out to determine the effectiveness of wastewater management programmes, or carried out in relation to specific uses, such as water abstraction, fisheries or recreation.

4. **Specific investigations**. These may involve determining the effects of a specific pollution incident and the subsequent recovery of the freshwater community, the effect of a new impoundment on water resources downstream, the development of communities within impoundments, and so on.

5. **Forecasting**. The forecasting of changes consequent on variation in the intensity of use of a resource or on altered pollution inputs is essential for the rational exploitation of water resources. The biologist can provide essential input for the development of predictive models for forecasting both in the short and long term.

For the tasks outlined above the biologist will ideally be working as a member of a multifunctional team, but there are two further areas in which the biologist has a dominant role:

6. **Fisheries**. The development and maintenance of commercial and recreational fisheries.

7. **Wildlife conservation**. Many aspects of water use, including pollution and various forms of development, are inimical to wildlife. The biologist's role is to assess wildlife resources, predict the impact of development and ensure that vulnerable species are conserved within wetland habitats.

Described below are some projects and problems which involve biologists.

Water quality standards

In Great Britain, releases of wastes from industry to land, air and water are controlled by Her Majesty's Inspectorate of Pollution (HMIP), which implements the provisions of the Environmental Protection Act 1990, as well as several other earlier Acts of Parliament. The mechanism of action is by way of integrated pollution control (IPC). Control of waste disposal is often considered in terms of the transfer of waste to the disposal route posing least environmental hazard.

The routes available for disposal are shown in Fig. 12.4. This concept of waste disposal is known as the *best practicable environmental option*. The approach aims to limit the potential waste disposal routes so that environmental

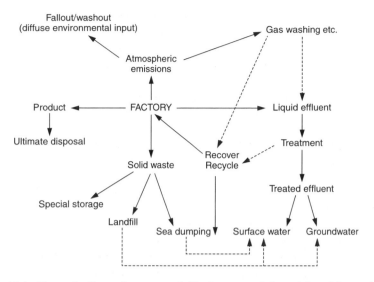

Fig. 12.4. The main disposal routes available for wastes (adapted from Mance, 1987).

protection is ensured in the long term. The integrated pollution control procedure is shown in Fig. 12.5. Before the release of a potentially harmful substance is authorized, it must be shown that attempts to minimize releases have been made under the BATNEEC principle (best available technology not entailing excessive cost).

Management decisions are based on the comparison of water quality data with criteria and standards. Such standards are reasonably well developed in the physical and chemical fields, but biological standards are at an earlier stage of development. The following definitions are generally accepted:

Criteria: scientific requirements on which a decision or judgement may be based concerning the suitability of water quality to support a designated use.
Objectives: a set of levels of water quality parameters to be attained in water quality management programmes, which also involve cost/benefit considerations.
Standards: legally prescribed limits of pollution which are established under statutory authority.

There are two different approaches to water pollution control. The first is to determine water quality objectives (WQOs), which are based on the uses to which surface water is to be put. Standards are then placed on individual discharges which will leave a river of the necessary quality defined by the WQO. This approach means that all discharges need not be of the same standard, and where the assimilative capacity of the river is high, relatively little expenditure may be required to treat some discharges. Opponents of this approach consider that individual quality standards are difficult to enforce and that they always tend towards the minimum values required to meet the classification, imposing

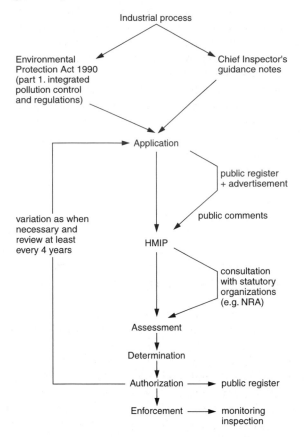

Fig. 12.5. Integrated pollution control procedure (adapted from Slater, 1992).

the minimum cost on the polluter. The alternative approach is the fixed uni-
form emission standards (UESs), which limit all polluting discharges irrespective of existing water quality, dilutions available or the future use of the
river. The philosophy here is that any discharge of pollutants is ultimately unacceptable and progress should be made towards their elimination by recovery
and recycling. The British approach has been towards the WQO while the
United States and major European countries have preferred the UES approach.

Under the Water Resources Act 1991, statutory water quality objectives
(SWQOs) are being developed for individual stretches of river in England and
Wales. WQOs are classifications of water defined by quantitative standards of
environmental quality related to the use of a watercourse, or its potential use
if pollution levels are reduced. The five uses are River Ecosystem, Special
Ecosystem (to protect aquatic ecosystems of high conservation value),
Abstraction for Potable Supply, Agricultural/Industrial Abstractions and
Watersports.

The River Ecosystem classification is the most developed and is based on the concentrations of seven common determinands which have effects on fish populations – dissolved oxygen, BOD, total ammonia, un-ionized ammonia, pH, copper and zinc. The WQO quality targets will consist of two parts, a target class and a target date by which compliance will be achieved. Further details are provided by Everard (1994).

A standard must be defined that allows for a clear assessment of performance and trend. Water quality management requires an integrated approach, which includes sampling of rivers and effluents, setting standards for effluent discharges, determining river quality targets, assessing compliance with standards and setting priorities for action (Miller and Tetlow, 1989). Quality standards must be defined as statistics whose values can be estimated sensibly from the results of sampling. The standards set for individual sewage and trade effluents are known as *consent conditions*, their aim being to achieve a particular quality standard for the surface water in question, in pursuit of the WQO. In Britain the 95 percentile approach to effluent quality is adopted, in which recorded values must be less than the stipulated consent for at least 95 per cent of the time and must exceed the consent no more than 5 per cent of the time, an agreed number of samples being taken over a year to assess performance. Warn (1989) describes modelling approaches to the fixing of water quality standards and the uncertainties surrounding them.

Figure 12.6 shows a scheme for deriving water quality standards to protect aquatic life. The initial stage is a review of published information on the effects of a pollutant on a specific water use to be protected. Laboratory and field studies are kept separate and both are evaluated for technical competence. The screened information on laboratory effects is used to assess the minimum adverse effect concentrations. This is then reduced by an arbitrary safety factor, which takes account of the severity of the effect and the length of the exposure

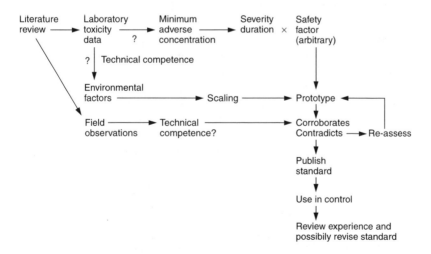

Fig. 12.6. A scheme for deriving water quality standards to protect aquatic life (from Mance, 1987).

period. This prototype standard is then further scaled to take account of toxicity, which may be influenced by environmental factors such as pH. The prototype standard is compared against available field observations, and if the prototype standard is corroborated it is accepted for use in controlling discharges from all sources. If the prototype conflicts with field observations, the data are reappraised to determine the reason for the conflict. If the conflict cannot be reconciled, the field observations should take precedence and a more stringent standard be adopted.

The detailed evaluations undertaken of a number of chemicals to derive water quality criteria to protect fisheries are presented by Howells (1994). This risk-assessment approach, developing criteria from toxicity tests, has been reasonably successful in producing standards to protect resources from toxic chemicals. It has been generally unsuccessful, however, in protecting river biota from episodic pollution (Seager and Maltby, 1989) or in protecting wildlife from bioaccumulating and biomagnifying contaminants (see Chapter 6). This latter problem has been approached in the Great Lakes region by determining the lowest adverse effect levels (LOAELs), either from field observations or from controlled laboratory studies in which domestic species were fed known quantities of Great Lakes fish. The standard, the no observable adverse effect level (NOAEL) was defined as 10 per cent of the LOAEL, that is allowing a safety factor of 10. It was considered that this would protect the most sensitive species (e.g. bald eagle) on which there were no data. The water quality criterion for total PCBs to protect the most sensitive species was calculated as $0.1 \, \mathrm{pg} \, l^{-1}$ (i.e. 1×10^{-16}) (Ludwig *et al.*, 1993).

Where there is reason to expect harmful effects but no conclusive scientific evidence to show a causal link, it is sensible to adopt the precautionary principle (Johnston and Simmonds, 1991). This is the avoidance or reduction of risks to the environment before specific environmental hazards are encountered. It can also be used to set stringent environmental standards, where for a variety of reasons causal links have not been proved, to protect a resource which is being damaged (see for example standards to protect otter populations in Mason, 1995).

The European Community has produced directives on quality requirements for bathing water, water abstracted for drinking, ecological water (e.g. supporting fish life), groundwater and dangerous substances. The requirements include fixed standards for a wide range of physical, chemical and microbiological determinands which must not be exceeded (Johnson and Corcelle, 1989). To meet these objectives will require considerable expenditure on the part of member states. The cost of meeting more stringent environmental requirements can be targeted at those causing the water quality problem by adopting the 'polluter pays' principle: those responsible for pollution should contribute towards the cost of monitoring, regulating and controlling the adverse effects of that pollution (McLoughlin and Bellinger, 1993).

Once standards have been set and consents given it is necessary to monitor their performance to ensure that water quality is maintained or improved.

Biologists within the water industry may spend much of their time in routine monitoring and some of the techniques applied were described in the previous two chapters. This monitoring may lead to a classification of surface water quality at the regional level. A review of the various classification schemes, both chemical and biological, adopted in the European Community is provided by Newman (1988).

The development and monitoring of water quality standards involves the biologist in objectives 1, 2 and 3 in the management scheme outlined on p. 271.

Catchment management plans

Evolving alongside water quality objectives are catchment management plans (CMPs). These involve a multifunctional and multi-use appraisal of the catchment, resulting in an agreed strategy for achieving the environmental potential of a catchment within prevailing economic and political constraints. The CMP:

1. provides for the implementation of functional strategies;
2. identifies present and future uses;
3. sets objectives and standards for each use;
4. identifies interactions and potential conflicts between uses;
5. sets out an action plan to achieve the agreed uses;
6. allocates responsibility for achieving actions together with an investment framework.

The plans include consideration of the uses required for the SWQOs described above and a range of non-statutory objectives related to activities which take place within a catchment.

The planning process begins with the formation of a multi-functional team under a project manager which identifies all of the current and future uses in the catchment. The required conditions to meet these uses are determined and compared with conditions which actually prevail. Shortfalls are identified and options to address them suggested. The provisional plan is then sent out for wide consultation amongst catchment user groups and the general public. The final plan includes the policy framework for the catchment and a series of action plans to achieve the policy.

For CMPs to be successful, the precise environmental requirements of each aspect requiring action will need to be determined, which may entail detailed scientific investigation at the local level. For example if the CMP for a particular river demands a good population of migratory salmonid fish, the water quality in the river will have to be good throughout, with high oxygen content and low levels of pollutants. It will require a flow regime to stimulate migration. There will need to be free access from the sea to the spawning grounds, which will require suitable gravel beds. For this one objective it is clear that a detailed scientific study, catchment-wide, is needed, especially if the re-establishment of a salmonid fishery, lost through environmental degradation in earlier years, is the aim.

More details of the CMP process are provided by Ayton (1994) and Gee and Jones (1995).

River regulation and inter-river transfer

In many parts of the world the use of water has risen dramatically. The population of the United States, for example, almost doubled between 1940 and 1980 but the use of water tripled, some $7500 \, l \, d^{-1}$ per person being withdrawn from freshwater supply. In many areas of the United States water shortages are critical. Over the past decade 80 000 ha of corn-producing land in the mid-west have been taken out of production because of shortages of water. In the republics of Uzbekistan and Kazakhstan in the Confederation of Independent States, large amounts of water have been diverted from rivers draining the inland Aral Sea to irrigate crops of cotton. The Aral Sea was once the world's fourth largest freshwater lake but the surface area has shrunk by 46 per cent, the volume by 69 per cent and the salinity has increased threefold. The fishery has collapsed, as has the shipping industry. The result is an ecological calamity (Williams and Aladin, 1991).

Much of the Mediterranean basin is suffering the effects of overabstraction. Shortfalls in water supply have also been regular events in recent years in England and Wales. Overabstraction has led to the headwaters of many streams running dry, even in winter (p. 197), having a major impact on the aquatic communities (Mantle and Mantle, 1992).

Many areas, of course, also suffer devastating floods, often with tremendous loss of life. Frequently floods are the result of major changes in the watershed, especially deforestation (p. 266). An engineering solution to both droughts and floods is river regulation. Almost every river in Europe and North America has been regulated, while there are few major rivers elsewhere which have not been regulated in some way (Petts, 1994).

River flows may be regulated by the addition of water from reservoirs, the water later being abstracted for use downstream, or water may be transferred from one river to another. Such schemes may have major impacts on the biota of rivers and may also influence levels of pollution (Petts, 1994). If waters are held back by a dam, for example, the reduced flow in the river downstream may lose much of its capacity to self-purify. Pollutants may be transferred between catchments. The biologist has an important role in investigating the influence of such major developments and in forecasting impacts (objectives 4 and 5, p. 272).

Edwards (1984) has summarized a large, interdisciplinary study, extending over five years, into the possible consequences of a river regulation scheme on the ecology of the River Wye in Wales. An enlargement of Craig Goch Reservoir, in the headwaters of the river, was planned in order to allow managed releases of water to compensate for abstractions downstream. Some of the impounded water was also to be used to regulate the adjacent catchment of the River Severn.

The predicted general and local effects of regulation on the River Wye are shown in Figs 12.7 and 12.8. Over the river as a whole it was thought that regulation would be beneficial to fisheries, because flow augmentation would avoid low oxygen concentrations developing during periods of macrophyte decay, a time at which considerable mortality of Atlantic salmon had occurred in the past. The food supply of salmonids should also be increased in summer by enhanced downstream drift of invertebrates, while oxygen-sensitive benthic invertebrates would be protected. The water in the river was likely to become softer and concentrations of iron and manganese were predicted to increase as a result of the release of hypolimnetic water from the reservoir, though it was unlikely to be sufficient to damage water quality.

The local effects were likely to be more varied and more difficult to predict. Increases in water velocity could influence the upstream movement of salmon as local temperature and oxygen regimes were modified. Plankton discharging from the reservoir could increase the food supply of filter-feeding animals, and indirectly the growth of salmonids. The softening of the water might lead to a reduction in the upstream distribution of molluscs and crustaceans, which require calcium. General reviews of the hydrological and ecological impacts of river regulation are provided by Petts (1984, 1994) and Petts *et al.* (1995).

Long-range transport of pollutants

In March 1973, severe damage to tomato crops grown in commercial glasshouses in Essex, eastern England, was reported, whole crops being

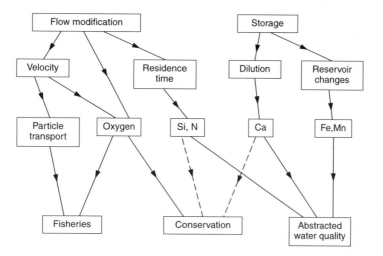

Fig. 12.7. Possible extensive effects of regulation on the aquatic resources of the River Wye; more doubtful effects are shown by dotted lines (from Edwards, 1984).

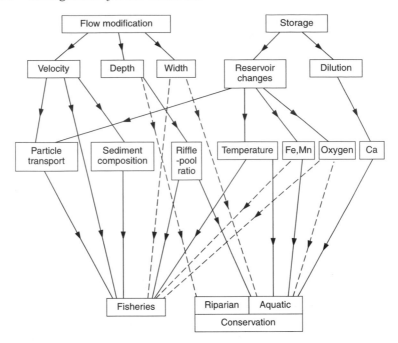

Fig. 12.8. Possible local effects of regulation on the aquatic resources of the River Wye; more doubtful effects are shown by dotted lines (from Edwards, 1984).

affected in some instances. Fifty tomato growers in the county reported problems. The damage was consistent with injury caused by growth regulatory herbicides and was eventually identified as due to 2,3,6-trichlorobenzoic acid (TBA), which was traced to the domestic water supply. By chemical analysis of river waters at various points and by plant bioassays, applying water to the cotyledons of tomato seedlings, the herbicide was eventually traced to a factory manufacturing herbicides (Williams *et. al.*, 1977).

The chemical factory began making TBA in 1971 and the herbicide became a component of its discharge to the River Cam above Cambridge. The effluent met the standards applied and monitoring of the river below the discharge detected very small and localized effects (Hawkes, 1978). Before 1973 the River Cam water was discharged, via the River Ouse, to the sea and none was abstracted for public water supply. In early 1973 the Ely Ouse–Essex water scheme became operational (Fig. 12.9). Water from the River Ouse is pumped at Denver into a Cut-off Channel which acts as an aqueduct to take water back southwards. From the channel it is pumped through tunnel and pipeline to be discharged into the upper reaches of the Rivers Stour and Blackwater to augment their flows. In the lower reaches of these rivers water is abstracted to Abberton and Hanningfield Reservoirs from where, after treatment, it is put into supply. TBA, the pollutant from the factory near Cambridge, was causing considerable losses to tomato growers who obtained water from different

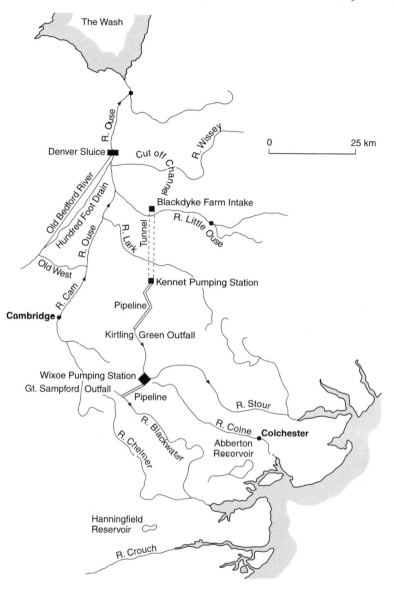

Fig. 12.9. The Ely Ouse–Essex water scheme.

catchments over 200 km away. Despite the dilution of the effluent during the inter-river transfer, the herbicide was still concentrated enough to kill tomatoes, which are sensitive to amounts as low as 0.0005 mg l^{-1} (Williams *et al.*, 1977). Charcoal towers were installed at the factory to treat the effluent as soon as it was implicated in the tomato failure and no further problems have been encountered.

Fisheries

Interest in recreational fishing has expanded rapidly over the past four decades and, in Britain for instance, angling has become the greatest single partici- pating sport, supporting a large service industry. On rivers with good runs of salmon and migratory trout the renting of fishing is extremely lucrative to riparian owners. The development and optimization of fisheries resources and their maintenance in relation to other conflicting uses of water requires a high level of expertise. In many cases fisheries management is kept separate from other biological work but ideally a multifunctional team of biologists, includ- ing those primarily involved in pollution assessment, is the most rational way of managing fisheries. A single pollution episode can destroy years of work in managing and improving a fishery (Fig. 12.10). The work carried out by fisheries biologists within the water industry includes:

1. Assessments of habitat and water quality to support particular species of fish.
2. Population, growth and production estimates of fish stocks in freshwater habitats.

Fig. 12.10. Major fish-kill caused by a saltwater intrusion, Norfolk Broads, England (photograph by Dr J. S. Wortley).

3. Investigations of the effects of land drainage and river channel modification works on fisheries.
4. Culturing of fish for stocking freshwaters.
5. Studies of populations and movements of migratory fish (especially salmonids).
6. Emergency investigations following pollution incidents.
7. Investigations of fish diseases and pathology.

As an example of the type of work which may be undertaken, fisheries biologists of the East Anglian region of the National Rivers Authority carry out a rolling programme to estimate the population density and biomass of fish stocks in the majority of their rivers, a total of over 3000 km. The results of such surveys for the River Waveney are shown in Fig. 12.11. The Waveney is in a popular tourist area, with much recreational angling. The river should support a fishery of at least Class B over much of its length, but in 1984–86 most of it was in Class D, with a biomass of fish of less than 5 g m^{-2}. The fisheries team then began a detailed investigation, in co-operation with water chemists, and it was found that much of the river regularly had ammonia concentrations in excess of those necessary to maintain coarse fisheries. The source of ammonia was piggeries, of which the catchment supports a large number. Efforts to control these discharges have been generally successful, with most of the fishery improving to Class B by 1994.

Models are frequently used to predict the effects of management activities on biological resources, including fisheries. One model which is widely used in the United States, and gaining currency elsewhere, is PHABSIM (Physical

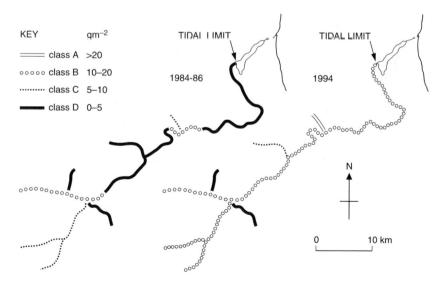

Fig. 12.11. Biomass of fish in the River Waveney, East Anglia, 1984–86 and 1994 (from data provided by the National Rivers Authority).

Habitat Simulation). This examines the effects of changes in discharge (for example caused by increases in resource use) on the habitat preferences of the biota. The underlying principles of PHABSIM are that individual species exhibit habitat preferences within a range of conditions that they can tolerate, that these preferences can be defined for each species and that the area of stream providing these conditions can be quantified as a function of discharge and channel structure (Petts and Maddock, 1994). Changing flow influences both current speed and water depth, primary variables used to predict the impact of altered flows on stream life. The model predicts changes in the physical habitat of the river and how this may influence ecological values. This then allows the biologist to recommend the flows required to maintain the ecological integrity of the site. For species of interest habitat suitability curves are derived (showing range of conditions tolerated, with optimum conditions) for velocity, width, depth and substratum type. This is done by field survey work, from the literature or by consulting experts. An example of the use of PHABSIM is provided by Petts *et al.* (1995).

Reviews of fisheries management include those of Elliott (1995), Mann (1995) and Cowx (1995), while Garciá de Jalón (1995) examines how habitats can be modified to benefit fish stocks.

Introduced species

For thousands of years human beings have introduced species to regions where they did not naturally occur, often for food or sport but also for cultural reasons, such as gardening. Anglers have been especially remiss at moving fish from water to water, country to country, with complete disregard for the ecological consequences (Maitland, 1995). When introduced species cause harm to ecological systems they can be considered as 'biological pollutants' under the definition given on p. 1.

One of the classic examples of the ecological consequences of an unplanned invasion is that of the sea lamprey (*Petromyzon marinus*), an ectoparasite of fish which spawns in fresh water. In 1829 the Welland Canal was opened to link the Great Lakes to the Atlantic Ocean, the Niagara Falls previously forming a barrier to navigation. Sea lampreys invaded slowly at first but there was an explosive population increase in the 1930s, causing the commercial catch of lake trout (*Salvelinus namaycush*) to decline more than thirty-fold and bankrupting fishing communities in Lakes Huron and Michigan. This apparently simple effect of predator on prey is probably not the whole story, however, for pollutants in the Great Lakes may have played a major role in the decline of the trout.

The zebra mussel (*Dreissena polymorpha*) has also invaded the Great Lakes from Europe, probably brought in with the ballast of a ship. First recorded in Lake St. Clair in 1985, it became a major pest in only three years, clogging water intakes and distribution networks, changing the ecology of the lakes

(p. 135) and fouling fishing gear and navigation buoys (Claudi and Mackie, 1994). Zebra mussels have, however, also become an important prey item of several species of diving ducks. Wildfowl populations could therefore benefit from this introduction (Hamilton and Ankney, 1994).

In Europe considerable concern has been expressed over four invasive alien plants. The swamp stonecrop (*Crassula helmsii*) is an amphibious aquatic, introduced from Australia as an oxygenation plant for ponds. Growing in both water and the damp margins it can form monocultures, replacing native species. Himalayan balsam (*Impatiens glandulifera*), Japanese knotweed (*Fallopia japonica*) and giant hogweed (*Heracleum mantegazzianum*) have all escaped from gardens to form dense stands on river banks, making access to the water difficult along extensive lengths of river in some cases. They suppress the native flora, including grasses and, when they die back in winter, they leave a bare riverbank, susceptible to erosion. The giant hogweed may reach 5 m in height, with leaves of 1 m in length. If its sap gets onto skin, which is then exposed to sunlight, an unpleasant dermatitis may develop. Aspects of the ecology and control of these species are discussed in De Waal *et al.* (1994).

In this 'ecologically correct' age it is regarded as inappropriate to introduce alien species and indeed it is illegal to do so in many countries. It has to be remembered that most introductions have had no obvious ecological impact, though island biota seem especially sensitive to alien predators, having evolved without these pressures. Some introductions can be seen as beneficial. The mandarin duck (*Aix galericulata*), deliberately introduced into England, has filled a vacant niche (a tree-hole nesting species of woodland pools) and this population is probably now of international significance for conservation.

The American mink (*Mustela vison*), which has escaped from fur-farms into many European countries, has been vilified for its supposed depredations on livestock, fisheries and wildlife, in Wales even being accused of attacking children! As the feral populations lose some of their acquired domestic traits, however, numbers seem to fall to a natural density. Existing on a wide prey base it is unlikely that their impact on native species will be measurable, except perhaps on water voles (*Arvicola terrestris*) when they first colonize a river (Mason, 1995).

The current bête-noir in Britain is the Canada goose (*Branta canadensis*), condemned by conservationists, on very flimsy evidence, for a variety of ills. In many parts of England, however, this is the only large, spectacular flocking species to delight birdwatchers, a benefit much greater than the alleged damage they do.

Wildlife conservation

The biological resources of immediate concern to the water industry are fishes and their food, together with those organisms of use in detecting changes in water quality. The work of the industry, however, affects a far greater range of habitats and their biota than merely the watercourse itself. Where a large-scale engineering scheme is planned an environmental impact assessment will take account of the biological changes which may result and the overall desirability of the scheme can be fully debated. In terms of wildlife conservation the small-scale destruction of habitat, over a wide area, will in the long term prove just as detrimental.

While habitat modification itself cannot be defined as pollution, it can be as damaging to the biota and hence should be the concern of the biologist within the water industry. Furthermore, habitat destruction, by reducing the carrying capacity, may render a population more susceptible to pollution, because its smaller size may be unable to make good any losses. The simplification of food chains may also promote the transfer of biomagnifying contaminants. For example a highly channelized stream may result in a fish community dominated by eels. Spending much of their time in the sediments, eels accumulate greater amounts of organochlorines and heavy metals than other fish (p. 254). An otter with only eels to eat may more rapidly accumulate contaminants harmful to its long-term survival.

The drainage of land for agriculture and channel improvements to aid flood prevention cause great concern to wildlife conservationists. In 1981, for example, 115 000 ha of land in France were drained, while in England and Wales in the mid 1970s drainage was running at 100 000 ha per year (Baldock, 1984). Within the European Commission the 'development agency' often gives very large sums of money for 'improvement' schemes to wetlands which have been declared of international importance by its own environmental agency; key wetlands are threatened in this way for example in Greece and Ireland.

Land drainage is not a new phenomenon. In the 17th century the Dutch engineer Cornelius Vermuyden constructed two massive parallel channels, 1 km apart and running for 30 km across the East Anglian fens to the sea, enabling 20 000 ha of wetland to be drained and turned into some of the richest arable land in the country. The land between the two channels acts as a safety valve when the incoming tide prevents the escape of water during spate conditions, especially in winter. The immense flood provides habitat for large numbers of waterfowl (Fig. 12.12), with peak winter numbers of over 54 000 recorded. The Washes also support good numbers of breeding waterfowl and waders. Outside the Washes the arable land for miles on either side is an ecological desert of flat, intensively cultivated land and this description is apt for the end point of most modern drainage schemes.

The drainage of wetlands results in the loss of species-rich grasslands, fens and marshes to species-poor, improved grasslands or ultimately to arable monocultures. Since the 1940s, land drainage, followed by ploughing, re-seeding and fertilization has left only 5 per cent of lowland neutral grasslands in England and Wales with any significant wildlife interest. Of plant species in Britain that have become extinct, rare or are rapidly declining, 22 per cent

Fig. 12.12. The Ouse Washes, eastern England. The grazing land between two channels is flooded each winter (a) to protect arable land and properties outside, which lie below the level of the channels. The floodlands are an internationally important wetland for birds and include a reserve of the Wildfowl and Wetlands Trust (b), where large populations of arctic-nesting Bewicks swans (*Cygnus columbianus*) (up to 6000), whooper swans (*C. Cygnus*) (up to 900) and ducks overwinter. The swans on this reserve are fed twice daily. (Photographs (a) by Dr S. M. Macdonald, (b) by the author.)

have been adversely affected by land drainage. The attractive snakeshead frit-illary (*Fritillaria meleagris*) (Fig. 12.13), once a widespread plant in English lowland meadows, has become restricted to a few nature reserves in the Thames Valley and Suffolk, while the snipe (*Gallinago gallinago*), a wading bird breeding in wetlands, has become extinct over much of lowland England (Marchant *et al.*, 1990). A localized drainage scheme may affect the water table over a considerably greater area so that a landowner, interested in pre-serving a wetland, may find his habitat deteriorating owing to the activities of his neighbours. The management of wet grasslands for birds (Fig. 12.14) is discussed by Ward (1994).

The costs of modern land drainage schemes can be measured in terms of en-vironmental damage, in terms of the over-production of food (which is bought

Fig. 12.13. Snakeshead fritillary (*Fritillaria meleagris*), once a common plant in alluvial meadows in England, now restricted to just a few sites (photograph by the author).

Fig. 12.14. Grazing marshes at Strumpshaw Fen, a reserve of the Royal Society for the Protection of Birds. Water levels are manipulated in spring to provide ideal nesting conditions for wading birds and ducks. Once breeding has finished cattle graze the plant-rich marshes through the summer (photograph by the author).

and stored at the tax-payers expense) and, in some areas, in a decline in the quality of agricultural land. The benefits are more difficult to determine. Calculations of benefit ignore the intangible environmental losses to the community and normally assume unrealistic yields of crops and a rapid take-up by farmers. The anticipated benefits are calculated on the 'farm gate' prices received by farmers for crops, which are unrealistically high because of the price protection which agriculture enjoys (Bowers, 1995). Most of the land drainage of ecologically valuable sites is undertaken to maximize the profits of individual landowners rather than to benefit the nation. Once destroyed, the original biodiversity of such wetlands can never be recreated so it would seem prudent, in the national interest, to conserve their ecological richness, at least until increased food production can be shown to be essential.

The other major area of conflict between wildlife conservationists and the water industry is river channelization for flood prevention. Some 44 000 km of rivers in England and Wales alone are subjected to regular maintenance (Brookes *et al.*, 1983). Management involves the straightening of river beds and the removal of natural obstructions, the destruction of aquatic macrophytes by hand, machines or herbicides and the clearance of bankside vegetation, especially of mature trees. Details can be found in Brookes (1988).

The majority of the river banks of the northern hemisphere were, if not swamp, once forested, but many are now devoid of riparian trees. In eastern England up to 70 per cent of bankside trees were removed between 1879 and

1970 (Fig. 12.15), 20 per cent of them in the period 1960 to 1970 (Mason and Macdonald, 1990). Figure 12.16 contrasts unmanaged and managed stretches. The level of management illustrated in Fig. 12.16b has a severe impact on both riparian and aquatic wildlife (Mason *et al.*, 1984).

The marked decrease in distribution and abundance of the otter due to contamination with organochlorines was described in Chapter 1 (p. 9). Habitat destruction has, however, also had a significant impact. Otters lie up and breed in bankside dens and, in some parts of their range, they favour the root systems of mature trees (Fig. 12.17), especially oak (*Quercus* spp.), ash (*Fraxinus excelsior*) and sycamore (*Acer pseudoplatanus*) (Macdonald and Mason, 1983). The roots of these trees dislike waterlogging so, although the trees readily establish and thrive on river banks, their major roots tend to grow shallowly and horizontally. In high waters the soil is easily eroded from amongst such roots, the trees gradually tilt and eventually fall into the river, where they may cause flooding. They were therefore removed by river engineers and such den sites for otters are now at a premium. Under pressure from conservationists, however, trees are now more usually pollarded or coppiced rather than ripped out, thereby preserving the otters' refuges. The two commonest trees on riverbanks, alder (*Alnus glutinosa*) and willows (*Salix* spp.), have water-seeking roots, the dense fibrous root mat suffering far less erosion but also not providing den sites.

Recent work has suggested that current management practices are excessive even for effective flood prevention. In England and Wales the annual cost of aquatic weed management alone is of the order of £45–75 million, and much of this may have become necessary because of the construction of drainage

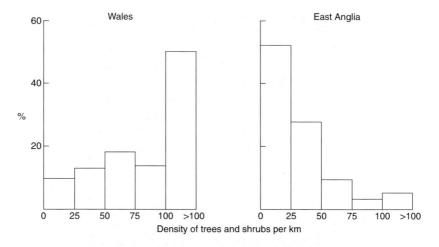

Fig. 12.15. The density of trees and shrubs (number per km) along the banks of rivers in Wales and East Anglia. Notice that heavy management in East Anglia has resulted in a very low density of woody vegetation, while many stretches in Wales have continuous tree lines. Vegetation along Welsh rivers has actually increased in recent years with a change to a more ecologically sensitive type of management.

Fig. 12.16. (a) Low-intensity management on the River Severn, Wales. (b) Typical stream management in eastern England (photographs by the author).

Fig. 12.17. An otter holt in the base of a sycamore tree, in Wales (photograph by the author).

schemes and the excessive removal of bankside trees, providing ideal conditions for macrophyte growth. Dawson (1989) has shown that the cutting of excess plant growth produces only a short-term improvement, but it may stimulate synchronized regrowth, accentuating problems in future years. Replanting river banks with trees, to provide a half-shade which reduces, but does not eliminate, macrophytes, would prove cost-effective and would benefit wildlife and fisheries. Dawson (1989) reviews options for macrophyte management. Sensitive and sensible approaches to river management are described in Brookes (1988), Purseglove (1988), Gore and Petts (1989) and Hey (1995).

Because of the regular conflicts between river engineers and wildlife conservationists, the water industry now employs a number of its own conservation officers. They are able to evaluate the possible impacts of engineering schemes *before* they are undertaken and suggest ways of alleviating the damage. Conservation staff can undertake, or commission, surveys to locate those stretches of river of especial conservation value. In England and Wales it is now general practice to carry out river corridor surveys using a standard methodology (National Rivers Authority, 1993). The river corridor can be defined as that width of river, bank and adjoining land which would be affected by major engineering works in the river. The surveyor walks slowly along the river bank and records features on a check sheet (mainly of geomorphology, habitat and vegetation). An annotated map is also produced (Harper *et al.*, 1995). An objective PC-based system for incorporating ecological information into broad-based descriptions of rivers is currently being developed so that the conservation value of rivers can be assessed: it is known as System for Evaluating Rivers for Conservation or SERCON (Boon, 1995).

One definite benefit to wildlife of the water industry has been the construction of some water supply reservoirs. Reservoirs in upland areas are normally steep-sided, deep and nutrient-poor, of little value to wildlife. They often drown important habitat and, where compensation below the dam is inadequately provided, the river may dry up to the detriment of aquatic life. In contrast, lowland reservoirs are a different proposition, for they are usually shallow, nutrient-rich and flood arable land or improved pasture. An example is Abberton Reservoir, in eastern England, completed in 1940 and covering 490 ha, with a mean depth of only 5 m. It supports a maximum average concentration of over 12 000 ducks, geese and swans and is a nature reserve of international importance. Rutland Water, in east midland England, was completed in 1976. A nature reserve was designed at its western end in the early stages of planning, covering 80 ha of water, with 160 ha of adjoining land. The rest of the 1260 ha of water is used for sailing and angling, while the reservoir perimeter has picnic sites, walks and cycle tracks – an excellent example of multiple use of a resource. The wildfowl numbers quickly became spectacular, with over 40 species recorded and peak totals of over 16 000 birds. Owens *et al.* (1986) provide detailed information on the value of reservoirs to wildlife in Great Britain, their importance being accentuated by the widespread destruction of natural wetlands over the past four decades.

The creation of wetlands for wildlife is not a new activity (Fig. 12.18) but there is currently much interest in developing new wetlands, as well as restoring damaged habitats. Wetland creation is described by Merritt (1994). Unfortunately very many of the world's key wetlands are threatened by development. Finlayson and Moser (1991) describe the types and values of wetlands throughout the world.

Conclusions

The management of water resources and the control of pollution is expensive and the benefits of a particular remedial action must be carefully weighed against the costs. Cut-backs in resources will lead to declines in the quality of our waters, as happened during the 1980s in England and Wales (Department of the Environment, 1992), following lack of investment in infrastructure and manpower in the water industry.

Politicians may argue that deaths caused by pollution have been rare, considerably less than mortality caused by accidents on the roads or in the home, or through self-inflicted damage caused by smoking or drinking. I have, however, emphasized several times in the preceding pages the extreme difficulty

Fig. 12.18. Bharatpur (Keoladeo Ghana National Park) in India is an artificial wetland created in the 1850s by the Maharajah of Bharatpur for hunting ducks and geese. It is now a much visited bird sanctuary of international importance, but there are current concerns that the excessive use of water in the surrounding farmland could cause it to dry out (photograph by the author).

of relating cause and effect in environmental pollution. Many materials, including newly synthesized ones, are constantly added to our waterways as traces in effluents and the long-term effects of pollutants, acting alone or in combination, are still largely unknown. That populations of animals have been severely reduced over wide areas, by pollutants previously not known to be environmental contaminants, should warn against complacency.

Politicians are also masters at prevarication, for pollution control costs money and may cut into profits. It took almost two decades, after a long rearguard action, for the British government to admit that acid pollution from its power stations may be partly responsible for the loss of Scandinavian fisheries, and for the government of the United States to accept that pollution from its industry was causing acidification in Canada (Pearce, 1987).

The 1980s and 1990s have seen a wave of public concern over the state of our environment. The late 1960s saw a similar mood, but as a popular movement it subsided with world recession. This period, however, saw the formation of environmental groups such as Greenpeace and Friends of the Earth, and over the past 25 years they have grown much in stature and expertise, both politically and scientifically. Their uncompromising position has done much to heighten public awareness, not least over water pollution, and the environmental movement is at last drawing political blood, which will hopefully lead to a greater commitment towards environmental protection. To protect our aquatic environment and to maintain constant vigilance over the state of our rivers and lakes will require the expertise of biologists, both within and outside the water industry.

Postscript

The Environment Act was granted Royal Assent on 19 July 1995. It establishes an Environment Agency in England and Wales and the Scottish Environment Protection Agency. The new agencies will begin operations on 1 April 1996. The Environment Agency will be formed by merging the National Rivers Authority, Her Majesty's Inspectorate of Pollution and the waste regulation authorities of the local councils. The Agency will employ around 9000 people.

The purpose of the Agency (and that in Scotland) is to produce high quality, integrated environmental protection, management and enhancement. The agencies will have the general functions of licensing, regulating and controlling activities relating to environmental protection. They will monitor the quality of the environment and will provide support and advisory services to other public authorities. They will be able to issue notices to prevent pollution from poorly maintained facilities, such as slurry pits, this being a distinct step forward in environmental management. The Environment Agency will have the duty to further conservation through all its functions except (unaccountably) water quality. This is a step backwards in terms of water resource management, though the conservation duty will be extended to other pollution functions (waste regulation, radioactive substances, integrated pollution control) which previously were not required to have regard to conservation. The agencies will also promote and coordinate environmental research and development and will liaise with the European Environmental Agency.

REFERENCES

ABDUL-HUSSEIN, M.M. (1985). Zooplankton and phytoplankton interactions in a eutrophic reservoir. Ph.D. thesis, University of Essex.

ABDUL-HUSSEIN, M.M. and MASON, C.F. (1988). The phytoplankton community of a eutrophic reservoir. *Hydrobiologia*, **169**, 265–277.

ABEL, P.D. (1989a). Pollutant toxicity to aquatic animals – methods of study and their applications. *Rev. Environ. Health*, **8**, 119–155.

ABEL, P.D. (1989b). *Water pollution biology*. Ellis Horwood, Chichester.

AHO, J.M., GIBBONS, J.W. and ESCH, G.W. (1976). Relationship between thermal loading and parasitism in the mosquitofish. In Esch, G.W. and McFarlane, R.W. (eds) *Thermal ecology II*, pp. 213–218. Technical Information Service, Springfield, VA.

ALABASTER, J.S. and LLOYD, R. (1982). *Water quality criteria for freshwater fish*, 2nd edn. Butterworths, London.

ALEXANDER, M.M., LONGABUCCO, P. and PHILLIPS, D.M. (1981). The impact of oil on marsh communities in the St. Lawrence River. In *Oil spill conference*, pp. 333–340. American Petroleum Institute, Washington, D.C.

ALVO, R., HUSSELL, D.J.T. and BERRILL, M. (1988). The breeding success of common loons (*Gavia immer*) in relation to alkalinity and other lake characteristics in Ontario. *Can. J. Zool.*, **66**, 746–752.

AMERICAN PUBLIC HEALTH ASSOCIATION (1989). *Standard methods for the examination of water and wastewater*, 17th edn. Amer. Public Health Assoc., Inc., New York.

ANDERSSON, T., BENGTSSON, B.-E., BERGQVIST, P.-A., ERIKSSON, T., LARSSON, A. and NORRGREN, L. (1993). Biochemical and physiological effects in farmed Baltic salmon fed lipids containing xenobiotics extracted from Baltic herring. *J. Aquat. Ecosystem Hlth.*, **2**, 185–196.

ANDRÉN, C., HENRIKSON, L., OLSSON, M. and NILSON, G. (1988). Effects of pH and aluminium on embryonic and early larval stages of Swedish brown frogs *Rana arvalis*, *R. temporaria* and *R. dalmatina*. *Hol. Ecol.*, **11**, 127–135.

APPELBERG, M., DEGERMAN, E. and NORRGREN, L. (1992). Effects of acidification and liming on fish in Sweden – a review. *Finnish Fish. Res.*, **13**, 77–91.

APSIMON, H.M., GUDIKSEN, P., KHITROV, L., ROHDE, H. and YOSHIKAWA, T. (1988). Modelling the dispersal and deposition of radionuclides. *Environment*, **30**(5), 17–20.

ARMITAGE, P.D., MOSS, D., WRIGHT, J.F. and FURSE, M.T. (1983). The performance of a new biological quality score system based on macroinvertebrates over a wide range of unpolluted running-water sites. *Water Res.*, **17**, 333–347.

ARTHUR, D.R. (1972). Katabolic and resource pollution in estuaries. In Cox, P.R. and Peel, J. (eds) *Population and pollution*, pp. 65–83. Academic Press, London.

ARTHUR, J.W., ZISCHKE, J.A. and ERICKSEN, G.L. (1982). Effect of elevated water temperature on macroinvertebrate communities in outdoor experimental channels. *Water Res.*, **16**, 1465–1477.

ASTON, R.J. (1973). Tubificids and water quality: a review. *Environ. Pollut.*, **5**, 1–10.

ASTON, R.J., BEATTIE, R.C. and MILNER, A.G.P. (1987). Characteristics of spawning sites of the common frog (*Rana temporaria*) with particular reference to acidity. *J. Zool. (Lond.)*, **213**, 233–242.

ATCHISON, G.J., HENRY, M.G. and SANDHEINRICH, M.B. (1987). Effects of metals on fish behaviour: a review. *Environ. Biol. Fish.*, **18**, 11–25.

AULERICH, R.J., RINGER, R.K. and SAFRONOFF, J. (1986). Assessment of primary vs. secondary toxicity of Aroclor 1254 to mink. *Arch. Environ. Contam. Toxicol.*, **15**, 393–399.

AYTON, W.J. (1994). Catchment management planning in the National Rivers Authority. *Wat. Sci. Tech.*, **3**, 351–355.

BALDOCK, D. (1984). *Wetland drainage in Europe*. International Institute for Environment and Development/Institute for European Environment Policy, London.

BALDWIN, I.G., HARMAN, M.M.I. and NEVILLE, D.A. (1994). Performance characteristics of a fish monitor for detection of toxic substances. 1. Laboratory trials. *Water Res.*, **28**, 2191–2199.

BALLS, H., MOSS, B. and IRVINE, K. (1989). The loss of submerged plants with eutrophication. 1. Experimental design, water chemistry, aquatic plant and phytoplankton biomass in experiments carried out in ponds in Norfolk Broadland. *Freshwat. Biol.*, **22**, 71–87.

BARAK, N.A.-E. and MASON, C.F. (1989). Heavy metals in water, sediment and invertebrates from rivers in eastern England. *Chemosphere*, **19**, 1709–1714.

BARAK, N.A.-E. and MASON, C.F. (1990a). Mercury, cadmium and lead in eels and roach: the effects of size, season and locality on metal concentrations in flesh and liver. *Sci. Total Environ.*, **92**, 249–256.

BARAK, N.A.-E. and MASON, C.F. (1990b). Mercury, cadmium and lead concentrations in five species of freshwater fish from eastern England. *Sci. Total Environ.*, **92**, 257–263.

BARAK, N.A.-E. and MASON, C.F. (1990c). A survey of heavy metal levels in eels (*Anguilla anguilla*) from some rivers in East Anglia, England: the use of eels as pollution indicators. *Int. Rev. ges. Hydrobiol.*, **75**, 827–833.

BARLAUP, B.T., ÅTLAND, Å. and KLEIVEN, E. (1994). Stocking of brown trout (*Salmo trutta* L.) cohorts after liming – effects on survival and growth during five years of reacidification. *Water Air Soil Pollut.*, **72**, 317–330.

BARR, J.F. (1986). Population dynamics of the common loon (*Gavia immer*) associated with mercury-contaminated waters in northwestern Ontario. Canadian Wildlife Service, Occasional Paper no. 56, 23 pp.

BARTHA, R. and ATLAS, R.M. (1977). The microbiology of aquatic oil spills. *Adv. appl. Microbiol.*, **22**, 225–266.

BATTARBEE, R.W. (1994). Diatoms, lake acidification and the Surface Water Acidification Programme (SWAP): a review. *Hydrobiologia*, **274**, 1–7.

BATTARBEE, R.W. and CARTER, C. (1993). The recent sediments of Lough Neagh, Part B: Diatom and chironomid analysis. In Wood, R.B. and Smith, R.V. (eds) *Lough Neagh*, pp. 133–147. Kluwer Academic, Dordrecht.

BATTARBEE, R.W. and CHARLES, D.F. (1994). Lake acidification and the role of paleolimnology. In Steinberg, C.F.W. and Wright, R.F. (eds) *Acidification of freshwater ecosystems; implications for the future*, pp. 51–65. Wiley, Chichester.

BAUDO, R. (1987). Ecotoxicological testing with *Daphnia*. *Mem. Ist. Ital. Idrabiol.*, **45**, 461–482.

BAVOR, H.J. and MITCHELL, D.S. (eds) (1994). Wetland systems in water pollution control. *Water Sci. Technol.*, **29**(4), 1–316.

BEATTIE, J.H. and PASCOE, D. (1978). Cadmium uptake by rainbow trout, *Salmo gairdneri*, Richardson, eggs and alevins. *J. Fish Biol.*, **13**, 631–637.

BÉLAND, P., DEGUISE, S., GIRARD, C., LAGACÉ, A., MARTINEAU, D., MICHAUD, R., MUIR, D.C.G., NORSTROM, R.J., PELLETIER, E., RAY, S. and SHUGART, L.R. (1993). Toxic compounds and health and reproductive effects in St. Lawrence beluga whales. *J. Great Lakes Res.*, **19**, 766–775.

BENNDORF, J. (1987). Foodweb manipulation without nutrient control: a useful strategy in lake restoration? *Schweiz. Z. Hydrol.*, **49**, 237–248.

BENNDORF, J. and HORN, W. (1985). Theoretical considerations on the relative importance of food limitation and predation in structuring zooplankton communities. *Arch. Hydrobiol.*, **21**, 383–396.

BERGMAN, A. and OLSSON, M. (1985). Pathology of Baltic grey seal and ringed seal females with special reference to adrenocortical hyperplasia: is environmental pollution the cause of a widely distributed disease syndrome? *Finn. Game Res.*, **44**, 43–62.

BERRY, D.R. and SENIOR, E. (1987). Applications of molecular biology. In Sidwick, J.M. and Holdom, R.S. (eds) *Biotechnology of waste treatment and exploitation*, pp. 209–233. Ellis Horwood, Chichester.

BERWICK, P.G. (1984). Physical and chemical conditions for microbial oil degradation. *Biotechnol. Bioeng.*, **24**, 1294–1305.

BEUMER, J.P. and BACHER, G.J. (1982). Species of *Anguilla* as indicators of mercury in the coastal rivers and lakes of Victoria, Australia. *J. Fish Biol.*, **21**, 87–94.

BEWLEY, R.J.F., SLEAT, R. and REES, J.F. (1991). Waste treatment and pollution clean-up. In Moses, V. and Cape, R.E. (eds) *Biotechnology: the science and the business*, pp. 507–519. Harwood Academic, Char.

BIANCHI, M. and COLWELL, R.R. (1986). Microbial indicators of environmental water quality: the role of microorganisms in the assessment and prediction of changes in the marine environment induced by human activities. In Salanki, J. (ed.) *Biological monitoring of the state of the environment: bioindicators*, pp. 5–15. IUBS, Paris.

BIELBY, G.H. (1988). The Lowermoor environmental report. South West Water Authority, Exeter.

BIGNERT, A., GÖTHBERG, A., JENSEN, S., LITZÉN, K., ODSJÖ, T., OLSSON, M. and REUTERGÅRDH, L. (1993). The need for adequate biological sampling in eco-toxicological investigations: a retrospective study of twenty years pollution monitoring. *Sci. Total Environ.*, **128**, 121–139.

BIRKHEAD, M. and PERRINS, C.M. (1986). *The mute swan*. Croom Helm, London.

BITTON, G. (ed) (1994). *Wastewater microbiology*. Wiley, New York.

BJÖRKLUND, I., BORG, H. and JOHANSSON, K. (1984). Mercury in Swedish lakes – its regional distribution and causes. *Ambio*, **13**, 118–121.

BLACK, J.J. and BAUMANN, P.C. (1991). Carcinogens and cancers in freshwater fishes. *Environ. Health Perspect.*, **90**, 27–33.

BLACK, P.E. (1987). *Conservation of water and related land resources*. Rownan and Littlefield, New Jersey.

BLEWETT, D.A., WRIGHT, S.E., CASEMORE, D.P., BOOTH, N.E. and JONES, C.E. (1993). Infective dose size studies on *Cryptosporidium parvum* using gnotobiotic lambs. *Water Sci. Technol.*, **27**, 61–64.

BOAR, R.R., CROOK, C.E. and MOSS, B. (1989). Regression of *Phragmites australis* reedswamps and recent changes of water chemistry in Norfolk Broadland, England. *Aquat. Bot.* **35**, 41–55.

BOERS, P., BALLEGOOIJEN, L. van and UUNK, J. (1991). Changes in phosphorus cycling in a shallow lake due to food web manipulations. *Freshwat. Biol.*, **25**, 9–20.

BOERS, P., DOES, J. van der, QUAAK, M. and VLUGT, J. van der (1994). Phosphorus fixation with iron (III) chloride: a new method to combat internal phosphorus loading in shallow lakes? *Arch. Hydrobiol.*, **129**, 339–351.

BOON, P.J. (1995). The relevance of ecology to the statutory protection of British rivers. In Harper, D.M. and Ferguson, A.J.D. (eds) *The ecological basis for river management*, pp. 239–250. Wiley, Chichester.

BOORMAN, L.A. and FULLER, R.M. (1981). The changing status of reedswamps in the Norfolk Broads. *J. appl. Ecol.*, **18**, 241–269.

BORYSLAWSKYJ, M., GARROOD, A.C., PEARSON, J.T. and WOODHEAD, D. (1987). Rates of accumulation of dieldrin by a freshwater filter feeder: *Sphaerium corneum. Environ. Pollut.*, **43**, 3–13.

BOSTRÖM, B. (1984). Potential mobility of phosphorus in different types of lake sediment. *Int. Rev. ges. Hydrobiol.*, **69**, 457–474.

BOWERS, J. (1995). The interface between ecology and economics in catchment management. In Harper, D.M. and Ferguson, A.J.D. (eds) *The ecological basis for river management*, pp. 515–523. Wiley, Chichester.

BOWKER, D.W. and MUIR, H.J. (1981). Algal bioassays in water pollution studies. *J. biol. Education*, **15**, 35–41.

BRAKKE, D.F., BAKER, J.P., BÖHMER, J. *et al.* (1994). Group report: physiological and ecological effects of acidification on aquatic biota. In Steinberg, C.E.W. and Wright, R.F. (eds) *Acidification of freshwaters: implications for the future*, pp. 275–312. Wiley, Chichester.

BRETT, M.T. (1989). Zooplankton communities and acidification processes (a review). *Water Air Soil Pollut.*, **44**, 387–414.

BREUKELAAR, A.W., LAMMENS, E.H.R.R., KLEIN BRETELER, J.G.P. and TATRAI, I. (1994). Effects of benthivorous bream (*Abramis brama*) and carp (*Cyprinus carpio*) on sediment resuspension and concentrations of nutrients and chlorophyll *a. Freshwat. Biol.*, **32**, 113–121.

BRIX, H. and SCHIERUP, H.-H. (1989). The use of aquatic macrophytes in water pollution control. *Ambio*, **18**, 100–107.

BROOK, A.J. (1994). Algae. In Maitland, P.S., Boon, P.J. and McLusky, D.S. (eds) *The fresh waters of Scotland*, pp. 131–146. Wiley, Chichester.

BROOKER, M.P. and EDWARDS, R.W. (1974). Effects of the herbicide paraquat on the ecology of a reservoir. III. Fauna and general discussion. *Freshwat. Biol.*, **4**, 311–335.

BROOKER, M.P. and EDWARDS, R.W. (1975). Aquatic herbicides and the control of water weeds. *Water Res.*, **9**, 1–15.

BROOKES, A. (1988). *Channelized rivers: perspectives for environmental management*. Wiley, Chichester.

BROOKES, A., GREGORY, K.J. and DAWSON, F.H. (1983). An assessment of river channelization in England and Wales. *Sci. Total Environ.*, **27**, 97–111.

BROWN, B.E. (1976). Observations on the tolerance of the isopod *Asellus meridianus* Rac. to copper and lead. *Wat. Res.*, **10**, 555–559.

BROWN, B.E. (1977). Uptake of copper and lead by a metal-tolerant isopod *Asellus meridianus* Rac. *Freshwat. Biol.*, **7**, 235–244.

BROWN, B.E. (1978). Lead detoxification by a copper-tolerant isopod. *Nature*, **276**, 388–390.

BROWN, D.J.A. and SADLER, K. (1989). Fish survival in acid waters. In Morris, R., Taylor, E.W., Brown, D.J.A. and Brown, J.A. (eds) *Acid toxicity and aquatic animals*, pp. 31–44. Cambridge University Press, Cambridge.

BROWN, L.R. (1987). Oil-degrading microorganisms. *Chem. Eng. Progr.*, **42**, 35–40.

BROWN, M. (1989a). Biodegradation of oil in freshwaters. In Green, J. and Trett, M. (eds) *The fate and effects of oil in freshwater*, pp. 197–213. Applied Science Publishers, London.

BROWN, M. (1989b). Clean-up technology. In Green, J. and Trett, M. (eds) *The fate and effects of oil in freshwater*, pp. 215–226. Applied Science Publishers, London.

BROWN, M.J. (1991). Metal recovery and processing. In Moses, V. and Cape, R.E. (eds) *Biotechnology: the science and the business*, pp. 567–580. Harwood Academic, Char.

BROWN, M.J., LINTON, E. and REES, E.C. (1992). Causes of mortality among wild swans in Britain. *Wildfowl*, **43**, 70–79.

BROWN, V.M., JORDAN, D.H.M. and TILLER, B.A. (1967). The effect of temperature on the acute toxicity of phenol to rainbow trout in hard water. *Water Res.*, **1**, 587–594.

BRUNBERG, A.-K. and BOSTRÖM, B. (1992). Coupling between benthic biomass of *Microcystis* and phosphorus release from the sediments of a highly eutrophic lake. *Hydrobiologia*, **235/236**, 375–385.

BRUSLE, J. (1991). The eel (*Anguilla* sp.) and organic chemical pollutants. *Sci. Total Environ.*, **102**, 1–19.

BUHL, K.J. and HAMILTON, S.J. (1991). Relative sensitivity of early life stages of Arctic grayling, coho salmon and rainbow trout to nine inorganics. *Ecotox. Environ. Safety*, **22**, 184–197.

BULL, K.R. (1991). The critical load/levels approach to gaseous pollutant emission control *Environ. Pollut.*, **69**, 105–123.

BURTON, T.M. and ALLAN, J.W. (1986). Influence of pH, aluminium, and organic matter on stream invertebrates. *Can. J. Fish Aquat. Sci.*, **43**, 1285–1289.

BURTON, T.M., STANFORD, R.M. and ALLAN, J.W. (1985). Acidification effects on stream biota and organic matter processing. *Can. J. Fish. Aquat. Sci.*, **42**, 669–675.

BURY, R.B. (1972). The effects of diesel fuel on a stream fauna. *Calif. Fish. Game*, **58**, 291 295.

BUTCHER, R.W. (1947). Studies on the ecology of rivers, VII. The algae of organically enriched waters. *J. Ecol.*, **35**, 186–191.

CABELLI, V.J., DUFOUR, A.P., MCCABE, L.J. and LEVIN, M.A. (1983). A marine recreational water quality criterion consistent with indicator concepts and risk analysis. *J. Water Pollut. Control Fed.*, **55**, 1306–1314.

CAIRNCROSS, S. and FEACHAM, R.G. (1983). *Environmental health engineering in the tropics*. Wiley, Chichester.

CAIRNS, J. (1984). Are single species toxicity tests alone adequate for estimating environmental hazard? *Environ. Monit. Assessment*, **4**, 259–273.

CAIRNS, J. and CHERRY, D.S. (1993). Freshwater multi-species test systems. In Calow, P. (ed) *Handbook of ecotoxicology*, vol. 1, pp. 101–116. Blackwell, Oxford.

CALAMARI, D. and MARCHETTI, R. (1973). The toxicity of mixtures of metals and surfactants to rainbow trout (*Salmo gairdneri* Rich.). *Water Res.*, **7**, 1453–1464.

CAMPBELL, P.G.C. and STOKES, P.M. (1985). Acidification and toxicity of metals to aquatic biota. *Can. J. Fish Aquat. Sci.*, **42**, 2034–3049.

CAMPBELL, R.N.B., MAITLAND, P.S. and LYLE, A.A. (1986). Brown trout deformities: an association with acidification? *Ambio*, **15**, 244–245.

CAMPLIN, W.C. and AARKROG, A. (1989). Radioactivity in north European waters: report of Working Group 2 of CEC Project MARINA. *Fisheries Research Data Report No. 20*. Ministry of Agriculture, Fisheries and Food, Lowestoft.

CANTON, J.H. and SLOOFF, W. (1982). Toxicity and accumulation studies of cadmium (Cd^{2+}) with freshwater organisms of different trophic levels. *Ecotoxicol. Environ. Safety*, **6**, 113–128.

CANTON, S.P. and CHADWICK, J.W. (1988). Variability in benthic invertebrate density estimates from stream samples. *J. Freshwat. Ecol.*, **4**, 291–297.

CARPENTER, S.R. (1989). Replication and treatment strength in whole-lake experiments. *Ecology*, **70**, 453–463.

CAVE, S. (1991). A green revolution down at the sewer ponds. *Our Planet*, **3**(5), 10–11.

CHAPMAN, D.V. (1987). Pesticides in the aquatic environment. MARC Report 39. Kings College, London.

CHRISTENSEN, G., HUNT, E. and FIANOT, J. (1977). The effect of methylmercuric chloride, cadmium chloride and lead nitrate on six biochemical factors of the brook trout (*Salvelinus fontinalis*). *Toxicol. appl. Pharm.*, **42**, 523–530.

CLARKE, A.G. (1992). The atmosphere. In Harrison, R.M. (ed) *Understanding our environment: an introduction to environmental chemistry and pollution*, 2nd edn, pp. 5–51. Royal Society of Chemistry, London.

CLAUDI, R. and MACKIE, G.L. (1994). *Practical manual for zebra mussel monitoring and control*. Lewis, Boca Raton.

CLESCERI, L.S. (1980). PCBs in the Hudson River. In Guthrie, F.E. and Perry, J.J. (eds) *Introduction to environmental toxicology*, pp. 227–235. Blackwell, Oxford.

COLBORN, T. and CLEMENT, C. (eds) (1992). *Chemically-induced alterations in sexual and functional development: the wildlife/human connection*. Princeton Sci. Publ. Co., New Jersey.

COLE, J.A., NORTON, R.L. and MONTGOMERY, H.A.C. (1994). Countering acute pollution events: procedures currently being adopted in the United Kingdom. *Wat. Sci. Tech.*, **29**, 203–205.

CONNELL, D.W. and MILLER, G.J. (1984). *The chemistry and ecotoxicology of pollution*. Wiley, New York.

COOKE, G.D. (1993). *Restoration and management of lakes and reservoirs*. Lewis, Boca Raton.

COONEY, J.J., SILVER, S.A. and BECK, E.A. (1985). Factors influencing hydrocarbon degradation in three freshwater lakes. *Microb. Ecol.*, **11**, 127–137.

COOPER, P.F. and FINDLATER, B.C. (1991). *Constructed wetlands for water pollution control*. Pergamon Press, Oxford.

COOPER, P.F., HOBSON, J.A. and JONES, S. (1989). Sewage treatment by reed bed systems. *J. IWEM*, **3**, 60–74.

CORREA, M. (1987). Physiological effects of metal toxicity on the tropical freshwater shrimp *Macrobrachium carcinus* (L., 1758). *Environ. Pollut.*, **45**, 149–155.

COSTE, M., BOSCA, C. and DAUTA, A. (1991). Use of algae for monitoring rivers in France. In Whitton, B.A., Rott, E. and Friedrich, G (eds) *Use of algae for monitoring rivers*, pp. 75–88. University of Innsbruck.

COUTANT, C.C., COX, D.K. and MOORED, K.W. (1976). Further studies of cold-shock effects on susceptibility of young channel catfish to predation. In Esch, G.W. and McFarlane, R.W. (eds) *Thermal Ecology II*, pp. 154–158. Technical Information Service, Springfield, VA.

COVER, E.C. and HARREL, R.C. (1978). Sequences of colonization, diversity, biomass and productivity of macroinvertebrates on artificial substrates in a freshwater canal. *Hydrobiologia*, **59**, 81–95.

COWX, I.G. (1995). Fish stock assessment – a biological basis for sound ecological management. In Harper, D.M. and Ferguson, A.J.D. (eds) *The ecological basis for river management*, pp. 375–388. Wiley, Chichester.

CRAIG, F. and CRAIG, P. (1989). *Britain's poisoned water*. Penguin, London.

CRIVELLI, A.J., FOCARDI, S., FOSSI, C., LEONZIO, C., MASSI, A. and RENZONI, A. (1989). Trace elements and chlorinated hydrocarbons in eggs of *Pelecanus crispus*, a world endangered bird species nesting at Lake Mikri Prespa, north-western Greece. *Environ. Pollut.*, **61**, 235–247.

CULLEN, P. and FORSBERG, C. (1988). Experiences with reducing point sources of phosphorus to lakes. *Hydrobiologia*, **170**, 321–336.

CURTIS, E.J.C. and CURDS, C.R. (1971). Sewage fungus in rivers in the United Kingdom: the slime community and its constituent organisms. *Water Res.*, **5**, 1147–1159.

CURTIS, E.J.C., DELVES-BROUGHTON, J. and HARRINGTON, D.W. (1971). Sewage fungus: studies of *Sphaerotilus* slimes using laboratory recirculating channels. *Water Res.*, **5**, 267–279.

CZARNEZKI, J.M. (1985). Accumulation of lead in fish from Missouri streams impacted by lead mining. *Bull. Environ. Contam. Toxicol.*, **34**, 736–745.

DALLINGER, R., PROSI, F., SEGNER, H. and BACK, H. (1987). Contaminated food and uptake of heavy metals by fish: a review and proposal for further research. *Oecologia*, **73**, 91–98.

DALY, H.B. (1993). Laboratory rat experiments show consumption of Lake Ontario salmon causes behavioural changes: support for wildlife and human research results. *J. Great Lakes Res.*, **19**, 784–788.

DAVIES, I.J. (1984). Sampling aquatic insect emergence. In Downing, J.A. and Rigler, F.H. (eds) *A manual on methods for the assessment of secondary productivity in fresh waters*, pp. 161–227. Blackwell, Oxford.

DAVIES, L.J. and HAWKES, H.A. (1981). Some effects of organic pollution on the distribution and seasonal incidence of Chironomidae in riffles in the River Cole. *Freshwat. Biol.*, **11**, 549–559.

DAVIES, P.H. and WOODLING, J.D. (1980). Importance of laboratory-derived metal toxicity results in predicting in-stream response of resident salmonids. In Eaton, J.G., Parish, P.R. and Hendricks, D.C. (eds) *Aquatic toxicology*, pp. 281–299. American Society for Testing and Materials, Philadelphia.

DAWSON, F.H. (1989). Ecology and management of water plants in lowland streams. *Freshwater Biol. Ass. Ann. Report*, **57**, 43–60.

DE BERNARDI, R. (1989). Biomanipulation of aquatic food chains to improve water quality in eutrophic lakes. In Ravera, O. (ed) *Ecological assessment of environmental degradation, pollution and recovery*, pp. 195–215. Elsevier, Amsterdam.

DE BOER, J. and HAGEL, P. (1994). Spatial differences and temporal trends of chlorobiphenyls in yellow eel (*Anguilla anguilla*) from inland waters of the Netherlands. *Sci. Total Environ.*, **141**, 155–174.

DEPARTMENT OF THE ENVIRONMENT (1992). *The UK environment*. HMSO, London.

DE PAUW, N. and VANHOOREN, G. (1983). Method for biological quality assessment of watercourses in Belgium. *Hydrobiologia*, **100**, 153–168.

DEPLEDGE, M.H. (1986). Global implications of Chernobyl. *Mar. Pollut. Bull.*, **17**, 281–282.

DEPLEDGE, M. (1989). The rational basis for detection of the early effects of marine pollutants using physiological indicators. *Ambio*, **18**, 301–302.

DE VOOGT, P., WELLS, D.E., REUTERGÅRDH, L. and BRINKMANN, U.A.T. (1990). Biological activity, determination and occurrence of planar, mono- and di-ortho PCBs. *Intern. J. Environ. Anal. Chem.*, **40**, 1–46.

DE WAAL, L.C., CHILD, L.E., WADE, P.M. and BROCK, J.H. (1994). *Ecology and management of invasive riverside plants*. Wiley, Chichester.

DHEER, J.M.S. (1988). Haematological, haematopoeitic and biochemical responses to thermal stress in an air-breathing freshwater fish, *Channa punctatus* Bloch. *J. Fish. Biol.*, **32**, 197–206.

DIAMOND, J., COLLINS, M. and GRUBER, D. (1988). An overview of automated biomonitoring – past developments and future needs. In Gruber, D. and Diamond, J. (eds) *Automated biomonitoring: living sensors as environmental monitors*, pp. 23–39. Ellis Horwood, Chichester.

DILLON, P.J. and RIGLER, F.H. (1975). A simple method for predicting the capacity of a lake for development based upon lake trophic status. *J. Fish. Res. Board Can.*, **32**, 1519–1531.

DODDS, W.K. and GUDDER, D.A. (1992). The ecology of *Cladophora. J. Phycol.*, **28**, 415–427.

DOTTO, L. (1986). *Planet earth in jeopardy*. Wiley, Chichester.

DOUBEN, P.E.T. (1989). Uptake and elimination of waterborne cadmium by the fish *Noemacheilus barbatulus* L. (Stone loach). *Arch. Environ. Contam. Toxicol.*, **18**, 576–586.

DOWNEY, D.M., FRENCH, C.R. and ODOM, M. (1994). Low cost limestone treatment of acid sensitive trout streams in the Appalachian Mountains of Virginia. *Water Air Soil Pollut.*, **77**, 49–77.

DOWNING, J.A. (1984). Sampling the benthos of standing waters. In Downing, J.A. and Rigler, F.H. (eds) *A manual on methods for the assessment of secondary productivity in fresh waters*, pp. 87–130. Blackwell, Oxford.

DUFFY, L.K., BOWYER, R.T., TESTA, J.W. and FARO, J.B. (1994). Chronic effects of the *Exxon Valdez* oil spill on blood and enzyme chemistry of river otters. *Environ. Toxicol. Chem.*, **13**, 643–647.

EDMONDSON, W.T. (1969). Eutrophication in North America. In *Eutrophication: causes, consequences, correctives*, pp. 124–149. National Academy of Sciences, Washington, DC.

EDMONDSON, W.T. (1970). Phosphorus, nitrogen and algae in Lake Washington after diversion of sewage. *Science*, **169**, 690–691.

EDMONDSON, W.T. (1971). Phytoplankton and nutrients in Lake Washington. In Likens, G. (ed.) *Nutrients and eutrophication: the limiting nutrient controversy*. American Society of Limnology and Oceanography, **1**, 172–193.

EDMONDSON, W.T. (1972a). The present condition of Lake Washington. *Verh. int. Verein. Limnol.*, **18**, 284–291.

EDMONDSON, W.T. (1972b). Lake Washington. In Goldman, C.R. (ed.) *Environmental quality and water development*, pp. 281–298. Freeman and Co., New York.

EDMONDSON, W.T. (1979). Lake Washington and the predictability of limnological events. *Ergebn. Limnol.*, **13**, 234–241.

EDMONDSON, W.T. (1991). *The uses of ecology: Lake Washington and beyond.* University of Washington Press, Seattle.

EDWARDS, R.W. (1975). A strategy for the prediction and detection of effects of pollution on natural communities. *Schweiz. Z. Hydrol.*, **37**, 135–143.

EDWARDS, R.W. (1984). Predicting the environmental impact of a major reservoir development. In Roberts, R.D. and Roberts, T.M. (eds) *Planning and ecology*, pp. 55–79. Chapman and Hall, London.

EDWARDS, R.W. (1989). Ecological assessment of degradation and recovery of rivers from pollution. In Ravera, O. (ed.) *Ecological assessment of environmental degradation, pollution and recovery*, pp. 159–194. Elsevier, Amsterdam.

EDWARDS, R.W., HUGHES, B.D. and READ, M.W. (1975). Biological survey in the detection and assessment of pollution. In Chadwick, M.J. and Goodman, G.T. (eds) *The ecology of resource degradation and renewal*, pp. 139–156. Blackwell, Oxford.

EDWARDS, R.W., ORMEROD, S.J. and TURNER, C. (1991). Field experiments to assess biological effects of pollution episodes in streams. *Verh. internat. Verein. Limnol.*, **24**, 1734–1737.

EDWARDS, R.W., WILLIAMS, P.F. and WILLIAMS, R. (1984). Ebbw. In Whitton, B. (ed.) *Ecology of European rivers*, pp. 83–111. Blackwell, Oxford.

EGGLISHAW, H., GARDINER, R. and FOSTER, J. (1986). Salmon catch decline and forestry in Scotland. *Scott. geogr. mag.*, **102**, 57–61.

EISENBUD, M. (1987). *Environmental radioactivity.* Academic Press, Orlando.

EKLUND, M.W. and DOWELL, V.R. (1987). *Avian botulism, an international perspective.* Charles C. Thomas, Springfield, IL.

ELLIOTT, J.M. (1977). *Some methods for the statistical analysis of benthic invertebrates. Scientific Publication No. 25.* Freshwater Biological Association, Windermere.

ELLIOTT, J.M. (1981). Some aspects of thermal stress in teleosts. In Pickering, A.D. (ed.) *Stress and fish*, pp. 209–245. Academic Press, London.

ELLIOTT, J.M. (1995). The ecological basis for management of fish stocks in rivers. In Harper, D.M. and Ferguson, A.J.D. (eds) *The ecological basis for river management*, pp. 323–337. Wiley, Chichester.

ELLIOTT, J.M. and DRAKE, C.M. (1981). A comparative study of seven grabs for sampling macroinvertebrates in rivers. *Freshwat. Biol.*, **11**, 99–120.

ELLIS, J.B., REVITT, D.M., SHUTES, R.B.E. and LANGLEY, J.M. (1994). The performance of vegetated biofilters for highway runoff control. *Sci. Total Environ.*, **146/147**, 543–550.

ELLIS, J.C. and HUNT, D.T.E. (1986). Surface water acidification: an assessment of historic water quality records. *Report TR 240*. Water Research Centre, Medmenham.

ELLIS, K.V. (1989). *Surface water pollution and its control*. Macmillan, London.

ENVIRONMENTAL DATA SERVICES (1988). Britain's eels extensively contaminated by organochlorines. *Report 165*, 7.

ENVIRONMENTAL DATA SERVICES (1989). Algal toxins implicated in human illness in U.K. *Report 177*, 5–6.

ERIKSSON, M.O. (1987). Some effects of freshwater acidification on birds in Sweden. In Diamond, A.W. and Filion, F. (eds) *The value of birds*, ICBP Tech. Publ. 6, 183–190. International Council for Bird Preservation, Cambridge.

ESSER, W. (1978). Über die Rolle sessiler Organismen auf die Selbstreinigungsgeschwindigkeit in Fliessgewässern. *Gas-u. Wasserfach.*, **119**, 582–586.

EVANS, G.P., JOHNSON, D. and WITHELL, C. (1986). Development of the WRC Mk III Fish Monitor: description of the system and its response to some commonly encountered pollutants. *Report TR 233*. Water Research Centre, Medmenham.

EVANS, G.P. and WALLWORK, J.F. (1988). The WRC Fish Monitor and other biomonitoring methods. In Gruber, D. and Diamond, J. (eds) *Automated biomonitoring: living sensors as environmental monitors*, pp. 75–90. Ellis Horwood, Chichester.

EVERARD, M. (1994). Water quality objectives as a management tool for sustainability. *Freshwater Forum*, **4**, 179–189.

EXTENCE, C.A., BATES, A.J., FORBES, W.J. and BARHAM, P.J. (1987). Biologically based water quality management. *Environ. Pollut.*, **45**, 221–236.

FALLOWFIELD, H.J., SVOBODA, I.F. and MARTIN, N.J. (1992). Aerobic and photosynthetic treatment of animal slurries. In Fry, J.C., Gadd, G.M., Herbert, R.A., Jones, C.W. and Watson-Craik, I.A. (eds) *Microbial control of pollution*, pp. 171–197. Cambridge University Press, Cambridge.

FEACHAM, R.G., BRADLEY, D.J., GARELICK, H. and MARA, D.D. (1983). *Sanitation and disease. Health aspects of excreta and wastewater management*, World Bank studies in water supply and sanitation 3. Wiley, Chichester.

FELTS, P.A. and HEATH, A.G. (1984). Interactions of temperature and sublethal environmental copper exposure on the energy metabolism of bluegill, *Lepomis macrochirus* Rafinesque. *J. Fish Biol.*, **25**, 445–453.

FENNESSY, M.S. and MITSCH, W.J. (1989). Design and use of wetlands for renovation of drainage from coal mines. In Mitsch, W.J. and Jørgensen, S.E. (eds) *Ecological engineering*, pp. 231–253. Wiley, New York.

FINLAYSON, M. and MOSER, M. (1991). *Wetlands*. Facts on File, Oxford.

FINNEMORE, E.J. and LYNARD, W.G. (1982). Management control technology for urban stormwater pollution. *J. Wat. Pollut. Fed.*, **54**, 1099–1111.

FISHLOCK, D. (1994). The dirtiest place on Earth. *New Scientist*, 19th February, 34–37.

FLANNAGAN, J.F. (1970). Efficiencies of various grabs and corers in sampling freshwater benthos. *J. Fish. Res. Bd. Can.*, **27**, 1691–1700.

FLANNAGAN, J.F. and ROSENBERG, D.M. (1982). Types of artificial substrates used for sampling freshwater benthic macroinvertebrates. In Cairns, J. (ed.) *Artificial substrates*, pp. 237–266. Ann Arbor Science Publishers, Ann Arbor.

FLEMING, W.J., CLARK, D.R. and HENNY, C.J. (1983). Organochlorine pesticides and PCBs: a continuing problem for the 1980s. *Trans. North Am. Wildlife Res. Conf.*, **48**, 186–199.

FLOWER, R.J. and BATTARBEE, R.W. (1983). Diatom evidence for recent acidification of two Scottish lochs. *Nature*, **305**, 130–133.

FLOWER, R.J., BATTARBEE, R.W. and APPLEBY, P.G. (1987). The recent palaeolimnology of acid lakes in Galloway, south-west Scotland: diatom analysis, pH trends, and the role of afforestation. *J. Ecol.*, **75**, 797–824.

FORBES, V.E. and FORBES, T.L. (1994). *Ecotoxicology in theory and practice*. Chapman and Hall, London.

FORSBERG, C. (1989). Importance of sediments in understanding nutrient cyclings in lakes. *Hydrobiologia*, **176/177**, 263–277.

FORSYTH, D.J., MARTIN, P.A., DE SMET, K.D. and RISHE, M.E. (1994). Organochlorine contaminants and eggshell thinning in grebes from Prairie Canada. *Environ. Pollut.*, **85**, 51–58.

FOSSI, C., FOCARDI, S., LEONZIO, C. and RENZONI, A. (1984). Trace-metals and chlorinated hydrocarbons in birds' eggs from the delta of the Danube. *Environ. Conserv.*, **11**, 345–350.

FOSTER, R.B. and BATES, J.M. (1978). Use of mussels to monitor point source industrial discharges. *Environ. Sci. Technol.*, **12**, 958–962.

FOX, G.A. (1993). What have biomarkers told us about the effects of contaminants on the health of fish-eating birds in the Great Lakes? The theory and a literature review. *J. Great Lakes Res.*, **19**, 722–736.

FOX, G.A., COLLINS, B., HAYAKAWA, H., WESELOH, D.V., LUDWIG, J.P., KUBIAK, T.J. and ERDMAN, T.C. (1991b). Reproductive outcomes in colonial fish-eating birds: a biomarker for development toxicants in Great Lakes food chains. II. Spatial variation in the occurrence and prevalence of bill defects in young double-crested cormorants in the Great Lakes, 1979–1987. *J. Great Lakes Res.*, **17**, 158–167.

FOX, G.A., KENNEDY, S.W., NORSTROM, R.J. and WIGFIELD, D.C. (1988). Porphyria in herring gulls: a biochemical response to chemical contamination of Great Lakes food chains. *Environ. Toxicol. Chem.*, **7**, 831–839.

FOX, G.A., WESELOH, D.V., KUBIAK, T.J. and ERDMAN, T.C. (1991a). Reproductive outcomes in colonial fish-eating birds: a biomarker for developmental toxicants in Great Lakes food chains. 1. Historical and ecotoxicological perspectives. *J. Great Lakes Res.*, **17**, 153–157.

FRANCE, R.L. and WELLBOURN, P.M. (1992). Influence of lake pH and macrograzers on the distribution and abundance of nuisance metaphytic algae in Ontario, Canada. *Can. J. Fish Aquat. Sci.*, **49**, 185–195.

FREDA, J. (1986). The influence of acidic pond water on amphibians: a review. *Water Air Soil Pollut.*, **30**, 439–450.

FRICK, K.G. and HERRMANN, J. (1990). Aluminium and pH effects on sodium ion regulation in mayflies. In Mason, B.J. (ed.) *The surface waters acidification programme*, pp. 409–412. Cambridge University Press, Cambridge.

FRIEDRICH, G. and MÜLLER, D. (1984). Rhine. In Whitton, B.A. (ed.) *Ecology of European rivers*, pp. 265–315. Blackwell, Oxford.

FRYER, G. (1980). Acidity and species diversity in freshwater crustacean faunas. *Freshwat. Biol.*, **10**, 41–45.

FULLER, G.B. and HOBSON, W.C. (1986). Effects of PCBs on reproduction in mammals. In Waid, J.S. (ed.) *PCBs and the environment*, Vol 2, pp. 101–125. CRC Press, Boca Raton.

FURSE, M.T., WRIGHT, J.F., ARMITAGE, P.D. and MOSS, D. (1981). An appraisal of pond-net samples for biological monitoring of lotic macro-invertebrates. *Water Res.*, **6**, 79–89.

GADD, G.M. (1992). Microbial control of heavy metal pollution. In Fry, J.C., Gadd, G.M., Herbert, R.A., Jones, C.W. and Watson-Craik, I.A. (eds) *Microbial control of pollution,* pp. 59–88. Cambridge University Press, Cambridge.

GARCIÁ DE JALÓN, D. (1995). Management of physical habitat for fish stocks. In Harper, D.M. and Ferguson, A.J.D. (eds) *The ecological basis for river management*, pp. 363–374. Wiley, Chichester.

GAUCH, H.G. (1982). *Multivariate analysis in community ecology*. Cambridge University Press, Cambridge.

GEE, A.S. and JONES, F.H. (1995). The use of biological techniques in catchment planning. In Harper, D.M. and Ferguson, A.J.D. (eds) *The ecological basis for river management*, pp. 475–489. Wiley, Chichester.

GEE, A.S. and STONER, J.H. (1988). The effects of afforestation and acid deposition on the water quality of upland Wales. In Usher, M.B. and Thompson, D.B.A. (eds) *Ecological change in the uplands*, pp. 273–287. Blackwell, Oxford.

GEORGE, M. (1992). *The land use, ecology and conservation of Broadland*. Packard, Chichester.

GERBA, C.P. and ROSE, J.B. (1990). Viruses in source and drinking water. In

McFeters, G.A. (ed.) *Drinking water microbiology*, pp. 386–396. Springer-Verlag, New York.

GERBA, C.P., WALTER, R. and FARRAH, S.R. (1991). *Water contamination by viruses: occurrence, detection, treatment*. Lewis Publishers, Boca Raton.

GERHARDT, A. (1993). Review of impact of heavy metals on stream invertebrates with special emphasis on acid conditions. *Water Air Soil Pollut.*, **66**, 289–314.

GIBSON, C.E. (1993). The phytoplankton populations of Lough Neagh. In Wood, R.B. and Smith, R.V. (eds) *Lough Neagh*, pp. 203–223. Kluwer Academic, Dordrecht.

GIBSON, D.T. (1983). Microbial degradations of hydrocarbons. In Albaiges, J., Frei, R.W. and Merian, E. (eds) *Current topics in environmental and toxicological chemistry*, pp. 177–190. Gordon and Breach, New York.

GIESY, J.P., LUDWIG, J.P and TILLITT, D.E. (1994). Deformities in birds of the Great Lakes region: assigning causality. *Environ. Sci. Technol.*, **28**, 128A–135A.

GILMOUR, C.C. and HENRY, E.A. (1991). Mercury methylation in aquatic systems affected by acid deposition. *Environ. Pollut.*, **71**, 131–169.

GOLTERMAN, H.L. (1991). Reflections on post-OECD eutrophication models. *Hydrobiologia*, **218**, 167–176.

GONCALVES, E.P.R., BOAVENTURA, R.A.R. and MOUVET, C. (1992). Sediments and aquatic mosses as pollution indicators for heavy metals in the Ave river basin (Portugal). *Sci. Total Environ.*, **114**, 7–24.

GORE, J.A. and PETTS, G.E. (1989). *Alternatives in regulated river management*. CRC Press, Boca Raton.

GOWER, A.M., MYERS, G., KENT, M. and FOULKES, M.E. (1994). Relationships between macroinvertebrate communities and environmental variables in metal-contaminated streams in south-west England. *Freshwat. Biol.*, **32**, 199–221.

GRAHN, O. (1986). Vegetation structure and primary production in acidified lakes in south-western Sweden. *Experentia*, **42**, 465–470.

GRAY, N.F. (1985). Heterotrophic slimes in flowing waters. *Biol. Rev.*, **60**, 499–548.

GRAY, N.F. (1992). *Biology of wastewater treatment*. Oxford University Press, Oxford.

GRAY, N.F. (1994). *Drinking water quality*. Wiley, Chichester.

GREEN, D.W.J., WILLIAMS, K.A., HUGHES, D.R.L., SHAIK, G.A.R. and PASCOE, D. (1988). Toxicity of phenol to *Asellus aquaticus* (L.) – effects of temperature and episodic exposure. *Water Res.*, **22**, 225–231.

GREEN, D.W.J., WILLIAMS, K.A. and PASCOE, D. (1985). Studies on the acute toxicity of pollutants to freshwater macroinvertebrates. 2. Phenol. *Arch. Hydrobiol.*, **103**, 75–82.

GREEN, J. and TRETT, M.W. (eds) (1989). *The fate and effects of oil in freshwater*. Applied Science Publishers, London.

GREEN, R.H. (1979). *Sampling design and statistical methods for environmental biologists*. Wiley, New York.

GROSCH, D.S. (1980). Radiation and radioisotopes. In Guthrie, F.E. and Perry, J.O. (eds) *Introduction to environmental toxicity,* pp. 44–61. Elsevier, New York.

GRUBER, D., FRAGO, C.H. and RASNAKE, W.J. (1994). Automated biomonitors – first line of defence. *J. Aquat. Ecosystem Health*, **3**, 87–92.

GUINEY, P.D., SYKORA, J.L. and KELETI, G. (1987). Environmental impact of an aviation kerosene spill on stream water quality in Cambria County, Pennsylvania. *Environ. Toxicol. Chem.*, **6**, 977–988.

GULATI, R.D. (1989). Concept of stress and recovery in aquatic ecosystems. In Ravera, O. (ed.) *Ecological assessment of environmental degradation, pollution and recovery*, pp. 81–119. Elsevier, Amsterdam.

HAINES, T.A. (1981). Acidic precipitation and its consequences for aquatic ecosystems: a review. *Trans. Am. Fish. Soc.*, **110**, 669–707.

HAINES, T.A. (1983). Organochlorine residues in brook trout from remote lakes in the northeastern United States. *Water Air Soil Pollut.*, **20**, 47–54.

HAINES, T.A. and BAKER, J.P. (1986). Evidence of fish population responses to acidification in the eastern United States. *Water Air Soil Pollut.*, **31**, 605–629.

HÅKANSON, L., ANDERSSON, T. and NILSSON, Å. (1989). Caesium-137 in perch in Swedish lakes after Chernobyl – present situation, relationships and trends. *Environ. Pollut.*, **58**, 195–212.

HÅKANSON, L., NILSSON, Å. and ANDERSSON, T. (1988). Mercury in fish in Swedish lakes. *Environ. Pollut.*, **49**, 145–162.

HAKKARI, L. (1992). Effects of pulp and paper mill effluents on fish populations in Finland. *Finn. Fish. Res.*, **13**, 93–106.

HÄKKINEN, I. and HÄSÄNEN, E. (1980). Mercury in eggs and nestlings of the osprey (*Pandion haliaetus*) in Finland and bioaccumulation from fish. *Ann. Zool. Fenn.*, **17**, 131–139.

HALL, R.J., DRISCOLL, C.T. and LIKENS, G.E. (1987). Importance of hydrogen ions and aluminium in regulating the structure and function of stream ecosystems: an experimental test. *Freshwat. Biol.*, **18**, 17–43.

HAMILTON, D.J. and ANKNEY, C.D. (1994). Consumption of zebra mussels *Dreissena polymorpha* by diving ducks in Lakes Erie and St. Clair. *Wildfowl*, **45**, 159–166.

HAMILTON, R.S. and HARRISON, R.M. (1991). *Highway pollution*. Elsevier, London.

HAMMER, D.A. (1989). *Constructed wetlands for wastewater treatment*. Lewis, Chelsea.

HAMMERTON, D. (1994). Domestic and industrial pollution. In Maitland, P.S., Boon, P.J. and McLusky, D.S. (eds) *The freshwaters of Scotland*, pp. 347–364. Wiley, Chichester.

HANCOCK, S.J. and BUDDHAVARAPU, L. (1993). Control of algae using duckweed

(*Lemna*) systems. In Moshiri, G.A. (ed.) *Constructed wetlands for water quality improvement*, pp. 399–406. Lewis, Boca Raton.

HANEL, K. (1988). *Biological treatment of sewage by the activated sludge process*. Ellis Horwood, Chichester.

HARDMAN, D.J., MCELDOWNEY, S. and WAITE, S. (1993). *Pollution: ecology and biotreatment*. Longman, Harlow.

HARE, L., SAOUTER, E., CAMPBELL, P.G.C., TESSIER, A., RIBEYRE, F. and BOUDOU, A. (1991). Dynamics of cadmium, lead and zinc exchange between nymphs of the burrowing mayfly *Hexagenia rigida* (Ephemeroptera) and the environment. *Can. J. Fish. Aquat. Sci.*, **48**, 39–47.

HARGREAVES, J.W., LLOYD, E.J.H. and WHITTON, B.A. (1975). Chemistry and vegetation of highly acidic streams. *Freshwat. Biol.*, **5**, 563–576.

HARPER, D. (1992). *Eutrophication of freshwaters*. Chapman and Hall, London.

HARPER, D., SMITH, C., BARHAM, P. and HOWELL, R. (1995). The ecological basis for the management of the natural river environment. In Harper, D.M. and Ferguson, A.J.D. (eds) *The ecological basis for river management*, pp. 219–238. Wiley, Chichester.

HARPER, D.B., SMITH, R.V. and GOTTO, D.M. (1977). BHC residues of domestic origin: a significant factor in pollution of freshwater in Northern Ireland. *Environ. Pollut.*, **12**, 223–233.

HARRAD, S.J., SEWART, A.P., ALCOCK, R., BOUMPHREY, R., BURNETT, V., DUARTE-DAVIDSON, R., HALSALL, C., SANDERS, G., WATERHOUSE, K., WILD, S.R. and JONES, K.C. (1994). Polychlorinated biphenyls (PCBs) in the British environment: sinks, sources and temporal trends. *Environ. Pollut.*, **85**, 131–146.

HARRIMAN, R., MORRISON, B.R.S., CAINES, L.A., CULLEN, P. and WATT, A.W. (1987). Long-term changes in fish populations of acid streams and lochs in Galloway, south west Scotland. *Water Air Soil Pollut.*, **32**, 89–112.

HARRIMAN, R. and WELLS, D.E. (1985). Causes and effects of surface water acidification in Scotland. *Wat. Pollut. Control*, **84**, 215–224.

HARRISON, A.D. (1984). The acidophilic thiobacilli and other acidophilic bacteria that share their habitat. *An. Rev. Microbiol.*, **38**, 265–292.

HARSHBARGER, J.C. and CLARK, J.B. (1990). Epizootiology of neoplasms in bony fish of North America. *Sci. Total Environ.*, **94**, 1–32.

HARTMANN, J. (1977). Fischereiliche Veranderungen in kulturbedingt eutrophierenden Seen. *Schweiz. Z. Hydrol.*, **39**, 243–254.

HASLAM, S.M. (1987). *River plants of Western Europe*. Cambridge University Press, Cambridge.

HAVAS, M., HUTCHINSON, T.C. and LIKENS, G.E. (1984). Red herrings in acid rain research. *Environ. Sci. Technol.*, **18**, 176A–186A.

HAVAS, M. and LIKENS, G.E. (1985). Changes in ^{22}Na influx in *Daphnia magna* (Straus) as a function of elevated Al concentrations in soft water at low pH. *Proc. natn. Acad. Sci. U.S.A.*, **82**, 7345–7349.

HAVENS, K. and DECOSTA, J. (1985). The effect of acidification in enclosures on the biomass and population size structure of *Bosmina longirostris*. *Hydrobiologia*, **122**, 153–158.

HAWKES, H.A. (1975). River zonation and classification. In Whitton, B. (ed.) *River ecology*, pp. 312–374. Blackwell, Oxford.

HAWKES, H.A. (1978). River bed animals tell tales of pollution. In Hughes, J.G. and Hawkes, H.A. (collators) *Biosurveillance of river water quality*, pp. 55–77. Proceedings of section K of the British Association for the Advancement of Science, Aston, 1977.

HAWKES, H.A. (1979). Invertebrates as indicators of water quality. In James, A. and Evison, L. (eds) *Biological indicators of water quality* pp. 2.1–2.45. Wiley, Chichester.

HAWKES, H.A. and DAVIES, L.J. (1971). Some effects of organic enrichment on benthic invertebrate communities in stream riffles. In Duffey, E.A. and Watt, A.S. (eds) *The scientific management of animal and plant communities for conservation*, pp. 271–293. Blackwell, Oxford.

HAYES, C.R., CLARK, R.G., STENT, R.F. and REDSHAW, C.J. (1984). The control of algae by chemical treatment in a eutrophic water supply reservoir. *J. Wat. Eng. Sci.*, **38**, 149–162.

HAYES, C.R. and GREENE, L.A. (1984). The evaluation of eutrophication impact in public water supply reservoirs in East Anglia. *Wat. Pollut. Control*, **83**, 45–51.

HAYES, J.T. (1991). Global climate change and water resources. In Wyman, R.L. (ed.) *Global climate change and life on earth*, pp. 18–42. Routledge, Chapman and Hall, New York.

HEANEY, S.I., CORREY, J.E. and LISHMAN, J.P. (1992). Changes of water quality and sediment phosphorus of a small productive lake following decreased phosphorus loading. In Sutcliffe, D.W. and Jones, J.G. (eds) *Eutrophication: research and application to water supply*, pp. 119–131. Freshwater Biological Association, Ambleside.

HEDTKE, S.F. and PUGLISI, F.A. (1982). Short-term toxicity of five oils to four freshwater species. *Arch. Environ. Contam. Toxicol.*, **11**, 425–430.

HEIT, M. and FINGERMAN, M. (1977). The influences of size, sex and temperature on the toxicity of mercury to two species of crayfishes. *Bull. Environ. Contam. Toxicol.*, **18**, 572–580.

HELLAWELL, J.M. (1986). *Biological indicators of freshwater pollution and environmental management*. Applied Science Publishers, London.

HENDERSON-SELLERS, B. and MARKLAND, H.R. (1987). *Decaying lakes*. Wiley, Chichester.

HENDREY, G.R. and VERTUCCI, F. (1980). Benthic plant communities in acid Lake Colden, New York: *Sphagnum* and the algal mat. In Drablos, D. and Tolan, A. (eds) *Ecological impact of acid precipitation*, pp. 314–315. SNFC Project, Oslo.

HENRIKSEN, A. (1989). Air pollution effects on aquatic ecosystems and their restoration. In Ravera, O. (ed.) *Ecological assessment of environmental degradation, pollution and recovery*, pp. 291–312. Elsevier, Amsterdam.

HENRIKSEN, A., LIEN, L., ROSSELAND, B.O., TRAAEN, T.S. and SEVELDRUD, I.S. (1989). Lake acidification in Norway: present and predicted fish status. *Ambio*, **18**, 314–321.

HENRIKSEN, A., SKOGHEIM, O.K. and ROSSELAND, B.O. (1984). Episodic changes in pH and aluminium-speciation kill fish in a Norwegian salmon river. *Vatten*, **40**, 255–260.

HENRIKSON, L., NYMAN, H.G., OSCARSON, H.G. and STENSON, J.A.E. (1980). Trophic changes without changes in the external nutrient loading. *Hydrobiologia*, **68**, 257–263.

HENRIKSSON, K., KARPPANEN, E. and HELMINEN, M. (1966). High residue of mercury in Finnish White-tailed Eagles. *Ornis fenn.*, **43**, 38–45.

HENRY, M.G. and ATCHISON, G.J. (1986). Behavioural changes in social groups of bluegills exposed to copper. *Trans. Am. Fish. Soc.*, **115**, 590–595.

HERBERT, D.W.M. (1961). Freshwater fisheries and pollution control. *Proc. Soc. Wat. Treat. J.*, **10**, 135–156.

HERNANDEZ, L.M., GONZALEZ, M.J., RICO, M.C., FERNANDEZ, M.A. and BALUJA, G. (1985). Presence and biomagnification of organochlorine pollutants and heavy metals in mammals of Doñana National Park (Spain) 1982–1983. *J. environ. Sci. Hlth.*, **B20**, 633–650.

HEY, R. (1995). River processes and management. In O'Riordan, T. (ed.) *Environmental science for environmental management*, pp. 131–150. Longman, Harlow.

HEYMAN, U. and LUNDGREN, A. (1988). Phytoplankton biomass and production in relation to phosphorus. *Hydrobiologia*, **170**, 211–227.

HIGGINS, I.J. and BURNS, R.E. (1975). *The chemistry and microbiology of pollution*. Academic Press, London.

HILL, I.R., HEIMBACH, F., LEEUWANGH, P. and MATHIESSEN, P. (eds) (1994). *Freshwater field tests for hazard assessment of chemicals*. Lewis, Boca Raton.

HILL, M.O. (1979a). *DECORANA – A Fortran program for detrended correspondence analysis and reciprocal averaging*. Ecology and Systematics, Cornell University, Ithaca, New York.

HILL, M.O. (1979b). *TWINSPAN – A Fortran program for arranging multivariate data in an ordered two-way table by classification of individuals and attributes*. Ecology and Systematics, Cornell University, Ithaca, New York.

HOFFMANN, D.J. (1979). Embryotoxic and teratogenic effects of crude oil on mallard embryos on day one of development. *Bull. Environ. Contam. Toxicol.*, **22**, 632–637.

HOLDGATE, M.W. (1979). *A perspective of environmental pollution*. Cambridge University Press, Cambridge.

HOLLAND, D.G. and HARDING, J.P.C. (1984). Mersey. In Whitton, B. (ed.) *Ecology of European rivers*, pp. 113–144. Blackwell, Oxford.

HOLMES, N. (1989). British rivers: a working classification. *British Wildlife*, **1**, 20–36.

HORAN, N.J. (1990). *Biological wastewater treatment systems.* Wiley, Chichester.

HOSPER, S.H. (1989). Biomanipulation, new perspectives for restoration of shallow, eutrophic lakes in The Netherlands. *Hydrobiol. Bull.*, **23**, 5–10.

HOUGH, R.A., FORNWALL, M.D., NEGELE, B.J., THOMPSON, R.L. and PUTT, D.A. (1989). Plant community dynamics in a chain of lakes: principal factors in the decline of rooted macrophytes with eutrophication. *Hydrobiologia*, **173**, 199–217.

HOUGHTON, J.T., CALLANDER, B.A. and VARNEY, S.K. (eds) (1992). *Climate change 1992. The supplementary report to the IPCC Scientific Assessment.* Cambridge University Press, Cambridge.

HOUSE, M.A. and NEWSOME, D.H. (1989). Water quality indices for the management of surface water quality. *Water Sci. Tech.*, **21**, 1137–1148.

HOWELLS, G. (1990). *Acid rain and acid waters.* Ellis Horwood, New York.

HOWELLS, G. (ed.) (1994). *Water quality for freshwater fish.* Gordon and Breach, Switzerland.

HOWELLS, G., DALZIEL, T.R.K. and TURNPENNY, A.W.H. (1992). Loch Fleet: liming to restore a brown trout fishery. *Environ. Pollut.*, **78**, 131–139.

HOWELLS, G.D. and GAMMON, K.M. (1984). Role of research in meeting environmental assessment needs for power station siting. In Roberts, R.D. and Roberts, T.M. (eds) *Planning and ecology*, pp. 310–330. Chapman and Hall, London.

HUCKLE, K.R. and MILLBURN, P. (1990). Metabolism, bioconcentration and toxicity of pesticides in fish. In Hutson, D.H. and Roberts, T.R. (eds) *Environmental fate of pesticides*, pp. 195–243. Wiley, Chichester.

HUGHES, B.D. (1975). A comparison of four samplers for benthic macroinvertebrates inhabiting coarse river deposits. *Water Res.*, **9**, 61–69.

HUGHES, B.D. (1978). The influence of factors other than pollution on the value of Shannon's diversity index for benthic macro-invertebrates in streams. *Water Res.*, **12**, 357–364.

HUGHES-CLARKE, S.A. and MASON, C.F. (1992). Ecological development of field corner tree plantations on arable land. *Landsc. Urban Plann.*, **22**, 59–72.

HUNN, J.B., CLEVELAND, L. and LITTLE, E.E. (1987). Influence of pH and aluminium on developing brook trout in low calcium water. *Environ. Pollut.*, **43**, 63–73.

HUNT, E.G. and BISCHOFF, A.I. (1960). Inimical effects on wildlife of DDD application to Clear Lake. *Calif. Fish. Game*, **46**, 91–106.

HUNT, G.J. (1987). Radioactivity in surface and coastal waters of the British Isles, 1986. *Aquatic Environment Monitoring Report 18*, 62 pp. Ministry of Agriculture, Fisheries and Food, Lowestoft.

HUTCHINSON, N.J., NEARY, B.P. and DILLON, P.J. (1991). Validation and use of Ontario's Trophic Status Model for establishing lake development guidelines. *Lake and Reserv. Manage.*, **7**, 13–23.

HYNES, H.B.N. (1960). *The biology of polluted waters.* Liverpool University Press, Liverpool.

IP, H.M.H. and PHILLIPS, D.J.H. (1989). Organochlorine chemicals in human breast milk in Hong Kong. *Arch. Environ. Contam. Toxicol.*, **18**, 490–494.

IRVINE, K., MOSS, B. and BALLS, H. (1989). The loss of submerged plants with eutrophication. II. Relationships between fish and zooplankton in a set of experimental ponds, and conclusions. *Freshwat. Biol.*, **22**, 89–107.

IRVINE, K., STANSFIELD, J. and MOSS, B. (1991). The use of enclosures to demonstrate the enhancement of *Daphnia* populations when isolated from fish predation in a shallow eutrophic lake. *Mem. Ist. ital. Idrobiol.*, **48**, 325–344.

JACKSON, D. (1992). Environmental impact of the United Kingdom nuclear fuel reprocessing industry. In Drake, J.A.G. (ed.) *The chemical industry – friend to the environment?*, pp. 126–148. Royal Society of Chemistry, London.

JACKSON, S.T. and CHARLES, D.F. (1988). Aquatic macrophytes in Adirondack (New York) lakes: patterns of species composition in relation to environment. *Can. J. Bot.*, **66**, 1449–1460.

JACOBSON, J.L. and JACOBSON, S.W. (1993). A 4-year followup study of children born to consumers of Lake Michigan fish. *J. Great Lakes Res.*, **19**, 776–783.

JAEGER, D. (1994). Effects of hypolimnetic water aeration and iron–phosphate precipitation on the trophic level of Lake Krupunder. *Hydrobiologia*, **275/276**, 433–444.

JANZEN, F. (1994). Climate change and temperature-dependent sex determination in reptiles. *Proc. Nat. Acad. Sci.*, **91**, 7487–7490.

JEFFRIES, M. and MILLS, D. (1990). *Freshwater ecology: principles and applications*. Belhaven Press, London.

JEPPESON, E., KRISTENSEN, P., JENSEN, J.P., SØNDERGAARD, M., MORTENSEN, E. and LAURIDSEN, T. (1991). Recovery resilience following a reduction in external phosphorus loading of shallow, eutrophic Danish lakes: duration, regulating factors and methods for overcoming resilience. *Mem. Ist. ital. Idrobiol.*, **48**, 127–148.

JOHNSON, S.P. and CORCELLE, G. (1989). *The environmental policy of the European Communities*. Graham and Trotman, London.

JOHNSTON, P. and SIMMONDS, M. (1991). Green light for precautionary science. *New Scientist*, 3rd August, 4.

JONES, K. and TELFORD, D. (1991). On the trail of a seasonal microbe. *New Scientist*, 6th April, 36–39.

JONES, K.C. (1985). Gold, silver and other elements in aquatic bryophytes from a mineralised area of North Wales, UK. *J. Geochem. Expl.*, **24**, 237–246.

JONES, K.C., PETERSON, P.J. and DAVIES, B.E. (1985). Silver and other metals in some aquatic bryophytes from streams in the lead mining district of mid-Wales, Great Britain. *Water Air Soil Pollut.*, **24**, 329–338.

JONES, K.C., SANDERS, G., WILD, S.R., BURNETT, V.B. and JOHNSTON, A.E. (1992). Evidence for the decline in PCBs and PAHs in rural vegetation and air. *Nature*, **356**, 137–140.

JØRGENSEN, S.E. (1980). *Lake management*. Pergamon, Oxford.

JOSÉ, P. (1989). Long-term nitrate trends in the River Trent and four major tributaries. *Regul. Rivers*, **4**, 43–57.

JOYCE, C. (1990). Lead poisoning lasts beyond childhood. *New Scientist*, 13 January, 26.

KARÅS, P., NEUMAN, E. and SANDSTRÖM, O. (1991). Effects of a pulp mill effluent on the population dynamics of perch *Perca fluviatilis*. *Can. J. Fish. Aquat. Sci.*, **48**, 28–34.

KAZAN, J., SINNOTT, D. and KAZAN, E.D.O. (1987). The toxicity of pyrene in the fish *Pimephales promelas*: synergism by piperonyl butoxide and by ultraviolet light. *Chemosphere*, **16**, 10–12.

KELLY, M. (1988). *Mining and the freshwater environment*. Elsevier, London.

KELLY, M.G., GIRTON, C. and WHITTON, B.A. (1987). Use of moss-bags for monitoring heavy metals in rivers. *Water Res.*, **21**, 1429–1435.

KHANGAROT, B.S. and RAY, P.K. (1987). Studies on the acute toxicity of copper and mercury alone and in combination to the common guppy *Poecilia reticulata* (Peters). *Arch. Hydrobiol.*, **110**, 303–314.

KIHLSTRÖM, J.E., OLSSON, M., JENSEN, S., JOHANSSON, J., AHLBOM, J. and BERGMAN, Å. (1992). Effects of PCB and different fractions of PCB on the reproduction of the mink (*Mustela vison*). *Ambio*, **21**, 563–569.

KING, J.M. and COLEY, K.S. (1985). Toxicity of aqueous extracts of natural and synthetic oils to three species of *Lemna*. In Bahner, R.C. and Hansen, D.J. (eds) *Aquatic toxicology and hazard assessment*. American Society for Testing and Materials, Philadelphia.

KINNERSLEY, D. (1994). *Coming clean: the politics of water and the environment*. Penguin, London.

KLAMER, J., LAANE, R.W.P.M. and MARQUENIE, J.M. (1991). Sources and fate of PCBs in the North Sea: a review of available data. *Wat. Sci. Technol.*, **24**, 77–85.

KLEIN, L. (1962). *River pollution II. Causes and effects*. Butterworths, London.

KLEINMANN, R.L.P. and HEDIN, R. (1990). Biological treatment of minewater: an update. In Chalkley, M.E., Conrad, B.R., Lakshmanan, V.I. and Wheeland, K.G. (eds) *Tailings and effluent management*. Pergamon, New York.

KRAAK, M.H.S., SCHOLTEN, M.C.T., PETERS, W.H.M. and DE KOCK, W.C. (1991). Biomonitoring of heavy metals in the western European Rivers Rhine and Meuse using the freshwater mussel *Dreissena polymorpha*. *Environ. Pollut.*, **74**, 101–114.

KRAMER, K.J.M. and BOTTERWEG, J. (1991). Aquatic biological early warning systems: an overview. In Jeffrey, D.W. and Madden, B. (eds) *Bioindicators and environmental management*, pp. 95–126. Academic Press, London.

KRAMER, K.J.M., JENNER, H.A. and DE ZWART, D. (1989). The valve movement response of mussels: a tool in biological monitoring. *Hydrobiologia*, **188/189**, 433–443.

KRATZ, K.W., COOPER, S.D. and MELACK, J.M. (1994). Effects of single and repeated experimental acid pulses on invertebrates in a high altitude Sierra Nevada stream. *Freshwat. Biol.*, **32**, 161–183.

KUBIAK, T.J., HARRIS, H.J., SMITH, L.M., STALLING, D.L., SCHWARTZ, T.R., TRICK, J.A., SILEO, L., DOCHERTY, D.E. and ERDMAN, T.C. (1989). Microcontaminants and reproductive success of the Forster's Tern on Green Bay, Lake Michigan – 1983. *Arch. Environ. Contam. Toxicol.*, **18**, 706–727.

LALIBERTÉ, G., PROULX, D., DE PAUW, N. and DE LA NOÜE, J. (1994). Algal technology in wastewater treatment. *Ergebn. Limnol.*, **42**, 283–302.

LAMMENS, E.H.R.R. (1989). Causes and consequences of the success of bream in Dutch eutrophic lakes. *Hydrobiol. Bull.*, **23**, 11–18.

LANGFORD, T.E. (1970). The temperature of a British river upstream and downstream of a heated discharge from a power station. *Hydrobiologia*, **35**, 353–375.

LANGFORD, T.E. (1975). The emergence of insects from a British river, warmed by power station cooling water. Part II. The emergence patterns of some species of Ephemeroptera, Trichoptera and Megaloptera in relation to water temperature and river flow, upstream and downstream of the cooling-water outfalls. *Hydrobiologia*, **47**, 91–133.

LANGFORD, T.E. (1983). *Electricity generation and the ecology of natural waters*. Liverpool University Press, Liverpool.

LARSSON, A., HAUX, C. and SJÖBECK, M.-L. (1985). Fish physiology and metal pollution: results and experiences from laboratory and field studies. *Ecotoxicol. Environ. Safety*, **9**, 250–281.

LARSSON, P. (1984a). Uptake of sediment-released PCBs by the eel *Anguilla anguilla* in static model systems. *Ecol. Bull.*, **36**, 62–67.

LARSSON, P. (1984b). Transport of PCBs from aquatic to terrestrial environments by emerging chironomids. *Environ. Pollut. (A)*, **34**, 283–289.

LARSSON, P. (1985). Contaminated sediments of lakes and oceans as sources of chlorinated hydrocarbons for release to water and atmosphere. *Nature*, **317**, 347–349.

LARSSON, P. (1986). Zooplankton and fish accumulate chlorinated hydrocarbons from contaminated sediments. *Can. J. Fish Aquat. Sci.*, **43**, 1463–1466.

LAURÉN, D.J. and MCDONALD, D.G. (1987a). Acclimation to copper by rainbow trout, *Salmo gairdneri*: physiology. *Can J. Fish. Aquat. Sci.*, **44**, 99–104.

LAURÉN, D.J. and MCDONALD, D.G. (1987b). Acclimation to copper by rainbow trout, *Salmo gairdneri*: biochemistry. *Can. J. Fish. Aquat. Sci.*, **44**, 105–111.

LAURIDSEN, T.L., JEPPESEN, E. and SØNDERGAARD, M. (1994). Colonization and succession of submerged macrophytes in shallow Lake Vaeng during the first five years following fish manipulation. *Hydrobiologia*, **275/276**, 233–242.

LAWS, E.A. (1993). *Aquatic pollution*. Wiley, New York.

LAWTON, L.A. and CODD, G.A. (1991). Cyanobacterial (blue-green algal) toxins and their significance in UK and European waters. *J. IWEM*, **5**, 460–465.

LEAN, G., HINRICHSEN, D. and MARKHAM, A. (1990). *Atlas of the environment.* Hutchinson, London.

LEARNER, M.A., DENSEM, J.W. and ILES, T.C. (1983). A comparison of some classification methods used to determine benthic macroinvertebrate species associations in river survey work based on data obtained from the River Eley, South Wales. *Freshwat. Biol.*, **13**, 13–36.

LEE, G.F., RAST, W. and JONES, R.A. (1978). Eutrophication of water bodies: insights for an age-old problem. *Environ. Sci. Technol.*, **12**, 900–908.

LEIVESTAD, H., JENSEN, E., KJARTANSSON, H. and XINGFU, L. (1987). Aqueous speciation of aluminium and toxic effects on Atlantic salmon. *Annl. Soc. r. zool. Belg.*, **117**, 387–398.

LELEK, A. and KÖHLER, C. (1990). Restoration of fish communities of the Rhine River two years after a heavy pollution wave. *Regul. Rivers*, **5**, 57–66.

LEMONICK, M.D. (1988). Nightmare on the Monongahela. *Time* 18 January, 34–35.

LETTERMAN, R.D. and MITSCH, W.J. (1978). Impact of mine drainage on a mountain stream in Pennsylvania. *Environ. Pollut.*, **17**, 53–73.

LEVIN, R. (1987). Reducing lead in drinking water: a benefit analysis. Office of Policy Planning and Evaluation, US Environmental Protection Agency, Report no. EPA-23-09-86-019, Washington, DC.

LEWIS, M.A. (1993). Freshwater primary producers. In Calow, P. (ed.) *Handbook of ecotoxicology, vol. 1*, pp. 28–50. Blackwell, Oxford.

LIERE, L. van, PARMA, S. and GULATI, R.D. (1992). Working group Water Quality Research Loosdrecht Lakes: its history, structure, research programme, and some results. *Hydrobiologia*, **233**, 1–9.

LIKENS, G.E., BORMANN, F.H., JOHNSON, N.M., FISHER, D.W. and PIERCE, R.S. (1970). Effects of forest cutting and herbicide treatment on nutrient budgets in the Hubbard Brook Watershed ecosystem *Ecol. Monogr.*, **40**, 23–47.

LILIUS, H., ISOMAA, B. and HOLMSTRÖM, T. (1994). A comparison of the toxicity of 50 reference chemicals to freshly isolated rainbow trout hepatocytes and *Daphnia magna*. *Aquat. Toxicol.*, **30**, 47–60.

LITTEN, S., MEAD, B. and HASSETT, J. (1993). Application of passive samplers (PISCES) to locating a source of PCBs in the Black River, New York. *Environ. Toxicol. Chem.*, **12**, 639–647.

LLOYD, R. (1960). The toxicity of zinc sulphate to rainbow trout. *Ann. appl. Biol.*, **48**, 84–94.

LLOYD, R. (1992). *Pollution and freshwater fish.* Fishing News Books, Oxford.

LLOYD, R. and ORR, L.D. (1969). The diuretic response by rainbow trout to sublethal concentrations of ammonia. *Water Res.*, **3**, 335–344.

LOCKHART, W.L., WAGEMANN, R., TRACEY, B., SUTHERLAND, D. and THOMAS, D.J. (1992). Presence and implications of chemical contaminants in the freshwaters of the Canadian Arctic. *Sci. Total Environ.*, **122**, 165–243.

LODENIUS, M., SEPPÄNEN, A. and HERRANEN, M. (1983). Accumulation of mercury in fish and man from reservoirs in northern Finland. *Water Air Soil Pollut.*, **19**, 237–246.

LOGANATHAN, B.G., TANABE, S., HIDAKA, Y., KAWANO, M., HIDAKA, H. and TATSUKAWA, R. (1993). Temporal trends of persistent organochlorine residues in human adipose tissue from Japan, 1928–85. *Environ. Pollut.*, **81**, 31–39.

LONG, S.P. and MASON, C.F. (1983). *Saltmarsh ecology.* Blackie, Glasgow.

LOPEZ, J., VAZQUEZ, M.D. and CARBALLEIRA, A. (1994). Stress responses and metal exchange kinetics following transplant of the aquatic moss *Fontinalis antipyretica. Freshwat. Biol.*, **32**, 185–198.

LOVETT DOUST, J., SCHMIDT, M. and LOVETT DOUST, L. (1994). Biological assessment of aquatic pollution; a review, with emphasis on plants as biomonitors. *Biol. Rev.*, **69**, 147–186.

LOVETT DOUST, L., LOVETT DOUST, J. and SCHMIDT, M. (1993). In praise of plants as biomonitors – send in the clones. *Functional Ecology*, **7**, 754–758.

LUDWIG, J.P., GIESY, J.P., SUMMER, C.L., BOWERMAN, W., AULERICH, R., BURSIAN, S., AUMAN, H.J., JONES, P.D., WILLIAMS, L.L., TILLITT, D.E. and GILBERTSON, M. (1993). A comparison of water quality criteria for the Great Lakes based on human and wildlife health. *J. Great Lakes Res.*, **19**, 789–807.

LÜKEWILLE, A. (1994). Billion dollar problem, billion dollar solution? Transboundary air pollution calls for transboundary solutions. In Steinberg, C.E.W. and Wright, R.F. (eds) *Acidification of freshwater ecosystems: implications for the future*, pp. 17–31. Wiley, Chichester.

LUND, J.W.G. (1978). Experiments with lake phytoplankton in large enclosures. *46th Annual Report of the Freshwater Biological Association*, pp. 32–39.

LUNDGREN, L. (1993). Alternative approaches to evaluating the impact of radionuclide release events – Chernobyl from the Swedish perspective. *Ambio*, **22**, 369–377.

LYNCH, J.A. and CORBETT, E.S. (1980). Acid precipitation – a threat to aquatic ecosystems. *Fisheries*, **5**, 8–12.

LYNCH, M. and SHAPIRO, J. (1980). Predation, enrichment and phytoplankton community structure. *Limnol. Oceanogr.*, **26**, 86–102.

MAC, M.J., SCHWARTZ, T.R., EDSALL, C.C. and FRANK, A.M. (1993). Polychlorinated biphenyls in Great Lakes trout and their eggs: relations to survival and congener composition 1979–1988. *J. Great Lakes Res.*, **19**, 752–765.

MACAN, T.T. (1959). *A guide to freshwater invertebrate animals.* Longman, London.

MACDONALD, S.M. and MASON, C.F. (1983). Some factors influencing the distribution of otters (*Lutra lutra*). *Mammal Rev.*, **13**, 1–10.

MACDONALD, S.M. and MASON, C.F. (1994). Status and conservation needs of the otter *(Lutra lutra)* in the western Palearctic. Council of Europe, Nature and Environment no. 67, pp. 1–54. Strasbourg.

MAGURRAN, A.E. (1988). *Ecological diversity and its measurement.* Croom Helm, London.

MAHANEY, P.A. (1994). Effects of freshwater petroleum contamination on amphibian hatching and metamorphosis. *Environ. Toxicol. Chem.*, **13**, 259–265.

MAITLAND, P.S. (1995). Ecological impact of angling. In Harper, D.M. and Ferguson, A.J.D. (eds) *The ecological basis for river management*, pp. 443–452. Wiley, Chichester.

MALINS, D.C. and OSTRANDER, G.K. (eds) (1994). *Aquatic toxicology: molecular, biochemical and cellular perspectives*. Lewis, Boca Raton.

MALLE, K.-G. (1994). Accidental spills – frequency, importance, control and countermeasures. *Wat. Sci. Technol.*, **29**, 149–163.

MALTBY, L. (1992). The use of physiological energetics of *Gammarus pulex* to assess toxicity: a study using artificial streams. *Environ. Toxicol. Chem.*, **11**, 79–85.

MALTBY, L. and CALOW, P. (1995). *Methods in ecotoxicology*. Blackwell, Oxford.

MANCE, G. (1987). *Pollution threat of heavy metals in aquatic environments*. Elsevier, London.

MANLY, B.F.J. (1986). *Multivariate statistical methods: a primer*. Chapman and Hall, London.

MANN, R.H.K. (1995). Natural factors influencing recruitment success in coarse fish populations. In Harper, D.M. and Ferguson, A.J.D. (eds) *The ecological basis for river management*, pp. 339–348. Wiley, Chichester.

MANTLE, A. and MANTLE, G. (1992). Impact of low flows on chalk streams and water meadows. *British Wildlife*, **4**(1), 4–14.

MARCHANT, J.H., HUDSON, R., CARTER, S.P. and WHITTINGTON, P. (1990). *Population trends in British breeding birds*. BTO, Tring.

MARSDEN, M.W. (1989). Lake restoration by reducing external phosphorus loading: the influence of sediment phosphorus release. *Freshwat. Biol.*, **21**, 139–162.

MASON, C.F. (1977a). Populations and production of benthic animals in two contrasting shallow lakes in Norfolk. *J. anim. Ecol.*, **46**, 147–172.

MASON, C.F. (1977b). The performance of a diversity index in describing the zoobenthos of two lakes. *J. appl. Ecol.*, **14**, 363–367.

MASON, C.F. (1978). Artificial oases in a lacustrine desert. *Oecologia*, **36**, 93–102.

MASON, C.F. (1987). A survey of mercury, lead and cadmium in muscle of British freshwater fish. *Chemosphere*, **16**, 901–906.

MASON, C.F. (1989). Water pollution and otter distribution: a review. *Lutra*, **32**, 97–131.

MASON, C.F. (1990). Biological aspects of freshwater pollution. In Harrison, R.M. (ed.) *Pollution: causes, effects and control*, pp. 99–125. Royal Society of Chemistry, London.

MASON, C.F. (1995). River management and mammal populations. In Harper, D.M. and Ferguson, A.J.D. (eds) *The ecological basis for river management*, pp. 289–305. Wiley, Chichester.

MASON, C.F. and ABDUL-HUSSEIN, M.M. (1991). Population dynamics and production of *Daphnia hyalina* and *Bosmina longirostris* in a shallow, eutrophic reservoir. *Freshwat. Biol.*, **25**, 243–260.

MASON, C.F. and BARAK, N.A.-E. (1990). A catchment survey for heavy metals using the eel (*Anguilla anguilla*). *Chemosphere*, **21**, 695–699.

MASON, C.F. and BRYANT, R.J. (1975). Changes in the ecology of the Norfolk Broads. *Freshwat. Biol.*, **5**, 257–270.

MASON, C.F. and MACDONALD, S.M. (1986). *Otters: ecology and conservation*. Cambridge University Press, Cambridge.

MASON, C.F. and MACDONALD, S.M. (1987). Acidification and otter (*Lutra lutra*) distribution on a British river. *Mammalia*, **51**, 81–87.

MASON, C.F. and MACDONALD, S.M. (1988a). Radioactivity in otter scats in Britain following the Chernobyl accident. *Water Air Soil Pollut.*, **37**, 131–137.

MASON, C.F. and MACDONALD, S.M. (1988b). Metal contamination in mosses and otter distribution in a rural Welsh river receiving mine drainage. *Chemosphere*, **17**, 1159–1166.

MASON, C.F. and MACDONALD, S.M. (1989). Acidification and otter (*Lutra lutra*) distribution in Scotland. *Water Air Soil Pollut.*, **43**, 365–374.

MASON, C.F. and MACDONALD, S.M. (1990). The riparian woody plant community of regulated rivers in eastern England. *Regul. Rivers*, **5**, 159–166.

MASON, C.F. and MACDONALD, S.M. (1993). Impact of organochlorine pesticide residues and PCBs on otters (*Lutra lutra*) in eastern England. *Sci. Total Environ.*, **138**, 147–160.

MASON, C.F., MACDONALD, S.M., BLAND, H.C. and RATFORD, J. (1992). Organochlorine pesticide and PCB contents in otter (*Lutra lutra*) scats from western Scotland. *Water Air Soil Pollut.*, **64**, 617–626.

MASON, C.F., MACDONALD, S.M. and HUSSEY, A. (1984). Structure, management and conservation value of the riparian woody plant community. *Biol. Conserv.*, **29**, 201–216.

MASON, J. (1989). The causes and consequences of surface water acidification. In Morris, R., Taylor, E.W., Brown, D.J.A. and Brown, J.A. (eds) *Acid toxicity and aquatic animals*, pp. 1–12. Cambridge University Press, Cambridge.

MAUND, S.J., PEITHER, A., TAYLOR, E.J., JÜTTNER, I., BEYERLE-PFNÜR, R., LAY, J.P. and PASCOE, D. (1992b). Toxicity of lindane to freshwater insect larvae in compartments of an experimental pond. *Ecotoxicol. Environ. Safety*, **23**, 76–88.

MAUND, S.J., TAYLOR, E.J. and PASCOE, D. (1992a). Population responses of the freshwater amphipod crustacean *Gammarus pulex* (L.) to copper. *Freshwat. Biol.*, **28**, 29–36.

MAYER, F.L. and ELLERSIECK, M.R. (1988). Experiences with single- species tests for acute toxic effects on freshwater animals. *Ambio*, **17**, 367–375.

MCCAHON, C.P. and PASCOE, D. (1988). Use of *Gammarus pulex* (L.) in safety evaluation tests: culture and selection of sensitive life stages. *Ecotoxicol. Environ. Safety*, **15**, 245–252.

MCCAHON, C.P. and PASCOE, D. (1990). Episodic pollution: causes, toxicological effects and ecological significance. *Functional Ecology*, **4**, 375–383.

MCCAHON, C.P. and POULTON, M.J. (1991). Lethal and sub-lethal effects of acid, aluminium and lime on *Gammarus pulex* during repeated simulated episodes in a Welsh stream. *Freshwat. Biol.*, **25**, 169–178.

MCCARTHY, J.F. and SHUGART, L.R. (1990). Biological markers for environmental contamination. In McCarthy, J.F. and Shugart, L.R. (eds) *Biomarkers of environmental contamination*, pp. 3–14. Lewis, Boca Raton.

MCFARLANE, R.W., MOORE, B.C. and WILLIAMS, S.E. (1976). Thermal tolerance of stream cyprinid minnows. In Esch, G.W. and McFarlane, R.W. (eds) *Thermal ecology II*, pp. 141–144. Technical Information Service, Springfield, VA.

MCLEAN, R.O. and JONES, A.K. (1975). Studies of tolerance to heavy metals in the flora of the rivers Ystwyth and Clarach, Wales. *Freshwat. Biol.*, **5**, 431–444.

MCLOUGHLIN, J. and BELLINGER, E.G. (1993). *Environmental pollution control*. Graham and Trotman, London.

MCMAHON, B.R. and STUART, S.A. (1989). The physiological problems of crayfish in acid waters. In Morris, R., Taylor, E.W., Brown, D.J.A. and Brown, J.A. (eds) *Acid toxicity and aquatic animals*, pp. 171–199. Cambridge University Press, Cambridge.

MEIER, P.G., PENROSE, D.L. and POLAK, L. (1979). The rate of colonization by macro-invertebrates on artificial substrate samplers. *Freshwat. Biol.*, **9**, 381–392.

MEIJER, M.L., JEPPESEN, E., van DONK, E. *et al.* (1994). Long-term responses to fish-stock reduction in small, shallow lakes: interpretation of five-year results of four biomanipulation cases in the Netherlands and Denmark. *Hydrobiologia*, **275/276**, 457–466.

MEIJER, M.L., RAAT, A.J.P. and DOEF, R.W. (1989). Restoration by biomanipulation of Lake Bleiswijkse Zoom (The Netherlands): first results. *Hydrobiol. Bull.*, **23**, 49–57.

MERRITT, A. (1994). *Wetlands, industry and wildlife: a manual of principles and practices*. The Wildfowl and Wetlands Trust, Slimbridge.

MES, J. (1990). Trends in the levels of some chlorinated hydrocarbon residues in adipose tissue of Canadians. *Environ. Pollut.*, **65**, 269–278.

METCALFE, C.D. (ed.) (1994). Chemical contaminants and fish tumours. *Sci. Total Environ.*, **154**, 1–167.

METCALFE, J.L. (1989). Biological water quality assessment of running waters based on macroinvertebrate communities: history and present status in Europe. *Environ. Pollut.*, **60**, 101–139.

MIERLE, G., CLARK, K. and FRANCE, R. (1986). The impact of acidification on aquatic biota in North America: a comparison of field and laboratory results. *Water Air Soil Pollut.*, **31**, 593–604.

MILLEMAN, R.E., BIRGE, W.J., BLACK, J.A., CUSHMAN, R.M., DANIELS, K.L., FRANCO, P.J., GIDDINGS, J.M., MCCARTHY, J.F. and STEWART, A.J. (1984). Comparative acute toxicity to aquatic organisms of components of coal derived synthetic fuels. *Trans. Am. Fish. Soc.*, **113**, 74–85.

MILLER, D.G. and TETLOW, J.A. (1989). The assessment and management of surface water quality in England and Wales. *Regul. Rivers*, **4**, 129–137.

MINISTRY OF AGRICULTURE, FISHERIES AND FOOD (MAFF) (1994). *Radionuclides in foods.* HMSO, London.

MOORE, M.J. and MYERS, M.S. (1994). Pathobiology of chemical-associated neoplasia in fish. In Malins, D.C. and Ostrander, G.K. (eds) *Aquatic toxicology*, pp. 327–386. Lewis, Boca Raton.

MOORE, M.M. (1986). Lead in humans. In Lansdown, R. and Yule, W. (eds) *The lead debate: the environment, toxicology and child health*, pp. 54–95. Croom Helm, London.

MORGAN, E.L., YOUNG, R.C. and WRIGHT, J.R. (1988). Developing portable computer-automated biomonitoring for a regional water quality surveillance network. In Gruber, D. and Diamond, J. (eds) *Automated biomonitoring: living sensors as environmental monitors*, pp. 127–141. Ellis Horwood, Chichester.

MORGAN, W.S.G. and KUHN, P.C. (1988). Effluent discharge control at a South African industrial site utilizing continuous automatic biological surveillance techniques. In Gruber, D. and Diamond, J. (eds) *Automated biomonitoring: living sensors as environmental monitors*, pp. 91–103. Ellis Horwood, Chichester.

MORRIS, R., TAYLOR, E.W., BROWN, D.J.A. and BROWN, J.A. (eds) (1989). *Acid toxicity and aquatic animals.* Cambridge University Press, Cambridge.

MOSHIRI, G.A. (1993). *Constructed wetlands for water quality improvement.* Lewis, Boca Raton.

MOSS, B. (1972). Studies on Gull Lake, Michigan. II. Eutrophication – evidence and prognosis. *Freshwat. Biol.*, **2**, 309–320.

MOSS, B. (1980). Further studies on the palaeolimnology and changes in the phosphorus budget of Barton Broad, Norfolk. *Freshwat. Biol.*, **10**, 261–279.

MOSS, B. (1983). The Norfolk Broadland: experiments in the restoration of a complex wetland. *Biol. Rev.*, **58**, 521–561.

MOSS, B. (1992). The scope for biomanipulation for improving water quality. In Sutcliffe, D.W. and Jones, J.G. (eds) *Eutrophication: research and application to water supply*, pp. 73–81. Freshwater Biological Association, Ambleside.

MOSS, B., BALLS, H., IRVINE, K. and STANSFIELD, J. (1986). Restoration of two lowland lakes by isolation from nutrient-rich water sources with and without removal of sediment. *J. appl. Ecol.*, **23**, 391–414.

MOSS, B. and LEAH, R.T. (1982). Changes in the ecosystem of a guanotrophic and brackish shallow lake in eastern England: potential problems in its restoration. *Int. Rev. ges. Hydrobiol.*, **67**, 625–639.

MOUVET, C. (1985). The use of aquatic bryophytes to monitor heavy metals pollution of freshwaters as illustrated by case studies. *Verh. int. verein. Limnol.*, **22**, 2420–2425.

MOUVET, C., MORHAIN, E., SUTTER, C. and COUTURIEUX, N. (1993). Aquatic mosses for the detection and follow-up of accidental discharges in surface waters. *Water Air Soil Pollut.*, **66**, 333–348.

MUDGE, G.P. (1983). The incidence and significance of ingested lead pellet poisoning in British wildfowl. *Biol. Conserv.*, **27**, 333–372.

MUIRHEAD-THOMSON, R.C. (1987). *Pesticide impact on stream fauna with special reference to macroinvertebrates*. Cambridge University Press, Cambridge.

MÜLLER, H. (1987). Hydrocarbons in the freshwater environment; a literature review. *Arch. Hydrobiol.*, **24**, 1–69.

MÜLLER, R. (1992). Bacterial degradation of xenobiotics. In Fry, J.C., Gadd, G.M., Herbert, R.A., Jones, C.W. and Watson-Craik, I.A. (eds) *Microbial control of pollution*, pp. 35–57. Cambridge University Press, Cambridge.

MUNIZ, J.P. (1991). Freshwater acidification: its effects on species and communities of freshwater microbes, plants and animals. *Proc. Roy. Soc. Edinburgh*, **97B**, 227–254.

MURDOCH, M.H. and HEBERT, P.D.N. (1994). Mitochondrial DNA diversity of brown bullhead from contaminated and relatively pristine sites in the Great Lakes. *Environ. Toxicol. Chem.*, **13**, 1281–1289.

MURPHY, P.M. (1978). The temporal variability in biotic indices. *Environ. Pollut.*, **17**, 227–236.

MYERS, M.S., STEHR, C.S., OLSON, O.P., JOHNSON, L.L., MCCAIN, B.B., CHAN, S.-L. and VARANASI, U. (1994). Relationships between toxicopathic hepatic lesions and exposure to chemical contaminants in English sole (*Pleuronectes vetulus*), starry flounder (*Platichthys stellatus*) and white croaker (*Genyohemus lineatus*) from selected marine sites on the Pacific coast, USA. *Env. Health Perspect.*, **102**, 200–215.

NALEWAJKO, C. and DUNSTALL, T.G. (1994). Miscellaneous pollutants: thermal effluents, halogens, organochlorines, radionuclides. *Ergeb. Limnol.*, **42**, 235–265.

NASU, Y., KUGIMOTO, M., TANAKA, O. and TAKIMOTO, A. (1984). *Lemna* as an indicator of water pollution and the absorption of heavy metals by *Lemna*. In Pascoe, D. and Edwards, R.W. (eds) *Freshwater biological monitoring*, pp. 113–120. Pergamon, Oxford.

NATIONAL RESEARCH COUNCIL (1987). Committee on biological markers. *Environ. Health Persp.*, **74**, 3–9.

NATIONAL RIVERS AUTHORITY (1991). *The quality of rivers, canals and estuaries in England and Wales*. NRA, Bristol.

NATIONAL RIVERS AUTHORITY (1993). *River corridor manual for surveyors*. NRA, Bristol.

NAYLOR, C., PINDAR, L. and CALOW, P. (1990). Inter- and intraspecific variation in sensitivity to toxins: the effects of acidity and zinc on the freshwater crustaceans *Asellus aquaticus* (L.) and *Gammarus pulex* (L.). *Water Res.*, **24**, 757–762.

NEEDLEMAN, H.L., SCHELL, A., BELLINGER, D., LEVITON, A. and ALLRED, E.N. (1990). The long-term effects of exposure to low doses of lead in childhood: an 11-year follow-up report. *New Eng. J. Med.*, **322**, 83–88.

NEWBOLD, C. (1975). Herbicides in aquatic systems. *Biol. Conserv.*, **7**, 97–118.

NEWMAN, M.C. and MCINTOSH, A.W. (eds) (1992). *Metal ecotoxicology: concepts and applications*. Lewis, Boca Raton.

NEWMAN, P.J. (1988). *Classification of surface water quality: review of schemes used in EC Member States*. Heinemann, Oxford.

NICHOLLS, K.H. and HOPKINS, G.J. (1993). Recent changes in Lake Erie (North Shore) phytoplankton: cumulative impacts of phosphorus loading reductions and the zebra mussel introduction. *J. Great Lakes Res.*, **19**, 637–647.

NRIAGU, J.O. (1988). A silent epidemic of environmental metal poisoning? *Environ. Pollut.*, **50**, 139–161.

NYHOLM, N. and KÄLLQVIST, T. (1989). Methods for growth inhibition toxicity tests with freshwater algae. *Environ. Toxicol. Chem.*, **8**, 689–703.

O'IIALLORAN, J., MEYERS, A.A. and DUGGAN, P.F. (1989). Some sub-lethal effects of lead on mute swan *Cygnus olor. J. Zool. Lond.*, **218**, 627–632.

ØKLAND, J. and ØKLAND, K.A. (1980). pH level and food organisms for fish: studies in 1000 lakes in Norway. In Drablos, D. and Tollan, A. (eds) *Ecological impact of acid precipitation*, pp. 326–327. SMSF, Oslo.

OLAVESON, M.M. and NALEWAJKO, C. (1994). Acid rain and freshwater algae. *Ergebn. Limnol.*, **42**, 99–123.

OLSON, R.K. and MARSHALL, K. (eds) (1992). The role of created and natural wetlands in controlling nonpoint source pollution. *Ecol. Eng.*, **1**, 1–170.

ULSSON, M. and REUTERGÅRDH, L. (1986). DDT and PCB pollution trends in the Swedish aquatic environment. *Ambio*, **15**, 103–109.

OPENSHAW, S. (1992). Radiation and the environment; types, sources, impacts and management. In Newson, M. (ed.) *Managing the human impact on the natural environment: patterns and processes*, pp. 213–231. Belhaven Press, London.

ORGANIZATION FOR ECONOMIC COOPERATION AND DEVELOPMENT (1977). *The OECD programme for long-term transport of air pollutants. Measurements and findings*. OECD, Paris.

ORGANIZATION FOR ECONOMIC COOPERATION AND DEVELOPMENT (1982). *Eutrophication of waters: monitoring, assessment and control*. OECD, Paris.

ORIS, J.T. and GIESY, J.P. (1987). The photo-induced toxicity of polycyclic aromatic hydrocarbons to larvae of the fathead minnow (*Pimephales promelas*). *Chemosphere*, **16**, 1395–1404.

ORMEROD, S.J., BOOLE, P., MCCAHON, C.P., WEATHERLEY, N.S., PASCOE, D. and EDWARDS, R.W. (1987). Short-term experimental acidification of a Welsh stream: comparing the biological effects of hydrogen ions and aluminium. *Freshwat. Biol.*, **17**, 341–356.

ORMEROD, S.J., DONALD, A.P. and BROWN, S.J. (1989). The influence of plantation forestry on the pH and aluminium concentration of upland Welsh streams: a re-examination. *Environ. Pollut.*, **62**, 47–62.

ORMEROD, S.J. and EDWARDS, R.W. (1985). Stream acidity in some areas of Wales in relation to historical trends in afforestation and the usage of agricultural limestone. *J. Environ. Management*, **20**, 189–197.

ORMEROD, S.J. and TYLER, S.J. (1993). Birds as indicators of change in water quality. In Furness, R.W. and Greenwood, J.J.D. (eds) *Birds as monitors of environmental change*, pp. 179–216. Chapman and Hall, London.

ORMEROD, S.J. and WADE, K.R. (1990). The role of acidity in the ecology of Welsh lakes and streams. In Edwards, R.W., Gee, A.S. and Stoner, J.H. (eds) *Acid waters in Wales*, pp. 93–119. Kluwer, Dordrecht.

ORMEROD, S.J., WADE, K.R. and GEE, A.S. (1987). Macro-floral assemblages in upland Welsh streams in relation to acidity, and their importance to invertebrates. *Freshwat. Biol.*, **18**, 545–557.

ORMEROD, S.J., WEATHERLEY, N.S. and GEE, A.S. (1990). Modelling the ecological impact of changing acidity in Welsh streams. In Edwards, R.W., Gee, A.S. and Stoner, J.H. (eds) *Acid waters in Wales*, pp. 279–298. Kluwer, Dordrecht.

OSBORNE, P.L. (1981). Phosphorus and nitrogen budgets of Barton Broad and predicted effects of a reduction in nutrient loading on phytoplankton biomass in Barton, Sutton and Stalham Broads, Norfolk, United Kingdom. *Int. Rev. ges. Hydrobiol.*, **66**, 171–202.

OSKAM, G. and BREEMAN, L. van (1992). Management of Biesbosch Reservoirs for quality control with special reference to eutrophication. In Sutcliffe, D.W. and Jones, J.G. (eds) *Eutrophication: research and application to water supply*, pp. 197–213. Freshwater Biological Association, Ambleside.

OSTENDORP, W. (1989). 'Die-back' of reeds in Europe – a critical review of literature. *Aquat. Bot.*, **35**, 5–26.

OWENS, M., ATKINSON-WILLES, G.L. and SALMON, D.G. (1986). *Wildfowl in Great Britain*, 2nd edn. Cambridge University Press, Cambridge.

PAASIVIRTA, J. (1991). *Chemical ecotoxicology*. Lewis, Boca Raton.

PAIN, S. (1993). The two faces of the Exxon disaster. *New Scientist* 22 May, 11–13.

PALMATEER, G.A., DUTKA, B.J., JANZEN, E.M., MEISSNER, S.M. and SAKELLARIS, M.G. (1991). Coliphage and bacteriophage as indicators of recreational water quality. *Water Res.*, **25**, 355–357.

PARNELL, J.F., SHIELDS, M.A. and FRIERSON, D. (1985). Hatching success of brown pelicans (*Pelicanus occidentalis*) eggs after contamination with oil. *Colon. Waterbirds*, **7**, 22–24.

PASCOE, D. and BEATTIE, J.H. (1979). Resistance to cadmium by pretreated rainbow trout alevins. *J. Fish Biol.*, **14**, 303–308.

PAYNE, A.G. (1975). Responses of the three test algae of the Algal Assay Procedure: Bottle Test. *Water Res.*, **9**, 437–445.

PEAKALL, D. (1992). *Animal biomarkers as pollution indicators*. Chapman and Hall, London.

PEARCE, F. (1987) *Acid rain*. Penguin, London.

PEARCE, F. (1993). The scandal of Siberia. *New Scientist* 27 November, 28–33.

PECKARSKY, B.L. (1984). Sampling the stream benthos. In Downing, J.A. and Rigler, F.H. (eds) *A manual on methods for the assessment of secondary productivity in freshwaters*, pp. 131–160. Blackwell, Oxford.

PERRINS, C.M. and SEARS, J. (1991). Collisions with overhead wires as a cause of mortality in Mute Swans *Cygnus olor. Wildfowl*, **42**, 5–11.

PERROW, M.R., MOSS, B. and STANSFIELD, J. (1994). Trophic interactions in a shallow lake following a reduction in nutrient loading: a long-term study. *Hydrobiologia*, **275/276**, 43–52.

PERRY, J.A., TROELSTRUP, N.H., NEWSOM, N. and SHELLEY, B. (1987). Results of a recent whole ecosystem manipulation: the search for generality. *Water Sci. Technol.*, **19**, 55–72.

PERSOONE, G. and JANSSEN, C.R. (1993). Freshwater invertebrate toxicity tests. In Calow, P. (ed.) *Handbook of ecotoxicology*, vol.1, pp. 51–65. Blackwell, Oxford.

PETERSEN, R.C., LANDNER, L. and BLANCK, H. (1986). Assessment of the impact of the Chernobyl reactor accident on the biota of Swedish streams and lakes. *Ambio*, **15**, 327–331.

PETTS, G.E. (1984). *Impounded rivers: perspectives for ecological management*. Wiley, Chichester.

PETTS, G.E. (1994). Large-scale river regulation. In Roberts, N. (ed.) *The changing global environment*, pp. 262–284. Blackwell, Oxford.

PETTS, G.E. and MADDOCK, I. (1994). Flow allocation for in-river needs. In Calow, P. and Petts, G.E. (eds) *The rivers handbook, vol. 2*, pp. 289–307. Blackwell, Oxford.

PETTS, G., MADDOCK, I., BICKERTON, M. and FERGUSON, A.J.D. (1995). Linking hydrology and ecology: the scientific basis for river management. In Harper, D.M. and Ferguson, A.J.D. (eds) *The ecological basis for river management*, pp. 1–16. Wiley, Chichester.

PFEIFFER, M.H. and FESTA, J.P. (1980). Acidity status of lakes in the Adirondack Region of New York in relation to fish resources. NYS-DEC Publ. FW-P168 (10/80).

PHILLIPS, D.J.H. (1993). Bioaccumulation. In Calow, P. (ed.) *Handbook of ecotoxicology, vol. 1*, pp. 378–396. Blackwell, Oxford.

PHILLIPS, D.J.H. and RAINBOW, P.S. (1993). *Biomonitoring of trace aquatic contaminants*. Elsevier, London.

PHILLIPS, G.L. (1992). A case study in restoration: shallow eutrophic lakes in the Norfolk Broads. In Harper, D. *Eutrophication of freshwaters*, pp. 251–278. Chapman and Hall, London.

PHILLIPS, G.L., JACKSON, R., BENNETT, C. and CHILVERS, A. (1994). The importance of sediment phosphorus release in the restoration of very shallow lakes (the Norfolk Broads, England) and implications for biomanipulation. *Hydrobiologia*, **275/276**, 445–456.

PHILLIPS, G.L. and KERRISON, P. (1991). The restoration of the Norfolk Broads: the role of biomanipulation. *Mem. Ist. ital. Idrobiol*, **48**, 75–97.

PIETERS, H. and HAGEL, P. (1992). Biomonitoring of mercury in European eel (*Anguilla anguilla* L.) in the Netherlands, compared with pike-perch (*Stizostedion lucioperca*): statistical analysis. In Vernet, J.P. (ed.) *Impact of heavy metals on the environment*, pp. 203–217. Elsevier, Amsterdam.

PILLINGER, J.M., COOPER, J.A., RIDGE, I. and BARRETT, P.R.F. (1992). Barley straw as an inhibitor of algal growth III: the role of fungal decomposition. *J. appl. Phycol.*, **4**, 353–355.

PINDER, L.C.V. (1989). Biological surveillance of chalk-streams. *Freshwater Biological Association Annual Report, 1989*, 81–92.

PINDER, L.C.V. and FARR, I.S. (1987a). Biological surveillance of water quality. 2. Temporal and spatial variation in the macroinvertebrate fauna of the River Frome, a Dorset chalk stream. *Arch. Hydrobiol.*, **109**, 321–331.

PINDER, L.C.V. and FARR, I.S. (1987b). Biological surveillance of water quality. 3. The influence of organic enrichment on the macroinvertebrate fauna of small chalk streams. *Arch. Hydrobiol.*, **109**, 619–637.

PINDER, L.C.V., LADLE, M., GLEDHILL, T., BASS, J.A.B. and MATTHEWS, A. (1987). Biological surveillance of water quality. 1. A comparison of macroinvertebrate surveillance methods in relation to assessment of water quality in a chalk stream. *Arch. Hydrobiol.*, **109**, 207–226.

PIOTROWSKI, J.K. and INSKIP, M.J. (1981). Health effects of mercury. *MARC Report 24*. Chelsea College, London.

PIPES, W.O. (1982). *Bacterial indicators of pollution*. CRC Press, Boca Raton.

POLMAN, H.J.G. and DE ZWART, D. (1994). The toxicity of organic concentrates to *Photobacterium phosphorium* of River Meuse water in the stretch between Remilly (France) and Keizersveer (The Netherlands). *Wat. Sci. Tech.*, **29**, 253–256.

POLPRASERT, C. (1989). *Organic waste recycling*. Wiley, Chichester.

PORTER, E. (1978). *Water management in England and Wales*. Cambridge University Press, Cambridge.

POULTON, M.J. and PASCOE, D. (1990). Disruption of pre-copula in *Gammarus pulex* L.: development of a behavioural bioassay for evaluating pollutant and parasite induced stress. *Chemosphere*, **20**, 403–415.

PRESCOTT, L.M., HARLEY, J.P. and KLEIN, D.A. (1993). *Microbiology*, 2nd edn. Brown, Dubuque-Iowa.

PRESTON, A. (1974). Application of critical path analysis techniques to the assessment of environmental capacity and the control of environmental waste disposal. In *Comparative studies of food and environmental contamination*, pp. 573–83. International Atomic Energy Agency, Vienna.

PRINCE, R.C. (1992). Bioremediation of oil spills with particular reference to the spill from the *Exxon Valdez*. In Fry, J.C., Gadd, G.M., Herbert, R.A., Jones, C.W. and Watson-Craik, I.A. (eds) *Microbial control of pollution*, pp. 19–34. Cambridge University Press, Cambridge.

PURDOM, C.E., HARDIMAN, P.A., BYE, V.J., ENO, N.C., TYLER, C.R. and SUMPTER, J.P. (1994). Estrogenic effects of effluents from sewage treatment works. *Chemistry and Ecology*, **8**, 275–285.

PURSEGLOVE, J. (1988). *Taming the flood*. Oxford University Press, Oxford.

RAINBOW, P.S. and DALLINGER, R. (1992). Metal uptake, regulation, and excretion in freshwater environments. In Dallinger, R. and Rainbow, P.S. (eds) *Ecotoxicology of metals in invertebrates*, pp. 119–131. Lewis, Boca Raton.

RAMADE, F. (1987) *Ecotoxicology*. Wiley, Chichester.

RASK, M. (1992). Effects of acidification and liming on fish populations in Finland. *Finnish Fish. Res.*, **13**, 107–117.

RAST, W. and HOLLAND, M. (1988). Eutrophication of lakes and reservoirs: a framework for making management decisions. *Ambio*, **17**, 2–12.

RAVEN, P.J. and GEORGE, J.J. (1989). Recovery by riffle macroinvertebrates in a river after a major accidental spillage of chlorpyrifos. *Environ. Pollut.*, **59**, 55–70.

RAVERA, O. (1989). Lake ecosystem degradation and recovery studied by the enclosure method. In Ravera, O. (ed.) *Ecological assessment of environmental degradation, pollution and recovery*, pp. 217–243. Elsevier, Amsterdam.

READ, M. (1989). Arrows v. arrogance. *BBC Wildlife*, **7**, 764–767.

REDSHAW, C.J. (1983). The effects of phosphorus control on reservoir eutrophication. Ph.D. thesis, University of Essex.

REDSHAW, C.J., MASON, C.F., HAYES, C.R. and ROBERTS, R.D. (1988) Nutrient budget for a hypertrophic reservoir. *Water Res.*, **4**, 413–419.

REEDERS, H.H., BIJ DE VAATE, A. and SLIM, F.J. (1989). The filtration rate of *Dreissena polymorpha* (Bivalvia) in three Dutch lakes with reference to biological water quality management. *Freshwat. Biol.*, **22**, 133–141.

REIMER, P. (1989). Concentrations of lead in aquatic macrophytes from Shoal Lake, Manitoba, Canada. *Environ. Pollut.*, **56**, 77–84.

REIMER, P. and DUTHIE, H.C. (1993). Concentrations of zinc and chromium in aquatic macrophytes from Sudbury and Muskoka regions of Ontario, Canada. *Environ. Pollut.*, **79**, 261–265.

RENBERG, I. and HEDBERG, T. (1982). The pH history of lakes in south-western Sweden as calculated from the subfossil diatom flora of the sediments. *Ambio*, **11**, 30–33.

RESH, V.H. and MCELRAVY, E.P. (1993). Contemporary quantitative approaches to biomonitoring using benthic macroinvertebrates. In Rosenberg, D.M. and Resh, V.H. (eds) *Freshwater biomonitoring and benthic macroinvertebrates*, pp. 159–194. Chapman and Hall, New York.

RICHARDS, W.N. and MOORE, M.R. (1982). Plumbosolvency in Scotland. The problem, remedial action taken and health benefits observed. *Proc. Ann. Conf. Amer. Water Works Assoc., Miami*, pp. 901–918.

RICHARDSON, C.J. and CRAFT, C.B. (1993). Effective phosphorus retention in wetlands: fact or fiction? In Moshiri, G.A. (ed.) *Constructed wetlands for water quality improvement*, pp. 271–282. Lewis, Boca Raton.

RIDGE, I. and BARRETT, P.R.F. (1992). Algal control with barley straw. *Aspects appl. Biol.*, **29**, 457–462.

ROBERTSON, L.A. and KUENEN, J.G. (1992). Nitrogen removal from water and waste. In Fry, J.C., Gadd, G.M., Herbert, R.A., Jones, C.W. and Watson-Craik, I.A. (eds) *Microbial control of pollution*, pp. 227–267. Cambridge University Press, Cambridge.

RODHE, W. (1969). Crystallization of eutrophication concepts in northern Europe. In *Eutrophication: causes, consequences, correctives*, pp. 50–64. National Academy of Sciences, Washington, DC.

ROSE, C. (1990). *The dirty man of Europe*. Simon and Schuster, London.

ROSENBERG, D.M. and RESH, V.H. (1982). The use of artificial substrates in the study of freshwater macroinvertebrates. In Cairns, J. (ed.) *Artificial substrates*, pp. 175–235. Ann Arbor Science Publishers, Michigan.

ROSSELAND, B.O. (1986). Ecological effects of acidification on tertiary consumers. Fish population responses. *Water Air Soil Pollut.*, **30**, 451–460.

ROSSELAND, B.O., SKOGHEIM, O.K., ABRAHAMSEN, H. and MATZOW, D. (1986). Limestone slurry reduces physiological stress and increases survival of Atlantic salmon (*Salmo salar*) in an acidic Norwegian river. *Can. J. Fish. Aquat. Sci.*, **43**, 1888–1893.

ROUND, F.E. (1991). Diatoms in river water-monitoring studies. *J. appl. Phycol.*, **3**, 129–145.

RUTT, G.P., PICKERING, T.D. and REYNOLDS, N.R.M. (1993). The impact of livestock farming on Welsh streams: the development and testing of a rapid biological method for use in the assessment and control of organic pollution from farms. *Environ. Pollut.*, **81**, 217–228.

RYDING, S-O. and RAST, W. (1989). *The control of eutrophication of lakes and reservoirs*. Parthenon, Paris.

SAFE, S. (1987). Determination of 2,3,7,8-TCDD toxic equivalence factors (TEFs). Support for the use of in vitro AHH induction assay. *Chemosphere*, **16**, 791–802.

SAFE, S. (1990). Polychlorinated biphenyls (PCBs), dibenzo-p-dioxins (PCDDs), dibenzofurans (PCDFs) and related compounds: environmental and mechanistic considerations which support the development of toxic equivalency factors (TEFs). *Crit. Rev. Toxicology*, **21**, 51–88.

SANDERS, G., JONES, K.C., HAMILTON-TAYLOR, J. and DÖRR, H. (1992). Historical inputs of polychlorinated biphenyls and other organochlorines to a dated lacustrine sediment core in rural England. *Environ. Sci. Technol.*, **26**, 1815–1821.

SÄRKKA, J., HATTULA, M.-L., PAASIVIRTA, J. and JANATUINEN, J. (1978). Mercury and chlorinated hydrocarbons in the food chain of Lake Päijänne, Finland. *Hol. Ecol.*, **1**, 326–332.

SAVCHENKO, V.K. (1995). *The ecology of the Chernobyl catastrophe.* Parthenon, London.

SAY, P.J., DIAZ, B.M. and WHITTON, B.A. (1977). Influence of zinc on lotic plants. 1. Tolerance of *Hormidium* species to zinc. *Freshwat. Biol.*, **7**, 357–376.

SAY, P.J., HARDING, J.P.C. and WHITTON, B.A. (1981). Aquatic mosses as monitors of heavy metal pollution in the River Etherow, Great Britain. *Environ. Pollut. B*, **2**, 295–307.

SAY, P.J. and WHITTON, B.A. (1984). Impact of heavy metal waste leachates on a stream in southwest France. *Conser. Recycling*, **7**, 321–327.

SCHARENBERG, W. GRAMANN, P. and PFEIFFER, W.H. (1994). Bioaccumulation of heavy metals and organochlorines in a lake ecosystem with special reference to bream (*Abramis brama* L.). *Sci. Total Environ.*, **155**, 187–197.

SCHEFFER, M., HOSPER, S.H., MEIJER, M.-L., MOSS, B. and JEPPESON, E. (1993). Alternative equilibria in shallow lakes. *TREE*, **8**, 275–279.

SCHINDLER, D.W. (1987a). Detecting ecosystem responses to anthropogenic stress. *Can. J. Fish. Aquat. Sci.*, **44**, 6–25.

SCHINDLER, D.W. (1987b). Recovery of Canadian lakes from acidification. In Barth, H. (ed.) *Reversibility of acidification*, pp. 2–13. Elsevier, London.

SCHINDLER, D.W. (1988a). Experimental studies of chemical stressors on whole lake ecosystems. *Verh. int. Verein. Limnol.*, **23**, 11–41.

SCHINDLER, D.W. (1988b). Effects of acid rain on freshwater ecosystems. *Science*, **239**, 149–157.

SCHINDLER, D.W., ARMSTRONG, F.A.J., HOLMGREN, S.K. and BRUNSKILL, G.J. (1971). Eutrophication of lake 227, Experimental Lakes Area, northwestern Ontario, by addition of phosphate and nitrate. *J. Fish. Res. Bd. Can.*, **28**, 1763–1782.

SCHINDLER, D.W., BEATTY, K.G. and FEE, E.J. (1990). Effects of climatic warming on lakes of the central boreal forest. *Science*, **250**, 967–970.

SCHINDLER, D.W. and FEE, E.J. (1973). Diurnal variation of dissolved inorganic carbon and its use in estimating primary production and CO_2 invasion in Lake 227. *J. Fish. Res. Bd. Can.*, **30**, 1501–1510.

SCHINDLER, D.W., KLING, H., SCHMIDT, R.V., PROKOPOWICH, J., FROST, V.E., REID, R.A. and CAPEL, M. (1973). Eutrophication of Lake 227 by addition of phosphate and nitrate: the second, third and fourth years of enrichment, 1970, 1971 and 1972. *J. Fish. Res. Bd. Can.*, **30**, 1415–1440.

SCHINDLER, D.W., MILLS, K.H., MALLEY, D.F., FINDLAY, D.L., SHEARER, J.A., DAVIES, I.J., TURNER, M.A., LINSEY, G.A. and CRUIKSHANK, D.R. (1985). Long-term ecosystem stress: the effects of years of experimental acidification on a small lake. *Science*, **228**, 1395–1401.

SCHÜLER, W., BRUNN, H. and MANZ, D. (1985). Pesticides and polychlorinated biphenyls in fish from the Lahn River. *Bull. Environ. Contam. Toxicol.*, **34**, 608–616.

SCHULTE-WÜLWER-LEIDIG, A. (1995). Ecological master plan for the Rhine catchment. In Harper, D.M. and Ferguson, A.J.D. (eds) *The ecological basis for river management*, pp. 505–514. Wiley, Chichester.

SCHÜPBACH, M.R. (1981). Halogenierte Kohlenwasserstoffe in der Nahrung. *Proc. Int. Conf. Chem. Environ. Man.*, pp. 105–124. Gottlieb Duttweiller Institute, Zurich.

SCULLION, J. and EDWARDS, R.W. (1980). The effects of coal industry pollutants on the macroinvertebrate fauna of a small river in the South Wales coalfield. *Fresh. Biol.*, **10**, 141–162.

SEAGER, J. and MALTBY, L. (1989). Assessing the impact of episodic pollution. *Hydrobiologia*, **188/189**, 633–640.

SEARS, J. (1989). A review of lead poisoning among the River Thames mute swan *Cygnus olor* population. *Wildfowl*, **40**, 151–152.

SELL, N.J. (1992). *Industrial pollution control*. Van Nostrand Reinhold, New York.

SHALES, S., THAKE, B.A., FRANKLAND, B., KHAN, D.H., HUTCHINSON, J.D. and MASON, C.F. (1989). Biological and ecological effects of oils. In Green, J. and Trett, M. (eds) *The fate and effects of oil in freshwater*, pp. 81–171. Applied Science Publishers, London.

SHAPIRO, J. and WRIGHT, D.I. (1984). Lake restoration by manipulation: Round Lake, Minnesota, the first two years. *Freshwat. Biol.*, **14**, 371–383.

SHARPE, R.M. and SKAKKEBAEK, N.E. (1993). Are oestrogens involved in falling sperm counts and disorders of the male reproductive tract? *The Lancet*, **341**, 1392–1395.

SHEATH, R.G., HAVAS, M., HELLEBUST, J.A. and HUTCHINSON, T.C. (1982). Effect of long-term acidification on the algal communities of tundra ponds at the Smoking Hills, N.W.T., Canada. *Can. J. Bot.*, **60**, 58–72.

SHUTES, B., ELLIS, B., REVITT, M. and BASCOMBE, A. (1992). The use of freshwater invertebrates for the assessment of metal pollution in urban receiving waters. In Dallinger, R. and Rainbow, P.S. (eds) *Ecotoxicology of metals in invertebrates*, pp. 201–222. Lewis, Boca Raton.

SIMPSON, K.W., BODE, R.W. and COLQUHOUN, J.R. (1985). The macroinvertebrate fauna of an acid-stressed headwater stream system in the Adirondack Mountains, New York. *Freshwat. Biol.*, **15**, 671–681.

SKIDMORE, J.F. (1970). Respiration and osmoregulation in rainbow trout with gills damaged by zinc sulphate. *J. exp. Biol.*, **52**, 484–494.

SLATER, D. (1992). Her Majesty's Inspectorate of Pollution's role in regulating industrial releases to the environment. In Drake, J.A.G. (ed.) *The chemical industry – friend to the environment?*, pp. 66–80. Royal Society of Chemistry, London.

SMART, G.R. (1981). Aspects of water quality producing stress in intensive fish culture. In Pickering, A.D. (ed.) *Stress and fish*, pp. 277–293. Academic Press, London.

SMITH, A.M. (1990). The ecophysiology of epilithic diatom communities of acid lakes in Galloway, southwest Scotland. *Phil. Trans. Roy. Soc. London*, **327B**, 25–30.

SMITH, D.I. (1993). Greenhouse climatic change and flood damages, the implications. *Climatic Change*, **25**, 319–333.

SMITH, M.A., GRANT, L.D. and SORS, A.I. (1989). *Lead exposure and child development: an international assessment*. Kluwer Academic, Lancaster.

SMITH, R.V. (1993). Phosphorus and nitrogen loadings to Lough Neagh and their management. In Wood, R.B. and Smith, R.V. (eds) *Lough Neagh*, pp. 149–169. Kluwer Academic, Dordrecht.

SMITH, S.R. (1994a). Effect of soil pH on availability to crops of metals in sewage sludge-treated soils. 1. Nickel, copper and zinc uptake and toxicity to ryegrass. *Environ. Pollut.*, **85**, 321–327.

SMITH, S.R. (1994b). Effect of soil pH on availability to crops of metals in sewage sludge-treated soils. II Cadmium uptake by crops and implications for human dietary intake. *Environ. Pollut.*, **86**, 5–13.

SOLBÉ, J.F. DE L.G. (1971). Aspects of the biology of the lumbricids *Eiseniella tetraedra* (Savigny) and *Dendrobaena rubida* (Savigny) f. *subrubicunda* (Eisen) in a percolating filter. *J. appl. Ecol.*, **8**, 845–867.

SOLBÉ, J.F. DE L.G. (1993). Freshwater fish. In Calow, P. (ed.) *Handbook of ecotoxicology*, pp. 66–82. Blackwell, Oxford.

SOLBÉ, J.F. DE L.G. and COOPER, V.A. (1976). Studies on the toxicity of copper sulphate to stone loach *Noemacheilus barbatulus* (L.) in hard water. *Water Res.*, **10**, 523–527.

SORENSEN, D.L., EBERL, S.G. and DICKSA, R.A. (1989). *Clostridium perfringens* as a point source indicator in non-point polluted streams. *Water Res.*, **23**, 191–197.

SPACIE, A. and HAMELINK, J.L. (1985). Bioaccumulation. In Rand, G.M. and Petrocelli, S.R. (eds) *Fundamentals of aquatic toxicology*, pp. 495–525. Hemisphere, Washington, D.C.

SPRAGUE, J.B. (1964). Lethal concentrations of copper and zinc for young Atlantic salmon. *J. Fish. Res. Bd Can.*, **21**, 17–26.

SPRAGUE, J.B. (1970). Measurement of pollutant toxicity to fish. II. Utilizing and applying bioassay results. *Water Res.*, **4**, 3–32.

SPRAGUE, J.B. (1971). Measurement of pollutant toxicity to fish. III. Sublethal effects and safe concentrations. *Water Res.*, **5**, 245–266.

SPRY, D.J. and WIENER, J.G. (1991). Metal bioavailability and toxicity to fish in low-alkalinity lakes: a critical review. *Environ. Pollut.*, **71**, 243–304.

STANSFIELD, J., MOSS, B. and IRVINE, K. (1989). The loss of submerged plants with eutrophication. III. Potential role of organochlorine pesticides: a palaeoecological study. *Freshwat. Biol.*, **22**, 109–132.

STEGEMAN, J.J. and HAHN, M.E. (1994). Biochemistry and molecular biology of monooxygenases: current perspectives on forms, functions, and regulation of cytochrome P450 in aquatic species. In Malins, D.C. and Ostrander, G.K. (eds) *Aquatic toxicology*, pp. 87–206. Lewis, Boca Raton.

STEINBERG, C.E. and HARTMANN, H.M. (1988). Planktonic bloom-forming cyanobacteria and the eutrophication of lakes and rivers. *Freshwat. Biol.*, **20**, 279–287.

STEINBERG, C.E.W. and WRIGHT, R.F. (eds) (1994). *Acidification of freshwater ecosystems: implications for the future*. Wiley, Chichester.

STEPHENSON, M. and MACKIE, G.L. (1986). Lake acidification as a limiting factor in the distribution of the freshwater amphipod *Hyalella azteca*. *Can. J. Fish. Aquat. Sci.*, **43**, 288–292.

STEPHENSON, M. and MACKIE, G.L. (1988). Multivariate analysis of correlations between environmental parameters and cadmium concentrations in *Hyalella azteca* (Crustacea: Amphipoda) from central Ontario lakes. *Can. J. Fish. Aquat. Sci.*, **45**, 1705–1710.

STOKES, P.M. (1975). Adaptations of green algae to high levels of copper and nickel in aquatic environments. In Hutchinson, T.C. (ed.) *International conference on heavy metals in the environment, Vol. 2. pathways and cycling*, pp. 137–154. Toronto.

STONER, J.H. and GEE, A.S. (1985). Effects of forestry on water quality and fish in Welsh rivers and lakes. *J. Inst. Water Eng. Sci.*, **39**, 27–45.

STONER, J.H., GEE, A.S. and WADE, K.R. (1984). The effects of acidification on the ecology of streams in the upper Tywi catchment in West Wales. *Environ. Pollut. (A)*, **35**, 125–157.

SUGIURA, K., WASHINO, T., HATTORI, M., SATO, E. and GOTO, M. (1978). Accumulation of organochlorine compounds in fishes. Differences of accumulation factors by fishes. *Chemosphere*, **9**, 359–364.

SUNS, K.R., HITCHIN, G.G. and TONER, D. (1993). Spatial and temporal trends of organochlorine contaminants in spottail shiners from selected sites in the Great Lakes (1975–1990). *J. Great Lakes Res.*, **19**, 703–714.

SUTCLIFFE, D.W. and HILDREW, A.G. (1989). Invertebrate communities in acid streams. In Morris, R., Taylor, E.W., Brown, D.J.A. and Brown, J.A. (eds) *Acid toxicity and aquatic animals*, pp. 13–29. Cambridge University Press, Cambridge.

SWAIN, W.R. (1988). Human health consequences of consumption of fish contaminated with organochlorine compounds. *Aquat. Toxicol.*, **11**, 357–377.

TANABE, S. (1988). PCB problems in the future: foresight from current knowledge. *Environ. Pollut.*, **50**, 5–28.

TANABE, S., TANAKE, H. and TATSUKAWA, R. (1984). Polychlorobiphenyls, total DDT and hexachlorohexane isomers in the western North Pacific ecosystem. *Arch. Environ. Contam. Toxicol.*, **13**, 731–738.

TAYLOR, E.J., MAUND, S.J. and PASCOE, D. (1991). Toxicity of four common pollutants to the freshwater macroinvertebrates *Chironomus riparius* Meigen (Insecta: Diptera) and *Gammarus pulex* (L.) (Crustacea: Amphipoda). *Arch. Environ. Contam. Toxicol.*, **21**, 371–376.

TAYLOR, L.R. (1978). Bates, Williams, Hutchinson – a variety of diversities. In Mound, L.A. and Waloff, N. (eds) *Insect faunas*, pp. 1–18. Blackwell, Oxford.

THOMPSON, P.-A. and RHEE, G.-Y. (1994). Phytoplankton responses to eutrophication. *Ergebn. Limnol.*, **42**, 125–166.

TIMMERMANS, K.K. (1992). Accumulation and effects of trace metals in freshwater invertebrates. In Dallinger, R. and Rainbow, P.S. (eds) *Ecotoxicology of metals in invertebrates*, pp. 134–148. Lewis, Boca Raton.

TIMMS, R.M. and MOSS, B. (1984). Prevention of growth of potentially dense phytoplankton populations by zooplankton grazing, in the presence of zooplanktivorous fish, in a shallow wetland ecosystem. *Limnol. Oceanogr.*, **29**, 472–486.

TITTIZER, T., SCHÖLL, F. and DOMMERMUTH, M. (1994). The development of the macrozoobenthos in the River Rhine in Germany during the 20th century. *Water Sci. Tech.*, **29**(3), 21–28.

TOLBA, M.R. and HOLDICH, D.M. (1981). The effect of water quality on the size and fecundity of *Asellus aquaticus* (Crustacea: Isopoda). *Aquat. Toxicol.*, **1**, 101–112.

TRAIN, R.E. (1979). *Quality criteria for water.* Castle House Publications, London.

TRANTER, M., DAVIES, T.D., WIGINGTON, P.J. and ESHLEMAN, K.N. (1994). Episodic acidification of freshwater systems in Canada – physical and geochemical processes. *Water Air Soil Pollut.*, **72**, 19–39.

TRIPP, B.W., FARRINGTON, J.W., GOLDBERG, E.D. and SERICANO, J. (1992). International mussel watch: the initial implementation phase. *Mar. Poll. Bull.*, **24**, 371–373.

TRIPPEL, E.A., ECKMANN, R. and HARTMANN, J. (1991). Potential effects of global warming on whitefish in Lake Constance, Germany. *Ambio*, **20**, 226–231.

TURNPENNY, A.W.H. (1989). Field studies on fisheries in acid waters in the United Kingdom. In Morris, R., Taylor, E.W., Brown, D.J.A. and Brown, J.A. (eds) *Acid toxicity and aquatic animals*, pp. 45–65. Cambridge University Press, Cambridge.

TURNPENNY, A.W.H. (1992). Fishery restoration after liming. In Howells, G. and Dalziel, T.R.K. (eds) *Restoring acid waters: Loch Fleet 1984–1990*, pp. 259–287. Elsevier Applied Science, London.

TURNPENNY, A.W.H., DEMPSEY, C.H., DAVIS, M.H. and FLEMING, J.M. (1988). Factors limiting fish populations in the Loch Fleet system, an acid drainage system in south-west Scotland. *J. Fish Biol.*, **32**, 101–118.

TWITCHEN, J.B. (1990). The physiological bases of resistance to low pH among aquatic insect larvae. In Mason, J.B. (ed) *The surface waters acidification programme*, pp. 413–419. Cambridge University Press, Cambridge.

TYLER, S.J. and ORMEROD, S.J. (1992). A review of the likely causal pathways relating to the reduced density of breeding dippers *Cinclus cinclus* to the acidification of upland streams. *Environ. Pollut.*, **78**, 49–55.

TYLER-JONES, R., BEATTIE, R.C. and ASTON, R.J. (1989). The effects of acid water and aluminium on the embryonic development of the common frog, *Rana temporaria*. *J. Zool., Lond.*, **219**, 355–372.

VANDERMEULEN, J.H. (1987). Toxicity and sublethal effects of petroleum hydrocarbons in freshwater biota. In Vandermeulen, J.H. and Hrudey, S.E. (eds). *Oil in freshwater: chemistry, biology, countermeasure technology*, pp. 267–303. Pergamon, New York.

VANDERMEULEN, J.H., FODA, A. and STUTTARD, C. (1985). Toxicity vs mutagenicity of some crude oils, distillates and their water-soluble fractions. *Water Res.*, **19**, 1283–1289.

VANDERMEULEN, J.H. and HRUDEY, S.E. (eds) (1987). *Oil in freshwater: chemistry, biology and countermeasure technology*. Pergamon, New York.

VAN HATTUM, B., DE VOOGT, P., VAN DEN BOSCH, L., VAN STRAALEN, N.M. and JOOSSE, E.N.G. (1989). Bioaccumulation of cadmium by the freshwater isopod *Asellus aquaticus* (L.) from aqueous and dietary sources. *Environ. Pollut.*, **62**, 129–151.

VAN HATTUM, B., KORTHALS, G., VAN STRAALEN, N.M., GOVERS, H.A.J. and JOOSSE, E.N.G. (1993). Accumulation patterns of trace metals in freshwater isopods in sediment bioassays – influence of substrate characteristics, temperature and pH. *Water Res.*, **27**, 669–684.

VAN HATTUM, B., TIMMERMANS, K.R. and GOVERS, H.A. (1991). Abiotic and biotic factors influencing *in situ* trace metal levels in macroinvertebrates in freshwater ecosystems. *Environ. Toxicol. Chem.*, **10**, 275–292.

VAN LIERE, L., PARMA, S., MUR, L.R., LEENTWAAR, P. and ENGELEN, G.B. (1984). Loosdrecht Lakes restoration project, an introduction. *Verh. int. verein. Limnol.*, **22**, 829–834.

VEITH, G.D. (1975). Baseline concentrations for polychlorinated biphenyls and DDT in Lake Michigan fish, 1971. *Pest. Monit. J.*, **9**, 21–29.

VEITH, G.D., DEFOE, D.L. and BERGSTEDT, B.V. (1979). Measuring and estimating the bioconcentration factor of chemicals in fish. *J. Fish. Res. Bd Can.*, **36**, 1040–1048.

VENANT, A. and CUMONT, G. (1987). Contamination des poissons du secteur français du Lac Léman par les composes organochlorés entre 1973 et 1981. *Environ. Pollut.*, **43**, 163–173.

VESELÝ, J. (1994). Effects of acidification on trace metal transport in fresh waters. In Steinberg, C.E.W. and Wright, R.F. (eds) *Acidification of freshwater ecosystems: implications for the future*, pp. 141–151. Wiley, Chichester.

VIESSMAN, W. and HAMMER, R.J. (1993). *Water supply and pollution control*, 5th edn. Harper Collins, New York.

VINCENT, W.F. (1989). Cyanobacterial growth and dominance in two eutrophopic lakes: a review and synthesis. *Arch. Hydrobiol.*, **32**, 239–254.

VOLLENWEIDER, R.A. (1969). Möglichkeite und Grenzen elementarer Modelle der Stoffbilanz von See. *Arch. Hydrobiol.*, **66**, 1–36.

VOLLENWEIDER, R.A. (1975). Input-output models with special reference to the phosphorus loading concept in limnology. *Schweiz. Z. Hydrol.*, **37**, 53–84.

VUORINEN, P.J. and VUORINEN, M. (1992). Acidification in Finland: a review of studies of fish physiology and toxicology. *Finnish Fish. Res.*, **13**, 119–132.

WADE, K.R., ORMEROD, S.J. and GEE, A.S. (1989). Classification and ordination of macroinvertebrate assemblages to predict stream acidity in upland Wales. *Hydrobiologia*, **171**, 59–78.

WAKAO, R., TACHIBANA, H., TANAKA, Y., SUKARAI, Y. and SHIOTA, H. (1985). Morphological and physiological characteristics of streamers in acid mine drainage waters from a pyrite mine. *J. gen. appl. Microbiol.*, **31**, 17–28.

WALESH, S.G. (1989). *Urban surface water management.* Wiley, Chichester.

WALSH, F. (1978). Biological control of mine drainage. In Mitchell, R. (ed.) *Water pollution microbiology* Vol. 2, pp. 377–389. Wiley-Interscience, New York.

WALTER, R., MACHT, W., DÜRKOP, J., HECHT, R., HORNIG, U. and SCHULZE, P. (1989). Virus levels in river waters. *Water Res.*, **23**, 133–138.

WANG, W. (1990). Literature review on duckweed toxicity testing. *Environ. Res.*, **52**, 7–22.

WARD, D. (1994). Management of lowland wet grassland for breeding waders. *British Wildlife*, **6**, 89–98.

WARN, A.E. (1989). River quality modelling. In Ellis, K.V. *Surface water pollution and its control*, pp. 322–343. Macmillan, London.

WARN, A.E. and PAGE, C. (1984). Estimating the effect of water quality on surface water supplies. *Water Res.*, **18**, 167–172.

WARNER, F. and HARRISON, R.M. (eds) (1993). *Radioecology after Chernobyl.* Wiley, Chichester.

WARREN, C.E. (1971). *Biology and water pollution control.* W.B. Saunders and Co., Philadelphia.

WASHINGTON, H.G. (1984). Diversity, biotic and similarity indices. A review with special relevance to aquatic ecosystems. *Water Res.*, **18**, 653–694.

WATANABE, T., ASAI, K. and HOUKI, P. (1988). Numerical index of water quality using diatom assemblages. In Yasumo, M. and Whitton, B.A. (eds) *Biological monitoring of environmental pollution*, pp. 179–192. Tokai University Press, Tokyo.

WATER AUTHORITIES ASSOCIATION (1989). *Water pollution from farm waste.* London.

WEATHERLEY, N.S., MCCAHON, C.P., PASCOE, D. and ORMEROD, S.J. (1990). Ecotoxicological studies of acidity in Welsh streams. In Edwards, R.W., Gee, A.S. and Stoner, J.H. (eds) *Acid waters in Wales*, pp. 159–172. Kluwer, Dordrecht.

WEATHERLEY, N.S. and ORMEROD, S.J. (1992). The biological response of acidic streams to catchment liming compared to the changes predicted from stream chemistry. *J. Environ. Management*, **34**, 105–115.

WEATHERLEY, N.S., ORMEROD, S.J., THOMAS, S.P. and EDWARDS, R.W. (1988). The response of macroinvertebrates to experimental episodes of low pH with different forms of aluminium, during a natural spate. *Hydrobiologia*, **169**, 225–232.

WEATHERLEY, N.S., RUTT, G.P. and ORMEROD, S.J. (1989a). Densities of benthic macroinvertebrates in upland Welsh streams of different acidity and land-use. *Arch. Hydrobiol.*, **115**, 417–431.

WEATHERLEY, N.S., THOMAS, S.P. and ORMEROD, S.J. (1989b). Chemical and biological effects of acid, aluminium and lime additions to a Welsh hill-stream. *Environ. Pollut.*, **56**, 283–297.

WEBER, D.N. and SPIELER, R.E. (1994). Behavioural mechanisms of metal toxicity in fishes. In Malins, D.C. and Ostrander, G.K. (eds) *Aquatic toxicology*, pp. 421–467. Lewis, Boca Raton.

WEHR, J.D., EMPAIN, A., MOUVET, C., SAY, P.J. and WHITTON, B.A. (1983). Methods for processing aquatic mosses used as monitors of heavy metals. *Water Res.*, **17**, 985–992.

WELCH, E.B. and SCHRIEVE, G.D. (1994). Alum treatment effectiveness and longevity in shallow lakes. *Hydrobiologia*, **275/276**, 423–431.

WELCH, I.M., BARRETT, P.R.F., GIBSON, M.T. and RIDGE, I. (1990). Barley straw as an inhibitor of algal growth I: studies in the Chesterfield Canal. *J. appl. Phycol.*, **2**, 231–239.

WELLBURN, A. (1994). *Air pollution and climate change: the biological impact*, 2nd edn. Longman, London.

WESTLAKE, G.F. and VAN DER SCHALIE, W.H. (1977). Evaluation of an automated biological monitoring system at an industrial site. In Cairns, J., Dickson, K.L. and Westlake, G.F. (eds) *Biological monitoring of water and effluent quality*, pp. 30–37. American Society for Testing and Materials, Philadelphia.

WHEATLEY, A.D. (1987). Recovery of by-products and raw materials and wastewater conversion. In Sidwick, J.M. and Holdom, R.S. (eds) *Biotechnology of waste treatment and exploitation*, pp. 173–208. Ellis Horwood, Chichester.

WHITEHEAD, P.G., NEAL, C. and NEALE, R. (1987). Modelling stream acidity in U.K. catchments. In Barth, H. (ed.) *Reversibility of acidification*, pp. 126–141. Elsevier, London.

WHITEHURST, I.T. (1991). The *Gammarus* : *Asellus* ratio as an index of organic pollution. *Water Res.*, **25**, 333–339.

WHITEN, A. and BARTON, R.A. (1988). Demise of the checksheet: using off-the-shelf miniature hand-held computers for remote fieldwork applications. *TREE*, **3**, 146–148.

WHITTON, B.A., BURROWS, I.G. and KELLY, M.G. (1989). Use of *Cladophora glomerata* to monitor heavy metals in rivers. *J. appl. Phycol.*, **1**, 293–299.

WHITTON, B.A. and DIAZ, B.M. (1981). Influence of environmental factors on photosynthetic species in highly acidic waters. *Verh. int. verein. Limnol.*, **21**, 1459–1465.

WHITTON, B.A., ROTT, E. and FRIEDRICH, G. (eds) (1991). *Use of algae for monitoring rivers*. University of Innsbruck.

WHITTON, B.A., SAY, P.J. and JUPP, B.P. (1982). Accumulation of zinc, cadmium and lead by the aquatic liverwort *Scapania*. *Environ. Pollut. B.*, **3**, 299–316.

WHITTON, B.A. and SHEHATA, F.H.A. (1982). Influence of cobalt, nickel, copper and cadmium on the blue-green alga *Anacystis nidulans. Environ. Pollut. A*, **27**, 275–281.

WHO WORKING GROUP (1986). Health impact of acidic deposition. *Sci. Total Environ.*, **52**, 157–187.

WICKHAM, P., VAN DE WALLE, E. and PLANAS, D. (1987). Comparative effects of mine wastes on the benthos of an acid and an alkaline pool. *Environ. Pollut.*, **44**, 83–89.

WIENER, J.G., FITZGERALD, W.F., WATRAS, C.J. and RADA, R.G. (1990). Partitioning and bioavailability of mercury in an experimentally acidified Wisconsin lake. *Environ. Toxicol. Chem.*, **9**, 909–918.

WILHM, J.L. and DORRIS, T.C. (1968). Biological parameters for water quality criteria. *BioScience*, **18**, 477–481.

WILLIAMS, D.D. and FELTMATE, B.W. (1992). *Aquatic insects*. CAB International, Wallingford.

WILLIAMS, J.H., KINGHAM, H.G., COOPER, B.J. and EAGLE, D.J. (1977). Growth regulator injury to tomatoes in Essex, England. *Environ. Pollut.*, **12**, 149–166.

WILLIAMS, K., GREEN, D. and PASCOE, D. (1984). Toxicity testing with freshwater macroinvertebrates: methods and application in environmental management. In Pascoe, D. and Edwards, R.W. (eds) *Freshwater biological monitoring*, pp. 81–91. Pergamon, Oxford.

WILLIAMS, W.D. and ALADIN, N.V. (1991). The Aral sea: recent limnological changes and their conservation significance. *Aquat. Conserv.*, **1**, 3–23.

WILLIS, M. (1985). A comparative survey of the *Erpobdella octoculata* (L.) populations in the Afon Crafnant, N. Wales, above and below an input of zinc from mine-waste. *Hydrobiologia*, **120**, 107–118.

WILLIS, M. (1989). Experimental studies on the effects of zinc on *Erpobdella octoculata* (L.) (Annelida: Hirudinea) from the Afon Crafnant, N. Wales. *Arch. Hydrobiol.*, **116**, 449–469.

WILSON, H.M., GIBSON, M.T. and O'SULLIVAN, P.E. (1993). Analysis of current policies and alternative strategies for the reduction of nutrient loads on eutrophicated lakes: the example of Slapton Ley, Devon. *Aquat. Conserv.*, **3**, 239–251.

WILSON, R.S. (1994). Monitoring the effect of sewage effluent on the Oxford Canal using chironomid pupal exuviae. *J. IWEM*, **8**, 171–182.

WILSON, R.S. and MCGILL, J.D. (1977). A new method of monitoring water quality in a stream receiving sewage effluent, using chironomid pupal exuviae. *Water Res.*, **11**, 959–962.

WINNER, R.W., VAN DYKE, J.S., CARIS, N. and FARREL, M.P. (1975). Response of the macro-invertebrate fauna to a copper gradient in an experimentally polluted stream. *Verh. int. verein. Limnol.*, **19**, 2121–2127.

WOLTERING, D.M. (1984). The growth response in fish chronic and early life stage toxicity tests: a critical review. *Aquat. Toxicol.*, **5**, 1–21.

WOOD, L.B. (1982). *The restoration of the tidal Thames*. Hilger, Bristol.

WOOD, R.B. and SMITH, R.V. (eds) (1993). *Lough Neagh.* Kluwer Academic, Dordrecht.

WORTLEY, J.S. and PHILLIPS, G.L. (1987). Fish mortalities and *Prymnesium* in the Norfolk Broads. In Wortley, J.S. (ed.) *Proc. 18th Study Course of Inst. Fish. Management, Cambridge*, pp. 152–162.

WREN, C.D. (1991). Cause-effect linkages between chemicals and populations of mink (*Mustela vison*) and otter (*Lutra canadensis*) in the Great Lakes basin. *J. Toxicol. Environ. Health*, **33**, 549–585.

WREN, C.D. and MACCRIMMON, H.R. (1986). Comparative bioaccumulation of mercury in two adjacent freshwater ecosystems. *Water Res.*, **20**, 763–769.

WREN, C.D. and STEPHENSON, G.L. (1991). The effect of acidification and toxicity of metals to freshwater invertebrates. *Environ. Pollut.*, **71**, 205–241.

WRIGHT, J.F., ARMITAGE, P.D., FURSE, M.T. and MOSS, D. (1989). Prediction of invertebrate communities using stream measurements. *Regul. Rivers*, **4**, 147–155.

WRIGHT, J.F., FURSE, M.T. and ARMITAGE, P.D. (1993). RIVPACS – a technique for evaluating the biological quality of rivers in the U.K. *European Wat. Pollut. Control*, **3**(4), 15–25.

WRIGHT, J.F., FURSE, M.T. and ARMITAGE, P.D. (1994). Use of macroinvertebrate communities to detect environmental stress in running waters. In Sutcliffe, D.W. (ed.) *Water quality stress indicators in marine and freshwater systems: linking levels of organization*, pp. 15–34. Freshwater Biological Association, Ambleside.

WRIGHT, R.F. (1985). Chemistry of Lake Hovvatn, Norway, following liming and reacidification. *Can. J. Fish. Aquat. Sci.*, **42**, 1103–1113.

WRIGHT, R.F. (1987). Rain project: results after two years of treatment. In Barth, H. (ed.) *Reversibility of acidification*, pp. 14–29. Elsevier, London.

WRIGHT, R.F., LOTSE, E. and SEMB, A. (1994). Experimental acidification of alpine catchments at Sogndal, Norway: results after eight years. *Water Air Soil Pollut.*, **72**, 297–315.

WRIGHT, R.M. and PHILLIPS, V.E. (1992). Changes in the aquatic vegetation of two gravel pit lakes after reducing the fish population density. *Aquat. Bot.*, **43**, 43–49.

XU, Q. and PASCOE, D. (1993). The bioconcentration of zinc by *Gammarus pulex* (L.) and the application of a kinetic model to determine bioconcentration factors. *Water Res.*, **27**, 1683–1688.

YAMASHITA, N., TANABE, S., LUDWIG, J.P., KURITA, H., LUDWIG, M.E. and TATSUKAWA, R. (1993). Embryonic abnormalities and organochlorine contamination in double-crested cormorants (*Phalacrocorax auritus*) and Caspian terns (*Hydroprogne caspia*) from the upper Great Lakes in 1988. *Environ. Pollut.*, **79**, 163–173.

ZAMORA-MUÑOZ, C. SAINZ-CANTERO, C.E., SÁNCHEZ-ORTEGA, A. and ALBA-

TERCEDOR, J. (1995). Are biological indices BMWP and ASPT and their significance regarding water quality seasonally dependent? Factors explaining their variations. *Water Res.*, **29**, 285–290.

ZILLIOUX, E.J., PORCELLA, D.B. and BENOIT, J.M. (1993). Mercury cycling and effects in freshwater wetland ecosystems. *Environ. Toxicol. Chem.*, **12**, 2245–2264.

INDEX TO GENERA AND SPECIES

GENERAL INDEX